Neuroscience
Intelligence
Unit

Sense and Senility: The Neuropathology of the Aged Human Brain

David M. A. Mann, PhD, FRCPath
University of Manchester
Manchester, U.K.

SPRINGER-SCIENCE+BUSINESS MEDIA, B.V.

Neuroscience Intelligence Unit

SENSE AND SENILITY: THE NEUROPATHOLOGY OF THE AGED HUMAN BRAIN

R.G. LANDES COMPANY
Austin, Texas, U.S.A.

Originally published by Chapman & Hall in 1997
Softcover reprint of the hardcover 1st edition 1997

Please address all inquiries to the Publishers:
R.G. Landes Company, 810 S. Church Street, Georgetown, Texas, U.S.A. 78626
Phone: 512/ 863 7762; FAX: 512/ 863 0081

Library of Congress Cataloging-in-Publication Data

Mann, David M.A.
Sense and senility: the neuropathology of the aged human brain / by David M.A. Mann
p. cm. — (Neuroscience intelligence unit)
Includes bibliographical references and index.
ISBN 978-1-4613-7749-8 ISBN 978-1-4615-6001-2 (eBook)
DOI 10.1007/978-1-4615-6001-2
1. Nervous system—Degeneration. 2. Nervous system—Aging.
I. Title. II. Series.
[DNLM: 1. Brain—physiology. 2. Aging—physiology. 3. Nerve Regeneration—physiology. 4. Alzheimer's Disease—physiopathology. 5. Dementia, Senile. WL 300 M281s 1997]
RC365.M36 1997
616.8'047—dc21
DNLM/DLC 96-29707
for Library of Congress CIP

Publisher's Note

R.G. Landes Company publishes six book series: *Medical Intelligence Unit, Molecular Biology Intelligence Unit, Neuroscience Intelligence Unit, Tissue Engineering Intelligence Unit, Biotechnology Intelligence Unit* and *Environmental Intelligence Unit.* The authors of our books are acknowledged leaders in their fields and the topics are unique. Almost without exception, no other similar books exist on these topics.

Our goal is to publish books in important and rapidly changing areas of bioscience and environment for sophisticated researchers and clinicians. To achieve this goal, we have accelerated our publishing program to conform to the fast pace in which information grows in bioscience. Most of our books are published within 90 to 120 days of receipt of the manuscript. We would like to thank our readers for their continuing interest and welcome any comments or suggestions they may have for future books.

Shyamali Ghosh
Publications Director
R.G. Landes Company

CONTENTS

CHAPTER 1

INTRODUCTION

Only two things are certain in life, one is that all of us will inevitably grow older, the other is that at some point during or at the end of this process we shall die. Inherent to the passage of time is a deterioration in the structural and functional integrity of our bodies, this progressing to such an extent that one or more organ systems will eventually begin to fail with the continued health and well-being of the individual coming under threat. Age-associated deficiencies in the musculo-skeletal, cardiovascular, or endocrine systems producing arthritis, hypertension, stroke or diabetes are all too apparent in our elderly population yet internally caused failures in the function of the nervous system provide the common, and mostly intractable, problems of memory and intellect or locomotion that face and frustrate clinicians.

Perhaps the most important factor which can decide the outcome of research studies professing to examine the effects of the passage of time (i.e. the 'process of aging') on the function of the nervous system, or indeed any other organ system, is the selection of appropriate or representative subjects for investigation. The heart of this problem lies in defining what might be considered as 'normal' aging as distinct from age-associated disease; setting the 'goal posts of normality' continues to be a matter of considerable debate.[1,2] Dictionary definitions of 'normal' imply findings that might be considered 'usual', 'customary' or even 'average' for the age of the individual. In statistical terms, normality lies within two standard deviations either side of a notional average. Katzman and Terry[3] have defined normal aging of the nervous system as "aging changes occurring in individuals free from overt neurological disease" yet, because the early stages of tissue damage occasioned by the common and insidious disorders of later life–Alzheimer's disease and Parkinson's disease–may not produce observable or distinctive clinical deficits, cases ascertained in this way might represent a mix of individuals, some truly free from disease, others yet to demonstrate an overt symptomatology but perhaps already showing subtle impairments that defy strict clinical classification, falling still within the normal range of expectations.

Many early studies dealing with the physiology of growing older, be it concerning the nervous or other organ system, compared elderly subjects in hospital or in nursing homes with younger subjects usually living in the community and, unsurprisingly, detected sizable decrements with age in the measure of physiological function under question (see Meneilly and Tuokka[2] for review). However, with increasing sophistication, subsequent studies excluded patients with tangible evidence of impairment due to disease in the organ system of interest and accordingly more modest age-related changes were determined. Nevertheless, because many such elderly subjects had common disease in organs other than that under study findings like these could be criticized on the grounds that pathologies elsewhere in the body might have secondarily influenced results producing a summation of age- and disease-related changes. Currently, a much higher stringency is employed with greater emphasis being placed upon physical and physiological screening of patients. For recruitment, older subjects, in many studies, are now required to have a 'normal' medical history and physical examination, a normal ECG, normal liver and kidney functions, taking no medication and so on; in some instances the passing of pulmonary function or stress exercise tests is needed. Hence, all levels of screening will refine the category of patients studied which in the extreme may result in a cohort of highly selected superfit individuals in whom, as might be anticipated, age-related changes become minimal or even non-existent. Thus, while early studies were subject to lack of rigor in patient selection many current studies may be too stringent, selecting out 'super performers' who might occupy perhaps only the upper levels of our statistically normal range. Such subjects are far removed from what might be thought of as 'usual' or 'average'. Further, it is arguable as to what relevance data from such 'superfit' persons might have for the majority of individuals who, despite being apparently free of overt disease in the system under study, fail to meet such rigorous general physical inclusion criteria.

In practice, most studies on aging of the nervous system meet minimal criteria that include subjects who show 'average' or 'usual' changes but who are free from overt evidence of neurological or psychiatric illness and free from medications known to influence nervous system function; such individuals may, or may not, have been suffering from the common physical ailments of later life, these perhaps requiring medication. Findings achieved in this way may therefore represent, and describe more closely, those which are likely to occur in most individuals. Such criteria might be equally applicable to studies of mental or neurological performance made in life, or to pathological investigations made following death. The study of 'superfit' individuals alone may have most value in providing insight into the pathogenesis of aging or of age-related disease that affects the wider elderly population or in the search for factors that might prevent these. They should not, however, be construed as providing a fair and full description of what might be average or expected for the nervous system as time passes.

The mode of study of individuals is also important. This can be 'cross-sectional' or 'longitudinal'. Both have benefits and limitations. In cross-sectional studies a wealth of information can be garnered over a short time by comparing an ascertained elderly group with younger individuals. However, such data may be subject to 'cohort effects' due, for example, to better educational attainment in later generations, or secular increases in body or organ size due to improved health and diet, which change the baseline values for young and older subjects. Longitudinal studies, where individuals are followed along their own time-span, are preferable since subjects act as their own 'controls' with later data being compared against previous achievement. However, because of the long life span of humans, observations of meaningful changes may require years or decades of individual study and in practice it is often necessary to follow several cohorts over different limited time spans of life, summating the 'slopes' of each to provide an overall index of change.

While on 'average' many physiological variables within the nervous system or other organ systems, may show a progressive decline in function with age, the rate or scale of change among individuals can vary greatly. Whether these reflect inherent genetic factors which determine such an outcome or whether significant variations in lifestyle (the environment) are to blame is not clear; both may contribute. It might even be the case that those failing at quickest rates represent individuals in whom the effects of disease, while silent at first, are beginning to gain momentum as the extent of tissue changes increase. Such distinctions are important since, although genetic variation is a matter that cannot be altered, lifestyle is amenable to change and in so doing (some of) the deleterious effects of growing older might at least be partially prevented or even reversed. Recognition of early signs of disease may allow for potential treatment and protection of tissue before the scale of damage has progressed beyond repair.

In the following section mental and neurological changes occurring in the 'normal' elderly with the passage of time will be briefly described. It is not within the purpose of this book to provide a comprehensive overview of clinical aspects of aging in the nervous system; these can be better gained elsewhere.[2,4-6] The changes described will usefully serve as a background to the comprehensive pathological and etiological sections that follow. The kinds of changes described represent those likely in the 'average' individual who is free from overt neurological or psychiatric disease, who is living at home and is taking no medication affecting the performance of the nervous system at time of examination. As such, the changes may be thought of as describing a 'hypothetical' average patient–most elderly persons will fare better or worse in each respect covered.

Neurological complaints are common, even in otherwise physically healthy old people. Forgetfulness is common and while this may equate with what has been termed 'benign senescent forgetfulness' it may represent an insidious forerunner of neurodegenerative disease. Mental status investigations show that, while vocabulary, attention and concentration are often well preserved into later life, semantic knowledge (object naming) and verbal fluency (word generation) decline. The ability to retain information over a brief period of time (short term memory) is compromised while recall of past events may be less affected. Visuospatial abilities diminish, as does the capability to conceptualize ideas and think in abstract terms. Dizziness, poor posture and impaired balance are frequent complaints, even in the absence of musculoskeletal deficiencies and these may represent decreases in visual, vestibular and sensory inputs. Impairments of smell, hearing, sight and taste are commonplace. Elderly individuals show a decrease in muscle strength, clumsiness especially for finicky locomotor tasks, and a slowing of movement. Sleep difficulty is frequent and nocturnal leg cramps or spasms common.

Whether such clinical changes are pathological or part of normal aging is difficult to ascertain; a (variable) contribution by both processes is likely with the balance between each differing much among individuals.

While such functional changes may in part reflect the 'efficiency' with which brain cells operate in later life and the extent to which their interconnections (pathways) are used, it is nonetheless an expectation that actual alterations in tissue organization or structure should occur in the elderly brain and that these will underlie many, or even most, of the clinical signs of a failing system. In the next chapter alterations in the structure of the brain which commonly occur in later life will be described.

References

1. Calne DB, Eisen A, Meneilly G. Normal aging of the nervous system. Ann Neurol 1991; 30: 206-207
2. Meneilly GS, Tuokko H. Normal aging of the nervous system. In: Calne D, ed.

Neurodegenerative Diseases. Philadelphia: WB Saunders Co, 1994: 383-397

3. Katzman R, Terry R. Normal aging of the nervous system. In: Katzman R, Terry R, eds. The Neurology of Aging. Philadelphia: FA Davis, 1983: 15-50.
4. Albert M. Cognition and aging. In: Hazzard WK, Andres R, Bierman EL, et al., eds. Principles of Geriatric Medicine and Gerontology. 2nd ed. New York: McGraw-Hill, 1990: 913-919
5. Calne DB. Normal aging of the nervous system. In: Andres R, Bierman EL, Hazzard WR, eds. Principles of Geriatric Medicine. New York: McGraw-Hill, 1985
6. Drachman DA, Long RR. Neurological evaluation of the elderly patient. In: Albert ML, ed. Clinical Neurology of Aging. New York: Oxford University Press, 1984

CHAPTER 2

Pathological Changes in the Elderly Human Brain

2.0. INTRODUCTION

Brain tissue is composed of two principal cell types, the nerve cells and the glial cells; the latter comprise three varieties, astrocytes, oligodendrocytes and microglial cells, each subserving different roles within neural tissue. Nerve cells are not uniformly similar in structure and even within a given region of brain much variation in shape and size occurs. For example, in the cerebellum the Purkinje cells are among the largest in the whole brain and the granule cells the smallest; minority cell types such as basket cells, Golgi cells and stellate cells are also present. In the cerebral cortex a laminar distribution of large pyramidal cells and small interneurones is seen. Cellular specialization is present in the hippocampus with each sub-division of the pyramidal cell layer possessing its own distinctive characteristics, these large cells contrasting with the small neurones of the granule cell layer. This diversity of form is endless throughout the brain reflecting the scale of purpose of each. The largest cells (i.e. pyramidal or Purkinje cells) have extensive dendritic trees with a vast receptive surface area appropriate to their role as principal effector cells of that region of brain, receiving and integrating a wealth of afferent messages into a single coordinated response. Smaller neurones spread information within areas, or between adjacent areas of brain.

Hence, given such a variety of form and function of brain cells, it is perhaps naive to expect all regions, let alone all cells, to respond equally to the challenges of time or the effects of pathogens. Changes with aging or disease are not uniformly seen throughout the brain. Nerve cells are especially vulnerable to the ravages of time or disease given their unique properties of having a long life and hence high potential for exposure to damaging agents, and because of their inability to undergo cell division; once lost through damage they are not replaced.

In this section of the book, the structural and chemical characteristics of pathological changes taking place in the elderly brain are presented. Histopathological features usually associated with the common diseases of old age, such as Alzheimer's disease (AD) or Parkinson's disease (PD), are also frequently seen, albeit usually to a lesser extent, in the brains of old persons apparently clinically free from such disease. At this point no useful purpose is served by drawing anatomical distinctions between such clinical diagnostic groups. For example, in broad terms, and on an individual basis, a plaque is a plaque whether this be in a patient with clinical AD or in an apparently mentally healthy centenarian. Hence, the descriptions provided are given without deference to the clinical setting in which the changes occur and will thus serve any particular structure equally well regardless of the state of health of the person bearing that change. Nevertheless, it has to be emphasized that the greater mass of knowledge regarding such pathological changes has been drawn from studies of the diseased, rather than the 'normal', state, though this in no way detracts from its applicability to all situations where that change might be encountered. It will be seen later that distinctions between clinical settings are based more upon the amount or distribution of change rather than the presence per se and as such may relate principally to etiological variations that determine whether pathology occurs and if so to what extent.

2.1. GROSS CHANGES IN THE BRAIN

2.1.1. Imaging Studies

Over the past decade, non-invasive neuroimaging techniques have been extensively used to investigate the structure and function of the brain in living patients, especially in those suffering from neurodegenerative disease. However, a fair number of studies have employed this methodology to look for evidence of structural or functional changes in the brains of elderly persons while they too are still alive.

Computerized tomography (CT) scans[1-4] indicate increases with age in the volume of cerebrospinal fluid present within and around the brain, in the size of the ventricles and the width of certain major cerebral cortical and cerebellar sulci. Yet, within this generality of change many (perhaps even 30-50%) individuals have CT scans akin to those seen in young(er) persons. When atrophy is present it is slight and usually diffusely spread; ventricular enlargement is even. However, in some instances, focal emphasis in frontal or parasagittal parietal sulci can occur and these may represent incipient AD. In addition about 30-50% of normal elderly subjects show white matter changes, periventricular translucencies or leukoaraiosis as these are called, which refer to local cerebrovascular insufficiencies. Similar structural changes to those seen on CT are evidenced on magnetic resonance imaging (MRI, NMR) in both grey[5] and white[6-9] matter, though more recent quantitative work [10] suggest a preferential atrophy of the medial temporal lobe (excluding the hippocampus) may occur whereas in other studies[11-14] hippocampal atrophy is emphasized.

Positron emission tomography (PET) can detect functional changes in brain tissue. Decreases in brain energy metabolism (as measured by fluorodeoxyglucose uptake) have been reported in some studies[15-19] whereas in others[20,21] no such decreases were seen, even in the presence of cerebral atrophy upon CT scanning of the same patients. Similarly, studies on the nigroneostriatal pathway, using fluorodopa uptake, have been conflictory. Martin et al[22] noted a 50% decrease in fluorodopa uptake in the corpus striatum over a 20-80 year age range whereas Sawle et al[23] found no such change. These discrepancies remain unresolved, though it is notable that in those studies[21,23] showing no age-related individuals were recruited who had scored maximally in Mini Mental Status Questionnaire or who had high (around 124) IQ scores. Hence, in these latter studies a high proportion of "superperformers" may have been investigated. A clear impression of what functional changes might "routinely" occur in the brains of elderly subjects has still to emerge.

2.1.2. Autopsy Changes

At autopsy, the brains of elderly persons often show a number of inconstant changes over their surface, these involving a thickening of the arachnoid by fibrosis, especially over the parasagittal cortex, and an increase in the size of the arachnoid granulations. Additionally it is often obvious from external inspection that the overall size of the brain may have changed and that shrinkage of the overlying folds (gyri) of the cerebral cortex may have taken place.

The actual weight of the fresh brain can often provide important clues as to its pathological state with deviations either side of an expected range suggesting either the presence of additional "tissue" (e.g. tumor, haemorrhage) or a loss of tissue (atrophy). Brain weight in apparently mentally normal individuals increases rapidly between birth and 3 years of age, attaining its maximal weight at around 20 years of age, then remains more or less steady until about 40-50 years of age, subsequently declining at a rate of about 2-3% per decade eventually reaching a value, some 10% below maximum, by the 9th decade of life.[24-27] It has been suggested[28,29] that atrophy might begin at an earlier age in women than in men.

As the brain shrinks in size with age, so the space surrounding it increases with the ratio between brain volume and skull volume (which remains constant throughout life) falling.[26] Not only does the subarachnoid space increase but so too does the intracerebral space–the ventricles. Ventricular enlargement is often most evident around the centrum semiovale, where a rounding of the angles of the ventricle is apparent, though the III ventricle, aqueduct and the IV ventricle may all become enlarged.[30-33]

It has been from such relatively simple analyses relating brain weight or brain volume to age at time of death, that the popular concept of the human brain losing 1 million nerve cells per day, every day of its life, has emerged. However, more recently, these data have been called into question. Careful secular analyses[34-36] have shown that the brain, in parallel with the rest of the body, has in fact over the past century undergone a real increase in size, this presumably being due to the influence of better health care, diet etc. A lower brain weight in elderly subjects may in part therefore reflect such an initially smaller brain. Correction for these secular effects shows that up to the age of 60 years there is little, if any, overall change in brain weight and it is only after this age that a gradual decline sets in leading eventually to a loss of substance amounting to about 5% of the original weight or some 60-70g.[36] Studies[26] relating brain volume to cranial capacity, a measure not influenced by secular effects, confirm these data.

While these changes in overall brain weight or the volume of the ventricles and amount of cerebrospinal fluid indicate a reduction in the size of the brain, it has been difficult to define the relative contributions made to this by cortical grey or white matter or by subcortical structures such as the basal ganglia, brainstem or cerebellum. Miller et al[34] noted, in a series of 130 autopsy brains aged between 20 and 98 years, that although both overall grey and white matter volumes decreased with age (even after accounting for secular effects), the ratio between the amount of grey and white matter followed an unusual pattern with age such that it decreased between the ages of 20-50 years, then increased thereafter. Harris et al[37] found a similar pattern of change in the grey/white matter ratio in 78 subjects between the ages of 19 and 77 years using MRI analysis. Others[2] have also observed a fall with aging in the ratio between the amounts of grey and white matter. Such data suggest that two separate processes are operating, one before, and one after, 50 years of age. In the first, a preferential loss or shrinkage of grey matter elements occurs towards later life. In the second, taking place later on, white matter lesions causing a loss of axons predominate; these are perhaps due to a loss of myelin or the effects of cerebrovascular disease.

Certainly, white matter hyperintensities indicative of vascular damage are widely seen on MRI and these would be consistent with this increase in ratio in later life.

It is only natural to assume that this loss of tissue, as evidenced by a shrinkage of the gyri and a widening of the sulci, is represented by an actual fall out of the brains' principal cells–the neurones. However, it is equally possible that cell shrinkage, with reductions in the amount of dendritic or axonal tissue (cell atrophy), in the absence of any actual change in cell number could also achieve this effect; indeed some combination of both processes might occur. It may also be the case that a loss of elements other than neurones (e.g. blood vessels) could influence brain weight and it is also plausible that age-related increases in, for example, glial cells might (partially) compensate for, and counterbalance, any weight loss due to changes in neuronal number or structure.

Hence, it is clear that gross measurements of brain weight or brain atrophy (volume) alone can at best provide only a rough guide as to what is really happening in the brain; microscopic examination of neuronal number and size and alterations in non-neuronal elements are necessary to comprehensively audit any relationship between structure and age and to determine the effects of the passage of time on the system, either as a whole or in respect of its component parts.

2.2. NERVE CELL NUMBERS IN AGING

It is another widely accepted belief that this loss of tissue means that the human brain must inevitably deteriorate in function with age and given that the neurones are those cells responsible for the receipt of incoming sensory information, its processing and its output in an appropriate form, it is assumed that changes in the brains nerve cell population should provide the substrate for this gradual failure. Yet, despite nearly a century of scientific effort, it is still far from clear as to what extent, or in what areas of brain, this age-related attrition might take place. The pioneering studies of Hodge[38] laid the foundations for the study of nerve cell loss with aging, but this challenge was not taken up until over half a century later by Brody.[39] Most of the present information available has been gathered within the last two decades and from this only now are some threads of consistency beginning to emerge.

2.2.1. Cerebral Cortex

The accurate quantification of nerve cell number in the cerebral cortex presents severe difficulties, partly because, out of necessity, only a small proportion of the total tissue can be sampled in any one patient and also because of the cytoarchitectural variations that occur within and between cortical regions. Nonetheless, Brody made the first study in 1955 addressing this question of neuronal loss from the human cerebral cortex as a function of increasing age.[39] He estimated a 57% decrease, between the ages of 18 and 95 years, in the density of nerve cells from the superior temporal cortex whereas other areas such as the pre- and post-central gyrus, striate cortex and inferior temporal gyrus showed reductions of 13-31%; most of the nerve cell loss was noted to occur after 30 years of age. In a later study Brody[40] noted a 51% decrease in nerve cell density within the superior frontal cortex in 18 patients between the ages of 41 and 87 years. Although no statistical tests were performed by Brody in these original studies[39,40] subsequent analysis of the data by Hanley,[41] using regression analysis and rank correlation, showed that most of the reported reductions were indeed age-related and statistically significant. In the later study[40] a reduction in cell density in the superior frontal cortex of 8% per decade was calculated.[41] While there was a decrease in nerve cell density in every layer in all cortical regions examined by Brody,[39,40] in general those neurones in layers II and IV showed the greatest loss of number.

About the same time, Colon[42] noted a reduction in nerve cell density in four cortical regions in two elderly persons (83 and 85 years of age) which, when compared with four younger individuals (18-39 years of age), amounted to an average loss of 44%. The distribution of this loss appeared to be 'random' over the cortex and was roughly equivalent in upper and lower cortical layers. Shefer[43] also reported decreases in density of the large cells in layer III of the cortex ranging from 12-28% in 15 patients older than 19 years of age. Affected areas included precentral gyrus, superior frontal and superior temporal gyri and temporal pole, inferior parietal cortex and occipital cortex.

Subsequent studies,[44-47] made in the same cortical regions examined in these early reports and in other cortical regions not previously investigated, have in general borne out these original conclusions of an age-related decline in neuronal density. Anderson et al[45] noted a decrease in neuronal density in 19 patients aged between 69 and 85 years of about 15% from the inferior frontal cortex (gyrus rectus) and superior temporal gyrus, this being equivalent to a reduction of about 10% per decade. Henderson et al[44] also found decreases of 12-41% in the density of 'small' neurones and 36-53% in 'large' neurones of the inferior frontal, inferior temporal, superior temporal and pre- and post-central gyri in 64 patients, ranging from 18-95 years of age. Mann et al[46] reported the density of large pyramidal cells of layers III and V of the middle temporal gyrus to decrease by 43-50% in 67 patients between 20 and 97 years of age. Lastly, in a cell suspension study, Devaney and Johnson[47] reported a 54% decrease in the number of neurones of the occipital cortex.

Hence, the broad conclusion to be drawn from these studies is that the density of nerve cells in many parts of the human cerebral cortex does indeed fall with age and that these changes in density probably reflect an actual loss of nerve cells. However, the extent of the losses seen with age are not uniform and may vary much from region to region. Furthermore, it is not by any means clear which (if any) particular nerve cell population is affected most. For example in some studies[43,44,46] emphasis is placed on the larger (presumably) pyramidal cells whereas in others[39,40] involvement of the small cells (interneurons) of layers II and IV is stressed; in yet others[42,45,47] no preferential cellular involvement was detected or mentioned. Moreover, it is also now clear that in most of these aforementioned studies, the actual decrease in neuronal density (cell number) occurring will, in all probability, be in excess of that quoted. This conclusion comes from observations that while the density of nerve cells during aging is falling, so too is the volume of cerebral cortex in which they are contained,[26,27,34,46,48] even allowing for secular increases in brain weight (and volume). Little attempt was made in previous studies to 'correct' such 'raw' density measurements for these age-related alterations in tissue volume which would produce an artefactually high numerical density value due to compacting of surviving tissue elements. Hence, a density value indicating no change with age may, in reality, mean that an actual loss of cells will have taken place but only up to a level that broadly matches the overall degree of tissue shrinkage.

However, not all studies, and particularly the most recent ones, have revealed a significant decrease in nerve cell density with aging. Hence, Terry et al,[27] examining the mid-frontal, superior temporal and inferior parietal regions of 51 patients aged from 24-100 years, noted that the overall neuronal density was unaltered with age. Although they detected a decrease in density of 'large' cells with age this was counterbalanced by an increase in the density of 'small' cells. From this finding Terry et al[27] concluded that nerve cells had simply shrunk in size with age rather than having been lost. However, it was acknowledged by the authors that because the cerebral cortex was thinner in the elderly patients, some actual neuronal loss was likely to have occurred (this being masked by tissue compaction (vide supra), but it was felt that the cell loss would be of a lesser magnitude than most previous studies had indicated. Haug and colleagues[35,36,49,50] examined the density of neurones in the inferior frontal cortex, superior frontal cortex and the occipital cortex of more than 50 individuals ranging in age from 30-110 years. When artefacts, peculiar to their own processing protocol, leading to differential tissue shrinkages between younger and older individuals were corrected for, these workers detected an overall increase in neuronal density with age in all three regions, this being roughly counterbalanced by a corresponding decrease in neuronal perikaryal size and cortical volume[51] within the same brain regions. Similar, though less pronounced, tendencies were seen within the temporal lobe and parietal cortex. From these data, Haug concluded that within these particular brain regions (and thence extrapolating to the whole cerebral cortex) the total number of nerve cells does not diminish with age; neurones merely shrink (by some 10-35% according to area) and that it is the cumulative effects of this shrinkage that produce the decline in cortical volume[34,35] and overall brain weight[36] after 60 years of age. Braak and Braak[52] also reported a shrinkage, rather than an actual loss, of neurones from the cerebral cortex in old age.

Hence, although the majority of early studies pointed towards a substantial age-related decline in nerve cell number from the cerebral cortex, this is apparently not borne out by later studies where cell shrinkage accompanied by an overall preservation of, or perhaps only a slight decrease in, cell number is emphasized. Undoubtedly, methodological problems arising from the examination of small amounts of tissue in relatively few individuals will account for much of these discrepancies. In most early studies the number of cases examined were small and thus more likely to be subject to the wide variations in neuronal number and density that occur between individuals, even ones of the same age. More doubt can be cast over the accuracy, and hence the validity, of the findings in these earlier studies when compared to later ones[27,35,36,49,50] where larger study groups were employed and technical factors were more rigorously controlled. Furthermore, problems of unrecognized or 'unscreened' pathology relating to the common degenerative diseases of old age may also bias data in elderly subjects and help to dictate an apparent cell loss with age; this will be commented upon later.

Therefore, at present, a definitive answer to the question as to whether, and to what extent, nerve cells are lost from different parts of the cerebral cortex as a function of aging is still awaited. On balance, some cell loss does seem likely in many old people but this probably does not take place to the extent formerly believed to be the case. Most recent studies have played down the actual loss of cells and have emphasized a shrinkage of the neuronal perikaryon (and presumably its processes) (see later) and this may be especially severe within the larger pyramidal cell population. This cellular atrophy will undoubtedly have functional consequences, both on an individual and on a cohort basis, at least as great as any actual loss of cell bodies, and may therefore contribute much to the decline in the weight of the cerebrum that occurs in later life.

2.2.2. Other (Non-Neocortical) Regions

Changes in nerve cell number or density with aging are often easier to determine in regions or structures outside of the human neocortex either because their anatomical boundaries are better defined or their component cells are more uniform in morphology and therefore more easily and consistently recognizable. Furthermore, because of their smaller size and more restricted neuronal population it becomes feasible to effectively sample more or even the whole of the region. Not surprisingly, therefore, there is a greater consensus of opinion concerning the changes that occur in nerve cell number with aging in these non-neocortical regions.

2.2.2.1. Hippocampus and amygdala

The possible changes with aging upon the nerve cell complement of the human hippocampus has been the subject of a good number of studies.[45,46,53-61] However, while in some studies only a specific part of the hippocampal formation (e.g. area CA1) was examined in others a more systematic approach sampling either the entire structure or different anatomical subdivisions was adopted. Ball [53] investigated all CA regions in coronal sections of the entire length of the hippocampus in 18 patients of age range 47-89 years and noted a negative correlation between pyramidal cell density and age, this being equivalent to a decrease of 5.4% per decade over this period. Shefer[54] reported the neuronal complement within the subiculum to decline by 29% in an unspecified number of elderly persons (mean age 77 years) when compared to five younger individuals (age range 19-27 years). Mouritzen Dam[55] estimated pyramidal neurone density in the 'H subfields' of the Ammons horn region in 30 individuals ranging from 21-91 years of age. Following a correction for the overall atrophy of the hippocampus taking place in later life, a reduction in neuronal density after 41 years of age, and ranging between 19 and 26% in the various subfields, was noted. This same author[55] also noted in these same patients a 15% decrease in the granule cell population of the dentate gyrus over this same period. Mann et al[46] again taking into account an overall atrophy of the hippocampus with age, observed, in 67 patients ranging from 10-97 years, a 6.2% fall in density per decade of pyramidal cells of area H1 (Sommer's sector) after 50 years of age. Miller et al[56] noted, again in this same part of the hippocampus, a decrease in pyramidal cell density of 3.6% per decade in 86 patients aged between 15 and 96 years. Anderson et al[45] reported a 14% loss of all neurones from a region they called the 'subiculum' (but which probably also contained parts of CA1 and prosubiculum) in 19 patients between the ages of 70 and 85 years. West[57] estimated a 52% loss of nerve cells from the subiculum and a 31% loss from the hilus of the dentate gyrus, in 32 males, across the age range 13-85 years; no changes with age were noted in areas CA1, CA2/3 or the dentate granule cells.

In contrast to these reports, other studies[58-60] have found only little, or a non-significant, fallout of neurones or that a regional specificity of change occurs, with age. Brown and Cassell[58] serially sectioned the whole hippocampus in six patients between 49 and 77 years of age and estimated the total number of pyramidal cells in areas CA1, CA2 and CA3/4; no correlations between pyramidal cell density and age in any region were found. Mani et al[59] investigated pyramidal cell density, within areas CA1 to CA4, in 23 patients aged between 4 and 98 years and although negative correlations between cell density and age were detected in all four regions, only in CA4 was the decrease statistically significant, this being equivalent to a reduction in cell number of 3.8% per decade. These same authors, however, suggested that the pattern of decrease in cell density with age might not be linear, becoming progressively greater after 65 years of age. In a recent study by Davies et al,[60] 12 patients (drawn from the original series of Mann et al)[46] aged between 6 and 87 years were investigated. Seven subfields, CA1-4, prosubiculum, subiculum and presubiculum, were separately analyzed and the possible affects of local atrophy (as opposed to an overall atrophy of the hippocampus) were accounted for; no significant reductions in pyramidal cell density with age were observed in any region. Finally, Devaney and Johnson,[61] using a cell suspension method, counted neurones from the entire hippocampus of 25 patients between 20 and 87 years of age and detected a slight, age-related, increase in neuronal density which they attributed to a greater loss of white matter rather than grey matter from the hippocampal formation during aging.

Therefore, while these latter four studies fail to confirm earlier reports of an age-related decline in pyramidal cells from (different parts of) the hippocampus, it has to be emphasized that in these the number of patients studied was usually lower (n=6 to 25) as compared to earlier reports in which, in most instances, the study groups exceeded 20 patients (range n=15 to 86). This limitation in the number of cases sampled combined with the wide age range investigated (usually a period of 80-90 years) and the considerable inter-individual variation that 'normally' occurs at any given age, would obviously militate against achieving a significant outcome in these latter studies.[58-61] Yet, it is also worth noting that in several of the earlier studies[46,54-56] a significant age-related decline in nerve cell number was accomplished only after making a correction for the overall atrophy of the hippocampus that had taken place rather than for that which might have taken place in the particular region where the nerve cells had been counted. In this latter context Davies et al[60] noted that the width of the pyramidal cell band did not alter with age in any subregion, indicating that "raw" cell density measures in the various parts of the pyramidal cell layer might not in fact require any correction for tissue shrinkage and that studies which had done this[46,54-56] might have done so erroneously. In these aforementioned studies raw cell density measures, i.e. uncorrected for atrophy, did not decrease with age. In the study by West,[57] an unbiased stereological technique (dissector) was employed and while this may avoid the technical limitations of manual counts on single (or representative) sections, the question of differential atrophy effects within tissue regions remains.

Hence, at present, it is strongly suspected that some cell loss from one or other parts of the hippocampus may indeed occur with aging, though a definitive study of all parts of this region that encompasses a sufficient number of individuals of a wide enough age range and one that also takes into account the possible effects of any local, rather than overall, atrophy is still awaited.

Only a single study[62] so far has investigated changes in nerve cell density in the amygdala. Small, but significant, decreases in both nerve cell density and regional volume, ranging from 3-9% and limited to the cortical, medial and central nuclei were seen in four elderly males (mean age, 73 years) when compared with four younger males (mean age, 45 years); no significant change in either of these measures was observed in other subregions of the amygdala.

2.2.2.2. Cerebellum

An apparently clear loss of Purkinje cells from the cerebellar cortex with aging has been detected in most of those studies[63-67] attempting to address this. Probably the most well-performed study so far was that of Hall et al[63] in which the total hemispheric volume of the cerebellum and the Purkinje cell density were both noted to decline with age. These data yielded an average reduction in Purkinje cell number amounting to about 2.5% per decade though it was remarked that even after the age of 80 years some individuals retained normal cell counts. These findings compared well with the much earlier studies of Ellis[64,65] who calculated in 23 individuals, carefully screened for the presence of overt pathological changes, a decrease in Purkinje cells of about 35% between the ages of 20 and 90 years. Delorenzi[66] studied the cerebella of six people ranging from 6 to 83 years of age but was unable to detect any age-related differences within these few individuals. More recently, Torvik et al[67] claimed a loss of Purkinje cell from different parts of the vermis ranging from 19 to 38% in 65 subjects aged between 49 and 89 years. However, in this particular study many patients with certain neurological disorders were included. Because parts of the cerebellum such as the vermis are particularly susceptible to the effects of damaging agents such as alcohol[67] it is likely that these quoted figures will have overestimated that component of cell loss due to age alone.

Only a single study has investigated changes in nerve cell number in the dentate nucleus of the cerebellum.[68] In 27 people aged between 6 and 99 years the total nerve cell number remained constant throughout life at about 820,000 cells.

2.2.2.3. Sub-cortical structures

Because of the relative ease of recognizing their anatomical boundaries and discriminating their nerve cell population, it has been possible to investigate age-related changes in many subcortical structures with a greater degree of accuracy and a higher level of consistency than has been the case for most cortical structures.

Hence, many regions, especially the cranial nerve nuclei, have been shown to be unaffected by age in terms of the size of their neuronal population. For example, Konigsmark and Murphy[69,70] showed no changes in nerve cell number in the ventral cochlear nucleus. Brody and colleagues reported no reductions in the number of nerve cells with aging in the motor nucleus of the trochlear[71] and abducens[72] nerves. Van Buskirk[73] also noted no nerve cell loss with aging from the facial nerve nucleus. Although one study[74] detected a 20% reduction in the number of cells of the inferior olivary nucleus, two others[75,76] found no change in number with age. Mann et al[77,78] noted no change in cell density in the dorsal raphe nucleus with age. However, among other catecholaminergic cell groups, the locus caeruleus complex has consistently been shown to suffer a substantial loss of cells with aging[77-85] and similar findings have been extended to the pigmented locus caeruleus pars cerebellaris[86] and the dorsal motor vagus.[87] Age-related decreases in the number of cells within the substantia nigra have reported by several workers[88-92] though other studies[93] have not confirmed these findings. Tomlinson and Irving[94] reported a stability in the number of anterior horn cells of the spinal cord up to 60 years followed by a loss averaging 30% by 90 years of age, though as with other areas of the brain (e.g. cerebellum) some very old people still retained a normal complement of neurones. In the hypothalamic region, Wilkinson and Davies[95] detected no loss of neurones with age from the mamillary bodies and both Goudsmit et al[96] and Fliers et al[97] reported that the density of nerve cells within the supraoptic and paraventricular nuclei also did not change. However, other studies from these latter workers, and mostly on the same patients, detected a loss of neurones, particularly the vasopressin containing ones, from both the suprachiasmatic nucleus and the sexually dimorphic nucleus of the medial preoptic area[98] whereas VIP containing cells were unchanged in number.[99]

Within the basal ganglia, one study[100] reported a decrease in neuronal density in the putamen, whereas in two other reports[50,101] the density of nerve cells in the putamen (and also the caudate nucleus[50]) actually increased with age, though because the volume of this region fell correspondingly no overall change in nerve cell number was recorded. Finally, within the cholinergic nucleus basalis complex in the basal forebrain, there have been three reports[102-104] showing a loss of cells with aging, while three others[105-107] have not confirmed this loss. McGeer et al[102] estimated the total cell number in 10 patients aged from 38 weeks to 95 years and detected a 60% loss by the age of 95 years. Mann et al[103] counted the number of neurones in representative sections of the nucleus basalis in 20 patients aged between 6 and 88 years and observed a 30% reduction in cell number per section by 90 years of age. LaCalle et al[104] noted a 50% reduction, by 100 years of age, in the total number of neurones in 60 patients aged 16 to 110 years. On the other hand, Whitehouse et al[105] found no change in cell counts in sections taken from the mid-portion of the nucleus basalis in 30 subjects from 44-88 years of age and Clark et al,[106] using the same approach, also detected no changes in cell number with age in seven individuals. Chui et al[107] counted cells in 17 patients aged between 23 and 87 years and likewise found no age-related decrease in cell number. In this latter study, however, many of the younger (i.e. under 60 years) patients died from alcohol-related disorders which are known to damage and reduce cell number in the nucleus basalis.[108,109] Hence, in this latter study baseline values in the young patients may have been lowered from normal by the effects of alcohol damage thereby obscuring any loss that might really have occurred in older subjects.

Sampling variations, combined with a high inter-individual variability in cell number within relatively small sample populations, is likely to account for these differences in findings in this rather dispersed brain region. Perhaps more emphasis should be placed on the findings from those two studies[102,104] in which the whole of the nucleus was serially sampled and total neuronal number derived.

2.3. REGRESSIVE CHANGES IN NEURONES WITH AGING

An actual loss of nerve cells with aging may not be the only mechanism whereby the function of a particular brain region may become compromised in later life. A loss of dendrites or synapses without perikaryal loss may just as drastically impair function as would the actual loss of the cell body; indeed loss of the cell body would, by necessity, involve a loss of its associated dendrites and synapses.

2.3.1. Dendritic Changes

Dendrites comprise some 95% of the total receptive surface that neurones offer for contact with other neurones[110] and hence, maintenance of the size and integrity of the dendritic tree is paramount in ensuring that the integrative capacity of individual nerve cells, brain regions or the system as a whole remains effective. Failure to maintain dendrites may lead to a deterioration in neuronal function and should this fall below a certain critical 'threshold', clinically observable dysfunction may occur.

In practice, in humans, it is only possible to demonstrate the whole of the neuronal dendritic tree at autopsy (and rarely at biopsy) by using Golgi (silver) type impregnation techniques (e.g. Golgi-Cox, rapid Golgi). These, although excellent for delineating the extent of dendrites in individual cells, are notoriously capricious to perform and are highly selective both in terms of the extent of impregnation that might occur between different individuals or within different tissues, or even within different parts of the same tissue, of any one individual. Hence, their degree of representativeness for both qualitative and quantitative studies has frequently been called into question and data obtained by such methods should be interpreted with appropriate caution.

Investigations concerning the extent of dendritic changes in human aging have been confined to studies on the cerebral cortex and hippocampus but even then widely differing results have been obtained by different workers. To what extent these variations stem from technical differences, the capricious nature of the impregnation method, or problems in selecting and quantifying the elaborate and often extensive dendritic systems of many neurones is not known; they may 'simply' reflect intrinsic intra-regional and inter-individual variations.

Hence, the early work of Scheibel and colleagues[111-113] based on the rapid Golgi technique, which impregnates only a small percentage of neurones in any given section, suggested that a reduction in the size of the dendritic tree occurred with aging in both the cerebral cortex and the hippocampus. Scheibel also described extensive and sequential morphological changes in the dendritic tree of the pyramidal cells of the frontal and temporal cortex, in the Betz cells of the precentral gyrus, and the hippocampal pyramidal cells. Such alterations included a loss of dendritic spines and a swelling and distortion of dendrites which eventually led to a loss of branches from both basal and apical dendrites and even a loss of the dendritic trunks themselves. These studies were not, however, substantiated by the later work of Buell and Coleman[114,115] who, using what they claimed to be a more reliable and artifactually free Golgi-Cox procedure[116] and applying computer analysis, demonstrated a net dendritic growth in pyramidal cells of layer II of the parahippocampal (entorhinal) cortex. These latter workers suggested that this increase was due to a compensatory response on the part of surviving cells to the loss of their neighbors. This concept was strengthened by other studies from this same group[117-118] who analyzed dendritic extent in hippocampal granule cells in middle-aged (52 years), old-aged (73 years) and very old (90 years) individuals. A significant increase in dendritic length occurred between middle and old age though this 'compensatory' response, within a region known[55] to become depleted of cells in old age, was lost in the very old in whom an actual net regression in dendritic extent (compared to the old-age group) was seen. Yet, in these very same patients, these same workers[119] detected no age-related changes whatsoever in the dendritic trees of CA2/3 pyramidal cells. Other studies[120] have suggested a continued regression throughout life, in the basal dendrites of layer III and V pyramidal cells of the human motor cortex. Coleman and Flood[121] also reported a similar change in layer III pyramidal cells of the middle frontal gyrus.

Therefore, the extent to which the dendritic system of nerve cells might regress, either individually or collectively, with age is still not clear. Nor is it apparent whether any such changes that might occur are uniformly distributed throughout the cortex or are confined to single or small clusters of cells. Local events (e.g. loss of afferent connections, glial cell relationships, vascular compromises) could impact upon the microenvironment of particular neurones and in so doing dictate these apparently varying and regionally specific patterns in dendritic change and plasticity during aging.

2.3.2. Axonal Changes

Even less is known concerning changes in axons with aging; only a few studies[122-125] have addressed this through quantification of synaptic density in post-mortem tissues. In one, Cragg[122] suggested an age-related decrease in synaptic density in the precentral gyrus, this being accompanied by a compensatory increase in the contact length of residual synapses. Adams[123] confirmed these findings, recording in this same brain region a 50% reduction in synaptic density in the 8th decade compared to that in the 5th and 6th decades; investigation of the post-central gyrus in the same patients revealed no age-related changes. Huttenlocher[124] and Gibson[125] both noted a loss of synapses with aging from the frontal cortex but no loss was found from the temporal cortex.[125]

Because of the difficulties in accurately estimating synaptic density at ultrastructural level, due to post-mortem tissue deterioration and the minute regions of tissue that can be effectively sampled, more recent studies have been based on the immunohistochemical detection of synapse-associated proteins such as synaptophysin. In this way Masliah et al[126] noted a 14-20% loss of synapses to occur after 60 years of age. Nonetheless, it is likely that any reduction in synapse number that might occur with aging is, at least partially, offset by compensatory changes in those that remain which include increasing their total pre-synaptic contact area (i.e. by a lengthening of the synaptic thickening) with post-synaptic sites[127] In this way it seems that even a substantial numerical loss of synapses, perhaps as much as 30%, may be tolerated without overall tissue dysfunction.

Thus, as with the loss of neurones and dendritic changes, synaptic alterations with age are likely to be regionally specific and compensatory changes may take place in remaining synapses in regions where actual loss has occurred.

2.3.3. Perikaryal Changes

In some of those studies that have addressed the question of cell loss the issue of possible changes in cell size with age has also been investigated[27,45] and in these a reduction in perikaryal size with aging has been detected. Although the structural and molecular counterparts of this perikaryal atrophy are not known, cell shrinkage may partly relate to reductions in the amount of internal membrane (e.g. rough endoplasmic reticulum). This inference is drawn from studies documenting a loss of ribonucleic acid with age in nerve cells from the cerebral cortex, hippocampus and cerebellum[128,129] and subcortical regions like the nucleus basalis of Meynert.[78,103] These changes, measured in postmortem tissues, may reflect alterations in the quantity of ribosomal RNA which, in turn, might represent either the overall number of ribosomes present or the actual extent of internal membrane (rough ER) upon which protein synthesis can occur, or some combination of both. Whether such changes in ribonucleic acid content are primary and relate perhaps to deficiencies within the genome or occur merely as secondary and adaptive responses to, for example, an anatomically reduced or functionally compromised dendritic tree or axonal synaptic plexus, is uncertain. Other perikaryal changes involving the formation and accumulation of inclusions, such as Lewy bodies or neurofibrillary tangles (NFT), or increases in the amount of neuropigments will be discussed later.

Hence, it is still not clear whether nerve cell atrophy and loss does indeed occur in aging and, if so, what the extent of such losses might be, and what regions of brain might be affected. Furthermore, it remains to be determined whether cell atrophy and loss is caused 'in-house' by programmed factors that regulate the life span of the nerve cell and dictate its point of demise (e.g. by apoptosis) or whether such losses are the secondary consequences of pathological insults occurring in later life (i.e. by necrosis). A lack of clear markers for these processes still hinders our understanding of the mechanisms underlying nerve cell death in later life.

2.4. SENILE PLAQUES

Senile plaques (SP) are complex foci of degeneration occurring within the brain parenchyma (also known as the neuropil) which are composed of both extracellular and cellular entities. They were originally observed towards the end of the last century, firstly by Blocq and Marinesco[130] in 1892, who noted their presence in the brain of a non-demented epileptic patient and termed them 'miliary foci'. Later, in 1898, Redlich[131] saw similar structures in the brains of two elderly and demented persons. Since then, their presence within the brains of elderly and demented subjects has been well documented;[132,133] indeed their numerical frequency within the tissue beyond a certain threshold value has formed part of some present day neuropathological criteria for the diagnosis of AD[134,135] However, it has also become equally clear that these very same structures commonly occur, albeit usually but not necessarily always, at a much lower numerical density, in the brains of many elderly but non-demented persons.[32,136-142]

It is still uncertain whether these pathological structures (plaques) reflect, whenever or wherever they occur, the tissue consequences of a time-dependent process that over many years progressively and adversely affects the structural and functional integrity of nerve cells (i.e. biological aging); a process which AD might represent the ultimate or exaggerated manifestation thereof. Alternatively, they may be the pathological signatures of a specific disease process of later life (i.e. pathological aging), which when they occur in high numbers is taken, by convention, to reflect the presence of AD, but when seen in lesser quantities, in normal individuals, might indicate a predestination to have eventually developed the full clinical and pathological features of that disease had the possessor lived longer.

2.4.1. Morphological Characteristics of Plaques

2.4.1.1. Histological features

SP are complex foci of damage to, and destruction of, brain tissue. Each contains a multiplicity of cellular (neuronal and glial) and non-cellular (amyloid and other proteins) elements. Consequently, they can present in the tissue with a spectrum of morphological appearances, sometimes with wide ranging numerical densities, according to which component is being sought and the particular staining procedure being used.

Routine histological staining by haematoxylin and eosin or cresyl violet (Nissl staining) fails, for the most part, to pick them out from the 'background' tissue though occasionally the 'cored' type of plaque can be discerned on account of the large quantity of compact amyloid present. The periodic-acid Schiff reaction can be used with more success, though again the efficacy of this can be highly variable from case to case. Early neuropathological descriptions relied upon silver impregnation methods and even by the time (1907) of Alzheimer's original description of the disease that was later to bear his name many variations on this methodology were in use for the detection of SP (and NFT). The 'classic' silver impregnation methods used today, which have been developed and perfected over many years, such as the Bielschowsky, Bodian and Palmgren techniques, reveal the 'classical' or 'neuritic' type of SP. (Fig. 2.1) This is so called because of the presence of filamentous structures within the plaque confines relating to altered (damaged) nerve terminals which surround the central core of proteinaceous material (amyloid). (Fig. 2.1a) Plaques with this appearance are roughly circular in profile and are perhaps even spherical in shape. Sometimes the plaque will lack this typical appearance, particularly when only the edges have been sectioned, and is seen as an irregular tangle of filamentous and granular material without an obvious central core. (Fig. 2.1b) When plaques are numerous they may form large, irregular conglomerates. Variations in this form where neurities are numerous but amyloid is sparse, and vice-versa, have been termed primitive and burnt-out (compact) plaques respectively.[143]

The amyloid within these neuritic plaques is present in a β-pleated sheet structure which is readily detected by dyes such as Congo red, where it appears as a red-green birefringence when viewed under polarized light, or Thioflavin S as a bright yellow fluorescence. Such dyes show many plaques to have a dense central core of amyloid, though in others the amyloid is more diffusely and evenly dispersed.

2.4.1.2. The size of plaques

Size frequency distributions of plaque diameter[144,145] show a positive skewed distribution with few deposits in the smallest (<10μm) size class and most deposits being in classes between 20μm and 40μm, with the frequency of larger deposits progressively declining as the diameter increases up to 200μm. This kind of distribution approximates a negative exponential model. Such a size frequency pattern is not however a reflection of sectioning artefacts of objects of fairly even size, nor is it likely to represent a preferential degradation and disappearance over time of the larger plaques which may have become 'overgrown' in size, even though it is known that the total amount of amyloid in the tissue represents a balance between competing rates of formation and dissolution.[146] Plaque size may be related to the size or frequency of those tissue elements (cells) responsible for its production;[145,147] the more cells present, the larger the plaque that is produced. In late stages of plaque evolution (i.e. compact or burnt out plaques) some condensation of amyloid and other elements might occur.

2.4.1.3. Amyloid plaques

In their 'classical' form, plaques thus contain a core of acellular material, now known as amyloid β protein (Aβ), surrounded by, and mixing in with, the neuritic elements and the processes and cell bodies of glial cells, both astrocytic and microglial. Immunohistochemistry, using antibodies produced against either the isolated Aβ protein or synthetic peptides of varying lengths within the overall Aβ sequence (Aβ protein was first purified from brain tissue and sequenced in 1985 by Masters et al,[148]see later) has now shown that, even in the cerebral cortex,[139,149-157] plaques with this classical description form only a minority type (Fig. 2.2a) and most appear as 'diffuse' plaques in which the Aβ is distributed in a more uniform and granular manner. (Fig. 2.2b) Such diffuse plaques occur in the cerebral cortex most commonly and at greatest densities within the association areas of the neocortex, particularly the frontal, temporal, cingulate, insular and occipital regions; plaques are less frequently in the primary motor and sensory areas. Archicortical areas such as the hippocampus (particularly the CA1 and CA4 regions and subiculum) and amygdala (cortical and medial nuclei) are also heavily involved). This latter type of

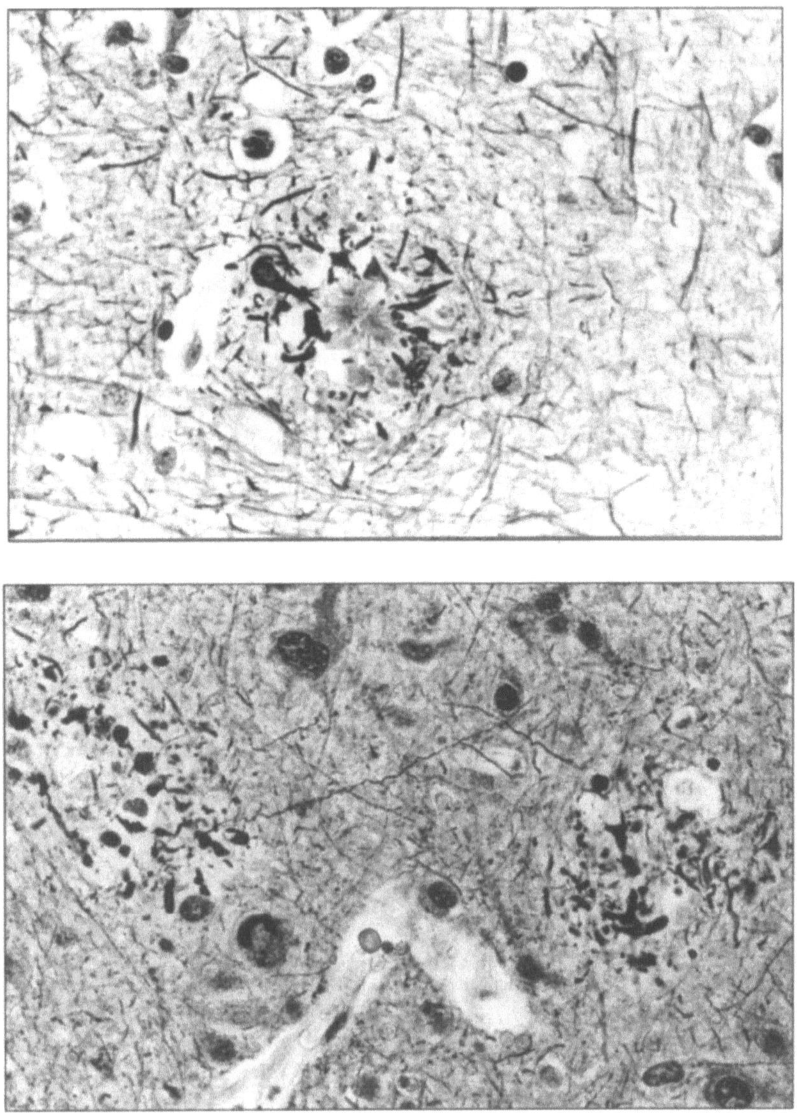

Fig. 2.1. Neuritic plaques in the cerebral cortex. In some plaques a distinct amyloid core is visible (a, top), whereas in others (b, bottom) the plaques appear as irregular masses of filamentous and granular material without a distinct core being present.

plaque is however not only common within these regions of the cerebral cortex[139,149-157] but is widespread, particularly in AD, throughout the cerebellum, (Fig. 2.3) corpus striatum, hypothalamus, thalamus and brain stem.[151,152,158-167] Such plaques do not occur in the deep grey matter nuclei of the cerebellum or brainstem nor in the spinal cord grey matter. They are likewise absent from white matter regions. Because the Aβ within these diffuse plaques is poorly organized, if at all, into the β-pleated sheet structure typical of Aβ within the classical plaques, they are not detectable with Congo red or Thioflavin S. Furthermore, such plaques are either minimally, or not at all, associated with neurites and hence do not stain up in standard Bielschowsky or Bodian-type preparations. It is probably for these reasons that the diffuse plaque has gone unrecognized for so long. Normal appearing axons often traverse such plaques and not infrequently the amyloid may enclose neuronal perikarya[147,168-172] and blood vessels,[173,174] particularly when present as a large conglomerate mass.

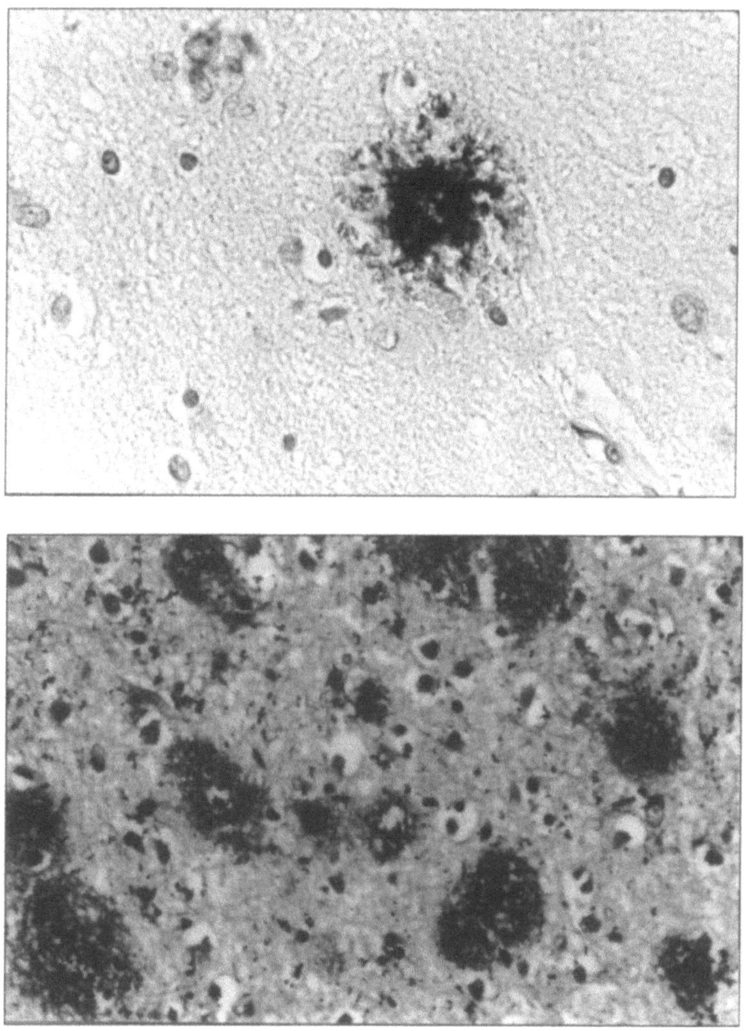

Fig. 2.2. Cored plaques (a, top) and diffuse plaques (b, bottom) in the cerebral cortex as shown by anti-Aβ immunostaining.

2.4.1.4. The prevalence of amyloid plaques in the elderly

This sort of deposition of Aβ in the form of diffuse plaques, principally within the grey matter of the cerebral cortex but also elsewhere in the brain, occurs in many old people with perhaps as many as 70% or more of those over 80 years of age showing some degree of plaque formation.[139,149-157] However, not all elderly people seem to accumulate such plaques in their brains no matter how long they live; an absence of amyloid deposits in many octogenarians and even in some centenarians is documented[142,156,157] though others[154] have suggested a universal involvement in subjects 100 years of age and older. Such observations imply that the potential to deposit Aβ in later life is commonplace and only small or subtle differences in biological factors that promote this deposition may be required to ensure that this strong inherent, and possibly inherited, tendency takes place.

2.4.1.5. Other plaque components

The amyloid protein is closely associated with many other molecules, this in general being so whether it is present in the morphological form of a cored, neuritic or diffuse type of plaque. Principal among these associated molecules are the apolipoproteins (Apo) E, A1 and J, heparan sulphate proteoglycan (HSPG), complement factors, amyloid P component and proteinase inhibitors like α-1 antichymotrypsin

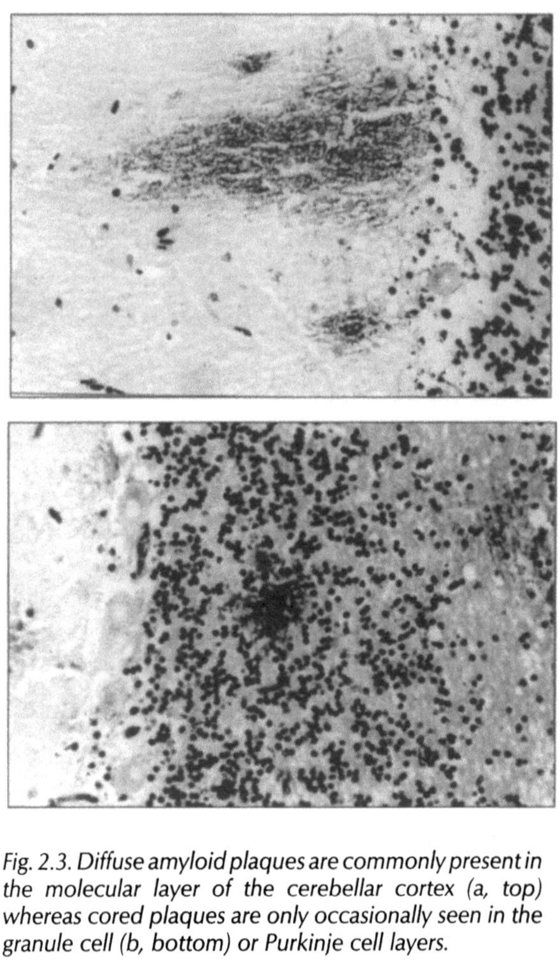

Fig. 2.3. Diffuse amyloid plaques are commonly present in the molecular layer of the cerebellar cortex (a, top) whereas cored plaques are only occasionally seen in the granule cell (b, bottom) or Purkinje cell layers.

(ACT) and (α_2-macroglobulin. These molecules appear to be tightly bound to the Aβ peptide, existing within the same anatomical locality as the amyloid fibrils, and their role in plaque formation will be discussed later. Another characteristic of plaques is the extracellular presence of the acid hydrolase enzymes, cathepsin D and β-hexosaminidase though the significance of these enzymes within plaques is not known. Some workers[148,175-177] have documented the presence, and highlighted the pathogenetic importance, of inorganic elements, such as aluminum and silicon (perhaps combined into aluminosilicates), within the amyloid cores of plaques. Others[178,179] however, have not been able to confirm these findings, explaining their presence as contaminants arising during tissue preparation and examination.

2.4.1.6. The ultrastructural appearance of plaques

Although early ultrastructural studies of diffuse plaques claimed[180-183] that the amyloid peptide existed as an amorphous non-fibrillary material termed "pre-amyloid" later studies on well-preserved biopsy or autopsy tissues[184,185] have shown that the Aβ is always present in fibrillar form. Although this is best seen within the cored regions of the so-called classical plaques, even within diffuse plaques the amyloid is still fibrillar with the fine bundles of amyloid or single fibrils being loosely interwoven between neuronal and neuritic elements and the cell bodies and processes of glial cells.[184,185] (Fig. 2.4) Although fibrillar, the Aβ within these diffuse plaques is presumably insufficiently organized into a β-pleated configuration as to make them detectable by Congo red or Thioflavin stains.

Each single unbranched amyloid fibril measures 5-10nm in diameter and longitudinally narrows every 30-40nm through a helical winding of the pair of filaments from which it is composed. The fibrils run in haphazard arrays sometimes radiating from an obvious central core of similar material. All such amyloid is present in the extracellular space. Surrounding, or often intermingled with, the amyloid fibrils are the dystrophic neurites (Fig. 2.4) and glial cell bodies and processes, particularly those of microglia (see later). The neurites are distended (being 5-10 times the normal diameter of nerve terminals) and contain numerous dense lamellar and multivesicular bodies, these probably originating from degenerate mitochondria. (Fig. 2.4) Many neurites also contain variable quantities of the paired helical filaments (PHF) which are present also within neuronal perikarya, as NFT, and in the proximal dendrites or dendritic sprouts, as neuropil threads.[186] Many of the altered nerve endings within plaques are considered to be axonal in origin, though a substantial dendritic contribution is also likely. The 'blackening' of these processes by silver impregnation (Fig. 2.1) is due to the presence of the paired helical filaments. PHF contain the microtubule associated protein, tau, and the 'stress protein', ubiquitin and as described later antibodies directed against these particular molecules can be used to equal effect to detect neuritic plaques by immunohistochemical procedures.

2.4.2. The Evolution of Plaques

It is now widely accepted that plaques are dynamic structures that undergo substantial structural changes over the many years they are present in the tissue and because diffuse Aβ plaques numerically far outweigh the neuritic type it has been assumed that the former may represent the morphological forerunners of the latter. Studies in Down's syndrome (DS), where development of the pathological changes of AD by middle age is entirely predictable,[187] substantiate this assumption. In most younger persons with DS, diffuse Aβ deposits are detectable within the cerebral (and especially the temporal) cortex by 30 years of age and increase in number and size thereafter.[150,166,172,187-192] These diffuse deposits subsequently acquire many of the non-cellular (Apo E,HSPG, complement factors, ACT,[191] and cellular (microglial cells)[172,189] characteristics of the later

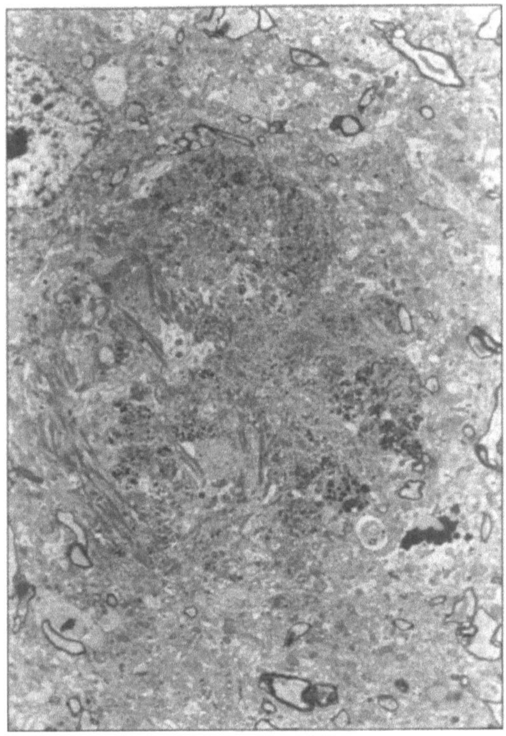

Fig. 2.4. Electron micrograph of a diffuse plaque showing bundles of amyloid fibrils intermingled with dystrophic neurites containing many electron dense lamellar and multivesicular bodies and some paired helical filaments. Astrocytic processes are sparsely present.

appearing cored plaques. Neuritic changes, NFT and astrocytes do not appear in DS until some 10-15 years after the initial Aβ deposition.[172,191] These kinds of data have challenged original postulates[143] that the abnormal neurite was the initial structural manifestation of plaque formation and that the earliest plaques (so-called primitive plaques) appeared as a small cluster of distended neurites, of axonal origin, with few or no associated amyloid fibrils. Instead, it is now widely believed that the process of plaque evolution begins with the deposition of Aβ within the tissue in the form of a diffuse plaque and that at least a proportion of these plaques evolve into the neuritic and cored types of plaques. It still however remains possible, though perhaps unlikely, that the diffuse and the neuritic types of plaques have separate origins and evolve independently. In this respect observations that the diffuse amyloid plaques of the cerebellum and the corpus striatum never evolve into congophilic cored or neuritic-type plaques,[166,167] nor do the diffuse-type plaques in the cerebral cortex of aged dogs[193,194] and primates,[195] may reflect important differences. Consequently, much effort has been exerted and progress made in recent times in characterizing the biochemical and molecular features of Aβ and its cellular sources and investigating factors which might influence its deposition within the brain tissue.

2.4.3. The Composition of Amyloid β Protein

2.4.3.1. Chemical analysis

About 25% of the dry weight of the plaque core is protein and of this 70% is Aβ.[196] These highly insoluble Aβ fibrils within plaque cores can be readily separated and purified from the rest of the brain tissue and their peptide structure analyzed following solubilization in harsh detergents. Initial studies[148] showed that the Aβ extractable from compacted plaque cores was composed of a peptide, of molecular mass around 4.2KDa whose actual length varied within a standard 39 to 42 or 43 (rarely) amino acids. Some amino-terminal heterogeneity was noted, though no carboxy-terminal species extending beyond amino acid 43 were noted. The name A4 peptide was given[148] in respect of its molecular mass. In 1986, Selkoe and colleagues[197] confirmed these findings, giving the name β-peptide in recognition of its physicochemical properties (i.e. β-pleating). Presently, these two names came to be combined into β/A4 peptide, this nowadays having been most commonly replaced by the abbreviation Aβ–the nomenclature adopted here.

Later analyses[198-206] substantiated these early findings and have confirmed that the insoluble Aβ within plaque cores is made of species of variable length, $A\beta_{x-42}$ (where x is mainly asp-1 or leu-17), though numerous other commencing amino acids are present including blocked peptides starting with glu-3 converted to pyroglutamate 3[202] or with isomerized or racemized asp-1.[201] Analysis of tissue regions rich in diffuse plaques[193,203] confirms the existence of Aβ as $A\beta_{x-42}$, but here $A\beta_{17-42}$ appears to be a prominent species. Interestingly, these chemical approaches have shown there to be little carboxyl-terminal heterogeneity with only relatively minor amounts of peptide terminating at either val-39 or val-40, rather than ala-42 or thr-43, being detected. Formerly, it was considered that the presence of species ending at val-40 might represent sample contamination by those Aβ deposits present within the walls of blood vessels, this vascular amyloid being known from the early work of Glenner and Wong [207] and others[198,206,208-210] to share the same initial 39 or 40 amino acid sequence as plaque amyloid. Later analyses of plaque Aβ[196,200,201,203,205] were mindful of this and excluded this possibility. It is of note that cerebrovascular amyloid, as well as containing much $A\beta_{1-40}$ or $A\beta_{1-39}$, also contains, but in lesser amounts, $A\beta_{1-42}$[196,200,206,209-211] Curiously much less amino-terminal heterogeneity seems to be present in vascular Aβ compared to plaque Aβ with most species commencing at asp1 or ala2.

2.4.3.2. Immunohistochemical analysis

Immunohistochemical studies,[189-191,212-220] using end-specific antibodies that detect amino acids val-40 and ala-42 respectively of the Aβ sequence, confirm these biochemical data and further show that in AD,[212-216,218-220] DS[189-191] and in the mentally able elderly[217,220] the major Aβ peptide species within cored (neuritic) type plaques of the cerebral cortex terminates at ala-42 (i.e. $A\beta_{42}$) (or thr-43 ($A\beta_{43}$) to a minor extent) with many fewer peptides terminating at val-40 ($A\beta_{40}$) being present. Furthermore, the diffuse type of plaque contains solely $A\beta_{42}$, irrespective of whether it occurs in AD,[212-216,218-220] DS[189-191] or in mentally able old persons,[217,220] or even in aged animals.[193] Nor does it matter whether the diffuse plaques are in the cerebral cortex, the cerebellum, the corpus striatum or the hypothalamus, in all these regions Aβ always and only occurs as $A\beta_{42}$[167] Hence, in every situation $A\beta_{42}$ is the predominant, and probably also the initial, Aβ species deposited in plaques of all morphological forms. $A\beta_{40}$ is present in only a subset of plaques, even in late stages of pathology, and then chiefly in those of the cored variety that occur mainly in the cerebral cortex. In such cored plaques it is notable that as the amount of $A\beta_{40}$ increases so that of $A\beta_{42}$ diminishes.[189-191] This suggests that $A\beta_{42}$ initiates plaque formation in a diffuse and non-aggregated form but later becomes aggregated and acts as a nidus for the deposition of $A\beta_{40}$. However, it is also possible that the apparent reduction in $A\beta_{42}$ may be due in part to a masking of its epitope by large amounts of deposited $A\beta_{40}$, or that some of the initial $A\beta_{42}$ deposited might undergo carboxy-terminal trimming to $A\beta_{40}$. Even more extensive proteolysis is perhaps possible leading eventually to a gradual removal of plaque amyloid. [146] Conversely, cerebrovascular amyloid is chiefly composed of $A\beta_{40}$, with lesser amounts of $A\beta_{42}$ being present,[190,212-214] but shows strong labeling for Aβ species starting at asp-1 and pyroglutamate-3. The amino-terminal heterogeneity in plaques apparent in protein chemistry is also seen by immunohistochemistry.[189,220,221] Both diffuse and cored plaques, in AD, DS and in the non-demented elderly show strong immunoreactivity for Aβ species commencing at asp-1, either isomerized or racemized, and pyroglutamate-3.[189,220,221] Such findings are again irrespective of whether the plaques occur in the cerebral cortex or in the cerebellum, corpus striatum or hypothalamus. Interestingly, even though chemical analyses[193,203] suggest the presence of much $A\beta_{17-42}$, particularly in diffuse plaques, immunohistochemical approaches[220] do not confirm this; findings are that Aβ species starting at pyroglutamate-11 and leu-17 are virtually absent from diffuse plaques and form only a minor component of cored plaques in any brain region.

Hence, immunohistochemical data suggest that Aβ species, that initially deposit as diffuse plaques, begin with asp-1, with or without structural modifications such as isomerization or racemization, as well as pyroglutamate-3 and preferentially terminate at $A\beta_{42}$ rather than $A\beta_{40}$. Since biochemical data[202,221] suggest that Aβ peptides beginning at pyroglutamate-3 predominate in amyloid plaques it is possible that the major initial species deposited is $A\beta_{3-42}$ with glu-3 being converted to pyroglutamate-3. Whether this particular Aβ species represents the earliest form of Aβ is not clear; such forms could be $A\beta_{1-42}$ which is then cleaved by two residues and subsequently 'pyroglutamated' to give this modified and depositable species. Alternatively, the initial Aβ peptide secreted in soluble form and then deposited could be $A\beta_{1-42}$ which might then undergo modifications involving isomerization, racemization and pyroglutamation following its deposition. In any case these amino-terminal modifications of Aβ will hamper proteolysis of Aβ by aminopeptidases and thereby contribute to its stability within the tissue.

2.4.3.3. The source of Aβ

Aβ deposits vary much in size and shape.[144,145] Diffuse plaques are often large, irregular and frequently merge into conglomerate deposits. Cored plaques are usually smaller, rounded in shape and form discrete deposits. The factors that determine these ultimate morphologies are unknown and may relate to the actual cellular source of the Aβ. It is well known that diffuse plaques are topographically closely associated with neuronal perikarya[147,168-172] as are cored plaques with microglial cells[172,189,222,223] or blood vessels,[173,174] though whether these represent more than chance associations between geographically common structures is not clear. Moreover, plaques are more common in the depths of sulci than in the crests of gyri[147,224,225] but again the reason for this is uncertain. It is possible that the initial deposition of Aβ, as a diffuse plaque relates to a release of this from nerve cells while development into what are widely thought of as more mature, cored plaques rests with the later presence of glial cells. Whether plaques originate and grow from a single point source, be this neuronal or vascular, is not known; certainly the rounded appearance of some plaques implies that this might be so, though the irregular appearance of others argues against this unless it is accepted that these represent fused multiple nearby point sources. Nonetheless, all of the earliest diffuse plaques seen in non-demented old people[217] or in young subjects with DS[188-190] seem to have a discrete outline and do not appear to be composed of multiple punctate deposits, suggesting that the plaque may indeed have originated and grown from a single point source in the neuropil. What this point source might be is unknown but a neuronal cell body or process, which may have been damaged or ruptured [226] allowing Aβ to 'escape' and become locally concentrated, is possible. Although a hematogenous source for Aβ was at one time a popular concept, this is now considered by most workers to be an unlikely origin not only of plaque amyloid but probably also that of cerebrovascular amyloid (see later).

2.4.4. Amyloid Precursor Protein

The sequencing of the Aβ peptide in 1985[148] quickly led to a cloning of the gene encoding this particular peptide. It turned out, as might have been anticipated, that Aβ is not itself a native genomic product but is derived through the proteolytic (catabolic) breakdown of a much larger precursor, this being known as the amyloid precursor protein (APP).

2.4.4.1. The APP gene

The gene encoding APP was discovered in 1987 and found to be located on the long arm of chromosome 21.[227-229] The open reading frame (coding region) consists of 19 exons and extends over some 400 kilobases (kB).[230-232] The region of the gene which encodes the Aβ sequence extends over parts of both exons 16 and 17. Hence, although alternative splicing of a single common mRNA transcript can produce multiple isoforms,[227,233-238] it is clear that because the Aβ sequence lies across two neighboring exons it cannot be produced through any aberrant or minor splicing mechanism. All APP isoforms are transcribed from the same promoter, but each is differentially expressed according to cell or tissue type. The APP gene is expressed in all major tissues with highest levels of expression occurring in brain and kidney and lowest, and barely detectable, levels of expression in liver. In brain, APP mRNA is abundant in neurones and endothelial cells[239-242] but is also present in glial cells, both in astrocytes[243,244] and microglial cells,[245,246] especially when activated.

Multiple protein isoforms of APP arise through complicated alternative splicings of the primary gene transcript. Two splicings result in the production of soluble secreted forms of APP, APP_{365},[239] and APP_{563} ,[237] that are carboxyl-truncated versions of APP_{695} and lack the whole of the Aβ sequence along with other parts of the molecule. Other alternative splicings yield eight isoforms, encoding proteins of 677-770 amino acids, according to the presence or absence of exons 7, 8 and 15. Hence, the predominantly transcribed isoforms APP_{695}, APP_{751}, and APP_{770} are all similar in basic structure but are produced by alternative splicing of exons 7 and 8. In APP_{751} and APP_{770} there is an additional 56 amino-acid insert (corresponding to exon 7) similar to a Kunitz-type (serine) protease inhibitor, and in APP_{770} alone there is a further insert (exon 8), of 19 amino acids, similar to the MRC OX2 surface antigen in thymus derived leukocytes.[233-235] More alternative splicing, affecting all three aforementioned principal isoforms, involves exon 15 in which an 18 amino acid sequence is either included or excluded.[236,247] Furthermore, Golde et al[238] have identified yet another splice variant, APP_{714}, which is like APP_{751} but truncated 37 amino acids from its carboxyl terminus. All isoforms are variously post-translationarily N- then O- glycosylated, tyrosine sulphated or phosphorylated[248,249] within the Golgi apparatus to produce a 'family' of secreted gene products of Mr 110-135. [250,251] Newly formed APP is then transported by axon flow[252] to the cell terminals where some is inserted as a full-length protein into the synaptic membrane[253] though other molecules undergo proteolytic processing along the way to produce secreted forms of APP[250,251] (see later). The longest form of Aβ ($Aβ_{1-42(43)}$) corresponds to amino acid residues 672 to 714(5) of the APP_{770} sequence.

In brain, APP_{695} is the principal isoform transcribed and contained within nerve cells, particular those of the hippocampus and association cortex,[234,238-241,254-256] whereas APP_{751} and APP_{770} are transcribed fairly uniformly across the brain and seemingly mostly by glial cells, especially microglial cells.[234,239,244-247,257] These latter species are also transcribed within peripheral tissues, particularly by cells of the kidney.[238] Non-neuronal cells within the brain and periphery produce large quantities of the KPI (exon 7) containing, exon 15-free, isoforms APP_{762} and APP_{733}[236,247] though the significance of this is not known.

It is now known that APP is not a 'unique' protein but belongs to a 'multigene' family of which related members are the amyloid precursor-like proteins APLP1 and APLP2. [258-261] APLP1 is encoded by a gene on chromosome 19 and APLP2 by one on chromosome 11; their mRNA transcripts undergo similar alternative splicing of a KPI domain and of an exon similar to APP exon 15.[247] The APLPs mainly differ from APP in lacking exons 16 and 17 of the APP gene. Although proteolysis of APLPs cannot result in the formation of Aβ they may still play a part in determining the extent of Aβ deposition by possibly competing with APP for similar processing enzymes, in similar cellular compartments, and in so doing they might influence how much APP is degraded along Aβ versus non-Aβ forming pathways.

Changes in APP expression favoring a relative increase in the expression of APP_{751} and APP_{770}[262,263] seem to occur with aging. Since these splice variants are mostly produced by glial cells this change may relate to age-dependent changes in the number, or activity, of astrocytes or microglia (see later), rather than indicating a change in the pattern of nerve cell expression in later life.

The absence (except for perhaps the skin)[264,265] of Aβ outside the brain, despite the almost universal tissue expression of the APP gene and the presence of APP itself, suggests a particular idiosyncrasy regarding either the way brain tissue catabolises APP or how its breakdown product, Aβ, is cleared from the brain.

2.4.4.2. The biological function of the amyloid precursor protein

Both APP and APLP can exist as transmembrane proteins with a large extracellular domain, a membrane-spanning domain and a short carboxy-terminal cytoplasmic domain. In APP the Aβ sequence lies partially across the membrane with the initial 28 amino acids of the sequence lying within the extracellular domain and the remaining 12-15 amino acids being embedded within the membrane. (Fig. 2.5) The biological function of membrane bound APP still remains uncertain, though roles as a G-protein receptor,[227,266] a cell adhesion molecule involved in tissue maintenance or repair through the modulation of cell-cell or cell-matrix interactions,[267-272] a growth factor,[273,274] a protease inhibitor,[235,275] or an agent in signal transduction[276,277] have all been proposed. Platelets produce much $APP_{751/770}$ and after processing secrete this into plasma,[278] in the form of the inhibitor of coagulation factor XIa,[279,280] following stimulation by thrombin or collagen-agents that induce platelet degranulation. This secreted APP is also identical to protease nexin II, a protease inhibitor secreted by fibroblasts.[275,280] Hence, platelet derived APP may function in terms of wound repair or in the regulation of the coagulation cascade.

The expression of APP, or its cellular distribution, changes following cell or tissue injury involving diverse etiologies such as ischemia/hypoxia,[281,282] trauma,[283] or toxins.[284,285] These changes in APP may reflect a response by pre-existing cells at that site (e.g. neurones), though reactive changes in cells migrating to, or proliferating at, the point of injury (e.g. astrocytes, microglia) might add to these. One unifying explanation for these observations could involve the secretion of interleukin-1 (IL-1) by damaged or reactive cells since this is known to increase APP expression or processing in sensitive cells like neurones[286-288] and alter the local population of reactive cells, like astrocytes, at the site of injury.[289] Correlations between IL-1 secretion and APP expression have been noted in other pathological situations such as temporal lobe epilepsy or head injury.[290] Hence, increased APP expression may form part of the cell or tissue response to "stresses" of diverse kinds and APP may play a vital role in adapting to, or compensating for, changes in tissue integrity.

Reductions in the amount of secreted APP occur at the expense of a preferential catabolism of APP along routes leading to Aβ formation (see later). This loss of secreted APP may itself be a pathogenic event which acts in concert with, or even instead of, Aβ deposition. Hence, because increases in Aβ formation occur at the

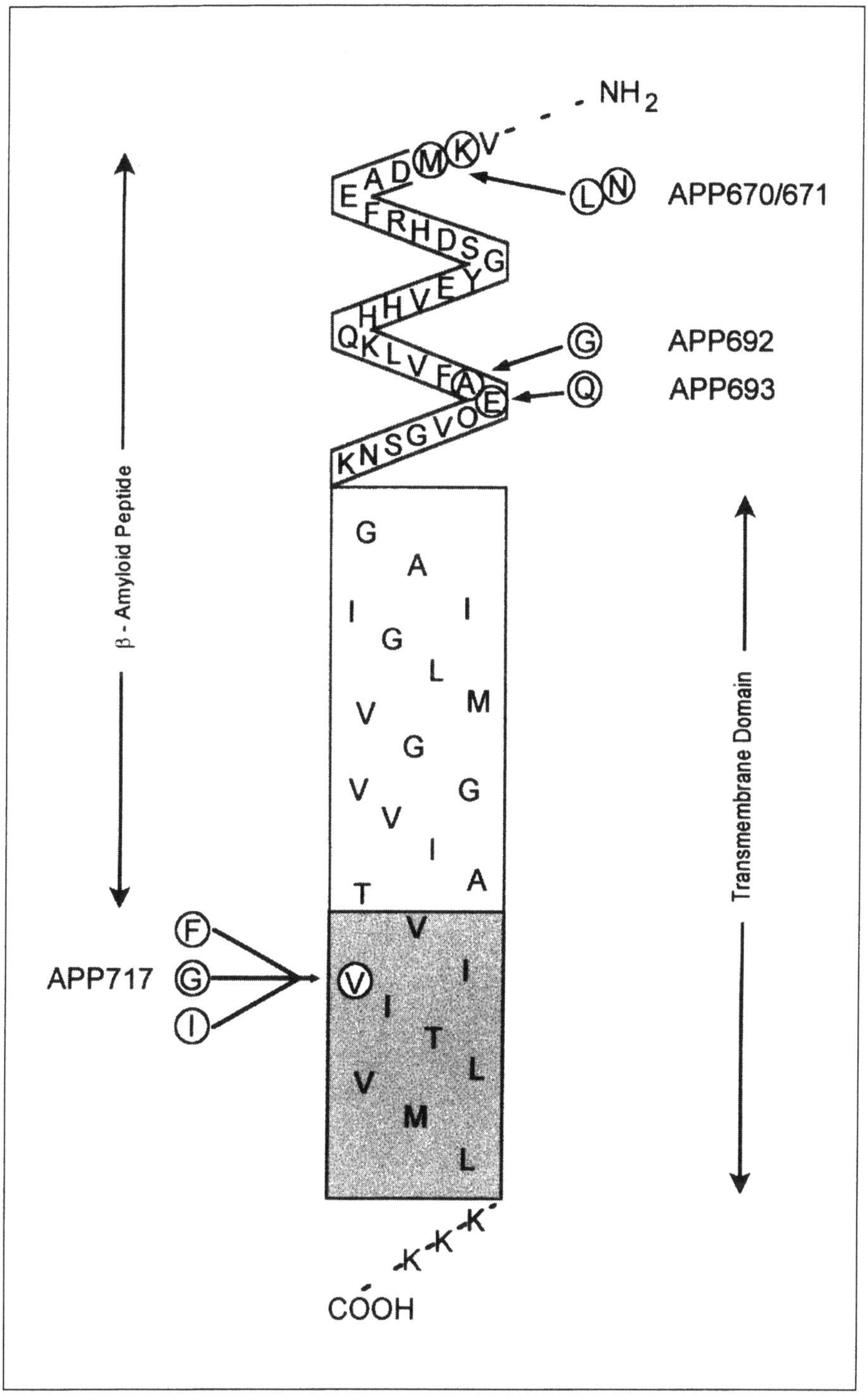

Fig. 2.5. Representation of the amyloid precursor protein showing the position of the Aβ peptide sequence lying within, and outside of, the cell membrane when the molecule is so inserted. The sites of the various point mutations associated with AD or hereditary cerebral hemorrhage with amyloidosis (Dutch variant) are indicated, together with their respective amino acid substitutions. Adapted from Clarke and Goate (1993). Arch Neurol 50: 1164-1172.

expense of secretion of APP, the deposition of Aβ may simply represent a marker of impaired APP metabolism or an APP function necessary for neuronal survival. In this way Aβ would act as a pathological 'hallmark' of disease, but would not necessarily be itself pathological.

2.4.4.3. The processing of amyloid precursor protein

As mentioned, Aβ is a proteolytic fragment of APP comprising the initial 28 amino acids immediately amino-terminal to the membrane surface (when inserted therein) plus the first 12-15 residues of the transmembrane domain. Hence, full-length Aβ can only be produced through the catabolism of APP and must be formed as a result of at least two separate proteolytic events. (Fig. 2.6) In one, cleavage of the APP molecule has to occur on the amino-side of the Aβ sequence; in the other, this must take place on the carboxyl-side. Neither of these sites appears to represent a consensus sequence recognized by any currently known protease.

However, the major route for the processing of APP, affecting at least half of the total newly produced full-length APP,[248] involves the (putative) enzyme designated α-secretase–a metallopeptidase that cleaves the APP molecule around, though principally between, lys-16 and leu-17 of the Aβ peptide sequence.[291-294] (Fig. 2.6) Following this, lys-16 is removed by an exopeptidase.[294] The effect of α–secretase enzyme action is to produce these amino-terminally truncated secreted forms of APP[248-251,275,278-280] leaving behind an intracellular carboxy-terminal fragment (CTF) of about 8 KDa consisting of the last 82 amino acids of the APP sequence.[295] The amino-terminal portion (Mr 105-124 kDa) is secreted into the extracellular space and is detectable in serum and or cerebrospinal fluid.[248-251,275,278-280] All KPI and non-KPI forms of APP can undergo this processing route, the secreted APP_{770} product having been identified in serum as protease nexin II, an antithrombin factor of platelets.[275,278-280] (Fig. 2.6)

The 8 KDa CTF left behind after α-secretase action is degraded by another putative enzyme, designated γ-secretase, which acts mainly at around amino acids 40 (γ_1-secretase) or 42 (γ_2-secretase) of the Aβ sequence but again can function at adjacent sites, to produce Aβ fragments starting at around $A\beta_{17}$ and terminating at $A\beta_{39-43}$, these having a molecular mass of about 3 KDa. Such fragments have been (collectively) termed P3 peptide.[296] (Fig. 2.6)

α-Secretase mediated secretion of APP seems to be regulated by the phosphorylative activity of phosphokinase C,[297-299] this in turn being driven by acetylcholine, via its action on muscarinic receptors,[288,300] or by IL-1.[286-288] Inhibitors of protein kinases have the opposite effect. Thus an increase in net phosphorylation will increase APP and P3 peptide secretion but decrease Aβ production, while a decrease in net phosphorylation increases Aβ production and lowers P3 formation by reducing the secretion of APP thereby diverting APP along competing catabolic pathways that favor the generation of Aβ instead of P3 peptide.

This competing mechanism involves a further enzyme, designated β-secretase. This also degrades APP but in this instance cleavage occurs close to the amino-terminus of the Aβ sequence (i.e. at or around amino acid 672 of the APP_{770} sequence). This enzyme acts principally upon APP_{695}, the neuronally produced isoform, and again produces a large (93KDa) secreted fragment but retains an approximately 11KDa CTF of 99 amino acids that starts with, and contains the whole of, the Aβ sequence.[301,302] (Fig. 2.6) Such a fragment clearly has amyloidogenic potential and this also acts as a substrate for α-secretase which again removes the carboxy-terminal portion to produce instead (up to full-length) Aβ molecules of molecular mass 4KDa. (Fig. 2.6) Like P3 peptide these Aβ molecules are then secreted from the cell into the extracellular fluid in **soluble** form.[296,303-307] Characterization of such secreted peptides, in cell cultures, shows

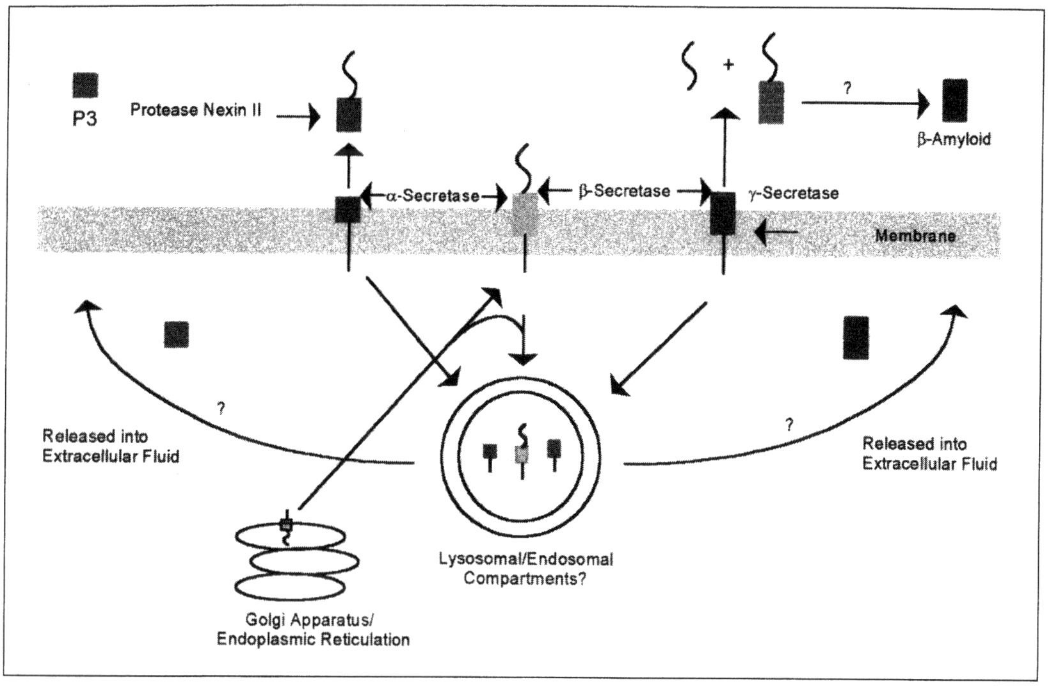

Fig. 2.6. Schematic representation of the intracellular processing of the amyloid precursor protein, APP. Newly formed APP is processed (glycosylated, phosphorylated, sulphated) in the Golgi and transported to the cell surface. On the way there, and at the cell surface, some APP is cleaved, by α-secretase, into secreted APP (protease nexin II in platelets) and P3 degradation product. Other APP molecules are processed via β and γ-secretases into Aβ. Membrane bound, carboxyl-terminal fragments of APP, and reinternalized full-length APP, can be degraded in lysosomes and this route may represent a further source of secreted Aβ. Adapted from Clarke and Goate (1993). Arch Neurol 50: 1164-1172.

them to be a heterogeneous mix of which the major Aβ peptide is $A\beta_{1-40}$, with lesser, but significant (approximately 17%) quantities of $A\beta_{1-42}$ being present along with minor species such as $A\beta_{1-28}$, $A\beta_{1-33}$, $A\beta_{3-34}$, $A\beta_{1-37}$, $A\beta_{1-38}$, and $A\beta_{1-39}$ and various other amino-terminally truncated forms commencing at val-3, phe-4 and glu-11. Similar secretory patterns seem to exist in vivo, since these same Aβ species are detectable within human cerebrospinal fluid.[302,308,309]

Analysis of the APP sequence shows that the cytoplasmic domain bears the consensus sequence, Asn-Pro-Thr-Tyr, for endocytosis via clathrin-coated pits. Hence, it seems that non-secretory pathways may also exist for APP breakdown, these probably occurring within the late Golgi or early endosomal compartment of the cell.[295,306,309-311] Here, full-length APP molecules, presumably re-internalized from the cell surface or extracellular space, are catabolized into a family of CTF of 8-12Kda, these corresponding to the last 82, 89, 95 and 99 amino acids of the APP sequence. The largest of these (at 11.4KDa) again contains the entire Aβ sequence and is hence potentially amyloidogenic. These fragments could be released from the cell directly or following further processing via γ-secretase activity into Aβ or P3 peptides. (Fig. 2.6)

Therefore, APP is cleaved in secretory pathways, either at the amino terminus (by β-secretase) or mid-sequence (by α-secretase) to produce CTF of 99 and 82 amino acids respectively; these are subsequently acted upon by γ-secretase to produce an array of secreted peptides of which P3 (e.g. $A\beta_{17-40}$, $A\beta_{17-42}$) and Aβ ($A\beta_{1-40}$,

$A\beta_{1-42}$) are the principals along with other minor species derived through alternative sites of α–, β– or γ–secretase action or by subsequent proteolytic "trimming" following release from the cell. Most of the peptides produced in this way terminate at val-40 (i.e. $A\beta_{17-40}$, $A\beta_{1-40}$) with lesser amounts terminating at ala-42 (i.e. $A\beta_{17-42}$, $A\beta_{1-42}$).

2.4.4.4. The cellular sites of APP processing

The intracellular sites of APP processing are not fully known and the routes for APP trafficking through the cell are complex. (Fig. 2.6) It seems likely that proteolysis, involving the various secretase enzymes, can occur at several distinct subcellular locations. Following translation of the spliced gene products APP molecules undergo post-translational modifications in the Golgi apparatus[248,249] while en route to the cell surface (cell terminals). Some APP molecules may proceed in full-length form for insertion as such into the cell membrane while others are cleaved, either in late Golgi compartments, or at the cell surface,[312] by α-secretase to produce secreted APP,[291-294] and P3 degradation product following γ-secretase action.[296] Also, within the late-Golgi,[312] β-secretase can compete with α-secretase for the same pool of APP molecules but in this instance Aβ will be formed as the degradation product. Finally, full-length, membrane bound APP, or extracellular APP that has escaped membrane insertion can be re-internalized into late endosomal, or even lysosomal, compartments for β– and γ-secretase action and potential formation of Aβ.[295,306,309-311]

Therefore, although the actual deposition of Aβ as plaques within the brain parenchyma or in blood vessel walls is considered to be a pathological event, the cellular mechanisms whereby Aβ is produced are not themselves pathological. The normal processing of APP will result in the production and release of the soluble peptide species Aβ and P3, the former having the potential to aggregate into amyloid plaques, the latter to become incorporated as a part of such tissue deposits. It is clear that P3 and Aβ are waste products of APP that are actively produced and secreted as such; they do not arise in the tissue simply as a result of a passive diffusion into the extracellular space following cell death and disruption of the cell membrane. They are secreted, in life, by brain cells and by cells from non-neural tissues alike; fibroblasts for example produce and release Aβ as part of their normal metabolism. Thus no abnormal pathways are necessary for Aβ production, neither is, as was once thought[227] pre-existing tissue damage a prerequisite. Present evidence favors Aβ being formed and released as a metabolic (waste) product without (known) biological action, eventually accumulating in the tissue as plaques because of its hydrophobicity and propensity to aggregate once a critical molecular concentration is reached. However, recent work [313] has demonstrated vasoconstrictive properties of Aβ on excised aorta, suggesting a possible role in blood vessel physiology.

It will not have gone unnoticed that the principal component of Aβ secreted into culture medium[296,303-307] or cerebrospinal fluid[302,308,309] is $A\beta_{1-40}$ (and $A\beta_{17-40}$) while that which is eventually deposited in plaque form is mostly $A\beta_{x-42}$,[148,189-191,197-206,212-221] where x = 1, 3 or other start points, but probably not 17.[220] Cerebrovascular amyloid, on the other hand, is a mixture of $A\beta_{40}$ (25-40%) and $A\beta_{42}$ (60-75%).[196,200,206,209-211] Thus while Aβ species ending at ala-42 are minor constituents of secreted Aβ they form the major and initial constituents of deposited Aβ. This may be because Aβ peptides terminating at ala-42 have a greater propensity to form fibrils than those ending at val-40[314,315] and in this context the precise site or balance (γ_1 or γ_2) of γ-secretase action may be critical in determining the relative proportions of $A\beta_{40}$ and $A\beta_{42}$ that are eventually secreted and deposited.

2.4.5. Plaque Formation

When first secreted 'in vivo' into the extracellular fluid of brain tissue, Aβ is in soluble form and must therefore be brought 'out of solution' for fibrillization and aggregation to take place. This conversion of soluble Aβ peptides into insoluble Aβ fibrils proceeds by a nucleation process,[314,316] similar to crystallization. During the 'lag-phase' soluble Aβ molecules form dimers and higher aggregates and eventually form a nucleus to "seed" fibril formation. Fibril formation can thus occur spontaneously when the local peptide concentration has reached a critical threshold or it may depend upon changes in the local microenvironment that 'persuade' such an event at the expense of its clearance from the tissue while still in soluble form. The former possibility will depend upon the rate at which soluble Aβ peptides are produced by the brain and the rate at which they are removed from the extracellular space, either 'passively' into the cerebrospinal fluid or 'actively' by extracellular proteolysis or uptake by other (glial) cells with internal catabolism. The rate of fibril formation will depend upon the prevailing soluble Aβ concentration, beginning only when levels of this exceed a threshold but gaining momentum once this is crossed. Also the amino- and particularly the carboxyl-terminal characteristics will dictate the rate of fibrillogenesis, particularly in respect of whether the peptide terminates at val-40 or ala-42.[314-316] Again the rate of accumulation (aggregation) of fibrils will depend upon the balance drawn between rates of production and removal. Local factors may influence aggregation of Aβ into fibrils or promote further seeding. For example, the Apo E E4 protein can reduce the seeding lag-time[316] and promote fibrillization[317] to a greater extent than the E3 isoform. Zinc,[318,319] aluminum and iron,[319] HSPG[320] and ACT[317] can all promote the conversion of Aβ peptides into fibrils. Changes in amyloid P component[321] inhibiting the proteolysis of Aβ may also promote fibrillization. When Aβ is first deposited in diffuse plaques it appears to be loosely bound either to any pre-existing plaque that may be present or to other tissue elements and can be readily extracted; such analysis[322,323] shows this to be exclusively $A\beta_{42}$ (perhaps even $A\beta_{1\text{-}42}$). Indeed solubilizable $A\beta_{1\text{-}42}$ seems to exist within the tissue very early in the course of plaque formation, even before Aβ fibrils become detectable by immunohistochemistry.[323]

2.4.5.1. Fibrillization of Aβ

Numerous early studies (see refs. 315,324 for examples) have demonstrated that many synthetic Aβ peptides of varying lengths with different amino- and carboxyl-termini can form amyloid fibrils in vitro though in such studies other important aspects of tissue amyloid such as its insolubility and fibril stability were not always present. Subsequently, more rigorous analyses have identified the hydrophobic carboxyl-terminal domain as the critical region conferring peptide aggregation,[314] insolubility[324] and β-pleating. Indeed, all of these physical characteristics can be faithfully recreated using a synthetic peptide $A\beta_{26\text{-}42}$[314,324] with a minimal sequence covering residues 25-35.[325]

2.4.5.2. Neurotoxicity of Aβ

Because large amounts of amyloid deposited extracellularly in the primary and secondary amyloidoses affecting the rest of the body induce organ dysfunction it has been argued that extracellular deposits of Aβ might similarly and adversely affect the brain. This argument is furthered by the presence of altered nerve terminals (dystrophic neurites) surrounding the amyloid masses of neuritic plaques and suggests that in some way Aβ might induce, or at least contribute to, the neural damage (neurofibrillary degeneration) taking place at and around the site of deposition.

Yankner[326] was the first, in 1989, to identify neurotoxic properties of externally applied Aβ fibrils upon cultured hippocampal neurones, using the peptide $A\beta_{1\text{-}40}$ at micromolar concentrations. Since then the potential neurotoxic effects of Aβ have been extensively studied both in vivo and in vitro with as many laboratories failing to demonstrate a neurotoxic effect of

Aβ as those that have. These early variations in findings were undoubtedly due to the many different methodologies that were employed at the time, a neurotoxic outcome depending upon variables such as type, strain and maturity of cell culture, type and age of animal species used, purity and concentration of Aβ sample employed, actual peptide fragment used, type of vehicle, presence of traumatic or inflammatory change at inoculation site (in animals), the contribution that other molecules such as HSPG, P component, complement factors, ACT or growth factors, etc might make, the criteria used for a positive detection of Alzheimer-associated pathology, and lastly and perhaps most importantly, the aggregation state of the Aβ fibril. It is now a consensus opinion that the neurotoxicity of Aβ in vitro depends upon its aggregation state and that amino acid residues 25-35 are not only critical for aggregation but also for neurotoxicity. (see ref. 325) However, the precise manner by which such a peptide is neurotoxic is still not understood. It has been suggested that Aβ might (i) induce apoptosis[327-330] or necrosis[331] following oxidative stress,[313,332,333] (ii) increase the vulnerability of neurones to excitotoxic damage mediated by Ca^{2+},[276] (iii) interfere with K^+ channels,[334] (iv) predispose towards injury following hyperglycemia,[335] or (v) induce microglial activation resulting in the production of neurotoxic substances.[336] Hence, at present, it remains uncertain whether extracellular Aβ deposits within human brain tissue are directly responsible for the death of neurones and if so how such a mechanism of cell death might be engendered. It may be that instead of being killed from outside by extracellular Aβ, the presence of this molecule in the brain parenchyma may be an innocent bystander simply representing a pathological marker or by-product of an abnormality of intracellular metabolism, perhaps involving APP, which is responsible for the neuronal degeneration. If this is so, extracellular Aβ may remain a pathological 'irrelevancy' or 'cul-de-sac' within the pathogenetic mechanism. These in vitro studies of Aβ fibrillization and neurotoxicity provide a certain paradox when applied to human disease. As mentioned earlier, the predominant, and indeed the initial, Aβ peptide species deposited as plaques is $A\beta_{42}$,[148,189-191,197-206,212-221] yet the major secreted species is $A\beta_{40}$.[296,302-309] These observations can however be reconciled according to the greater tendency of $A\beta_{1-42}$ to form fibrils than that of $A\beta_{1-40}$.[314-316] Nonetheless, most of these early cerebral cortical deposits are free from obvious neuritic changes and many apparently remain so during their entire life history; for example those in the cerebellum and corpus striatum never acquire an additional pathology.[166,167,172] These latter observations suggest either that $A\beta_{42}$ is not neurotoxic or that such brain regions are "insensitive" to its deposition or can resist its effects. On the other hand, neuritic plaques, in which neural damage is presumed to take place, always contain $A\beta_{40}$. Hence, although a minor and late deposited Aβ species, $A\beta_{40}$, and not $A\beta_{42}$, may be the peptide responsible, in vivo, for neurotoxicity when it is deposited in a suitably aggregated state upon pre-existing $A\beta_{42}$ containing plaques. The actual source of $A\beta_{40}$ in plaques is not clear but it may be produced by microglial cells and not by neurones (see later).

2.4.6. Amyloid Associated Proteins

While most Aβ deposits remain throughout their lifetime as diffuse plaques lacking a significant glial or neuronal dystrophy, others, and particularly those in the hippocampus and amygdala, become increasingly dense and fibrillar and form stable complexes with many extracellular proteins the most important among which appear to be Apo E and J, HSPG, ACT, complement factors and amyloid P component. These latter molecules have been termed amyloid associated proteins or 'chaperone proteins' and are of great potential importance since they may influence (i) how much Aβ remains soluble and free within the tissue, (ii) the rate of removal of soluble Aβ from the tissue, (iii) the rate of fibril formation and (iv) the rate of fibril removal. Hence, these associated proteins may promote amyloidogenesis or favor the "remodeling" of Aβ over the many years the plaque exists, this resulting in a 'maturation' (evolution) of plaques especially in areas such as the hippocampus, amygdala and association cortex.

2.4.6.1. Apolipoproteins

Apolipoprotein E (ApoE) is a polymorphic protein which facilitates the movement of cholesterol in or out of cells, via its interaction with the LDL receptor and LDL receptor-related protein and in so doing it modulates the catabolism of triglyceride rich lipoproteins.[337] In humans, ApoE is a 34KDa protein encoded by a 4 exon gene (APO E), 3.6Kb in length, situated on the long arm of chromosome 19. Two common polymorphisms at residues 112 and 158 determine the three major allelic forms, E2, E3 and E4. The ancestral E3 allele has a cysteine at residue 112 and an arginine at residue 158, whereas in the E4 allele an arginine is present at both sites and in the E2 allele, a cysteine at both. These polymorphisms result in six major genotypes, E2/E2, E2/E3, E2/E4, E3/E3, E3/E4 and E4/E4, of which the latter three are the most common. In the nervous system ApoE mRNA, and the protein itself, is principally expressed by, and contained within, glial cells, mostly astrocytes.[338-340] Although nerve cells do not seemingly express the ApoE message[338,340] they do contain the protein[339] this presumably being taken up, following release by glial cells, via the LDL receptor and LDL receptor-related protein sited on their cell surface.[341]

Two observations firmly implicate the apolipoproteins, and particularly ApoE in the pathogenesis of plaque formation. Firstly, numerous immunohistochemical studies[191,341-346] have shown that ApoE can bind to the Aβ within plaques, probably after this has already been deposited in the tissue. In the systemic amyloidoses the deposited protein again binds ApoE. Secondly, possession of the E4 allelic variant of the APO E gene is a major genetic determinant of some cases of early onset and most cases of late onset AD where the allelic frequency is raised from a norm of around 15% to about 30-45%.[347-353] Hence, within the AD population about 50% of sufferers possess at least one copy of the E4 allele.

How ApoE might influence Aβ deposition is not clear. Several studies have shown that 'in vitro' "purified" ApoE E4 protein binds more avidly to synthetic Aβ than does the E3 protein[317,348,354,355] and may thus reduce the lag time to fibrillization.[316] The ApoE E4 protein might exert its effect in plaque formation (and thus in AD) by increasing the amount of Aβ that is deposited within the brain.[341,344,356-359] However, other in vitro studies[360,361] have suggested that the purification procedures employed in previous studies[317,348,354,355] may have caused a relative failure of E3 binding to occur since native (i.e. unpurified) E3 and E4 isoforms actually bind to Aβ with at least equal propensity; indeed E3 may actually bind better than E4.[360,361] Furthermore, much other recent work has also failed to substantiate the suggestion that overall Aβ deposition might be increased in possessors of an E4 allele.[362-367] Nonetheless, it has been noted[363,365] that the proportion of $A\beta_{40}$ (relative to $A\beta_{42}$) does increase with ApoE allele copy number with possessors of one E4 allele having about twice as much $A\beta_{40}$ in their brains as those with no E4 alleles and those with two E4 alleles having more than three times as much. Since $A\beta_{40}$ is present mostly within the cored rather than the diffuse plaques,[190,191,212,217] it is possible that the ApoE E4 protein influences the fate of Aβ only once it has been deposited in the tissue, perhaps by facilitating its aggregation into cored deposits or increasing its "attractiveness" to microglial cells. In this latter context it is notable that the diffuse plaques of the cerebellum and striatum[346] contain little ApoE and are also deficient in microglial cells.[172] Co-localization studies[191] showing that not all (early) diffuse Aβ deposits contain ApoE, and that no ApoE immunoreactivity is present extracellularly in the absence of Aβ, reinforce the view that ApoE progressively accrues onto pre-existing $A\beta_{42}$ containing plaques. Hence, the ApoE isotype may affect how Aβ is "handled" once it has been deposited in fibrillar form within the tissue, though it has no influence over the total amount that is initially deposited (as $A\beta_{42}$). ApoE however is not the only apolipoprotein present in plaques; ApoJ (also known as clusterin or SP40, 40)[368-371] and ApoA1[348] are also located in plaques and these together with ApoE may act in concert to regulate the solubility of Aβ. Changes in the relative balance of these may favor its deposition in fibrillary form.[372-375]

2.4.6.2. Heparan sulphate proteoglycan (HSPG)

As in all forms of systemic amyloid, HSPG is present, bound to the Aβ, in both the diffuse and the cored plaques of the cerebral cortex[376-380] but is notably absent from the diffuse plaques of the cerebellum.[381] Binding of HSPG to Aβ can be inhibited, in vitro, by anionic sulphates and sulphonates,[382-384] including sulphonate dyes such as Congo red,[385] which reduce fibril formation and the extent of neurotoxicity.[382-384] This latter effect may be achieved by preventing aggregation of Aβ peptides or by favoring their aggregation in a form different from the neurotoxic β-pleated structure. Alternatively, these kinds of molecules may "coat" aggregated Aβ in such a way as to prevent its toxic interaction with cells.

The source of HSPG within plaques is not known though because microglial cells are known to secrete HSPG[386,387] they may therefore be responsible for this. The (relative) absence of microglial cells in the diffuse, non-β pleated cerebellar plaques[172] accords with such a view. The role of HSPG in plaque formation is not known, though it may favor the adoption by Aβ of a β-pleated sheet configuration (rather than an α-helix) characteristic of all amyloids. The absence of β-pleating within cerebellar diffuse plaques (as shown by the lack of staining with Congo red or Thioflavin S) would be consistent with this role, though many of the diffuse plaques of the cerebral cortex contain HSPG[376,387] yet these remain non-Congophilic and Thioflavin negative; presumably in these plaques the degree of β-pleating is minimal and insufficient for visualization by such methods. Nonetheless, observations [320] that co-perfusion of rat brain with a synthetic $A\beta_{1-40}$ and perlecan (a form of HSPG) can produce cored amyloid plaques within brain tissue indicates an important role for HSPG in plaque evolution.

In addition to HSPG, other proteoglycans, like dermatan sulphate proteoglycan,[388] chondroitin sulphate proteoglycan[389] and keratan sulphate proteoglycan[390] are also present in plaques though usually within the dystrophic neurites rather than bound to Aβ. Furthermore, other extracellular matrix proteins such as fibronectin,[391] laminin and collagen type IV[392] and basic fibroblast growth factor all occur in cerebral cortical plaques.[393,394] The intercellular adhesion molecule (ICAM) has been reported to be present in the cored plaques of the cerebral cortex[395] and cerebellum,[396] but not in the diffuse deposits of these regions,[395,396] though others have not found either this particular molecule nor the nerve cell adhesion molecule (NCAM) within amyloid deposits of any morphological kind.[397] These latter substances are all neurite inducing molecules and as such may be responsible, at least in part, for the neuritic changes ongoing in plaques. Their absence from the diffuse plaques of the cerebellum may help to explain the lack of neuritic changes in the Aβ deposits of this region.

2.4.6.3. Complement factors

Complement factors are a group of at least 20 proteins that become activated during the immune response and inflammation. Activation of the classical pathway arises through the binding of the Fc region of Ig to Clq, the first component in the cascade. This is followed by a sequence of autocatalytic reactions, involving C4, C2,

C3 and C5 components, culminating in the membrane attack complex (MAC) C5b-9. This can then insert into the target membrane, disturbing its continuity and promoting cell damage.

Most of the complement factors and complement inhibitors characteristic of the activation cycle of the classical, but not the alternative, pathway are present in all amyloid plaques of the cerebral cortex, especially those of the cored type.[368-371,398-409] This reactivity is only present to a much lesser extent, and sometimes not at all, in the diffuse plaques of the cerebellar cortex.[398-401] C1 will bind directly to Aβ, and especially to $A\beta_{1-42}$,[402,410] but only when in an aggregated form.[411] This binding may be the signal for engagement of the activation cycle, thereby avoiding the need for immunoglobulin. Since binding of C1 depends upon the aggregation state of Aβ,[411] this may partially explain the lack of C1 and other complement factors in diffuse cerebellar deposits[398-401] where the Aβ exists in a non-congophilic, non-β pleated conformation. Nonetheless, the diffuse plaques of the cerebral cortex still contain complement factors even though here again much of the Aβ seems to exist in a non-β pleated form.

The complement factors present in plaques do not originate from blood and are probably synthesized within the brain itself. Glial cells, particularly microglial cells, as well as neurones can produce mRNA for C1, C3 and C4,[409,412-414] especially when extracellular Aβ is present. Microglial cells express complement (Fc) receptors[403] and are presumably sensitive to the presence of complement proteins within brain tissue. Since complement reactivity within diffuse plaques appears to predate the presence of microglial cells[415] this may be the "trigger" to migration and activation of microglial cells rather than the presence of Aβ, (or even $A\beta_{40}$), per se. The relative absence of microglial cells,[172] complement factors[398-401] and $A\beta_{40}$[167] within the diffuse plaques of the cerebellum would all accord with this view.

In addition to these early cycle complement proteins immunoreactivity for the terminal membrane attack complex (MAC) is seen, but apparently only in neuritic plaques and in NFT[369,402,404] suggesting that in these plaques alone the full classical pathway may have been activated. Activation of the full pathway is meant to result in lysis of "foreign tissues", though in doing so unwanted damage to host tissue (bystander lysis) can occur. Complement inhibitors such as vitronectin, clusterin (also known as ApoJ, or SP 40,40), protectin (CD59) can restrict bystander lysis and all of these have been noted in plaques of both neuritic and diffuse types.[368-371,405,407] These molecules too may be synthesized in the brain[405,410] and while their expression may represent an attempt to minimize bystander lysis via MAC, the close structural and functional similarities between ApoE and ApoJ suggests that the combined presence of these, from a very early stage, in plaques[368,369] may relate more to Aβ deposition or stabilization. Indeed, specific ionic interactions between Aβ and ApoJ[374] would support such a role, perhaps by maintaining the solubility of Aβ in extracellular fluids and preventing its aggregation. The purpose of ApoA1 in plaques[348] is unknown, though because ApoA1 can, like ApoE, bind to Aβ[376] and also to ApoJ,[375] it may function in concert with these other two apolipoproteins in regulating the solubility of Aβ or favoring its deposition in fibrillary form.

2.4.6.4. α-1 antichymotrypsin

The diffuse and the neuritic (cored) plaques of the cerebral cortex and the diffuse plaques of the cerebellum all contain α-1 antichymotrypsin (ACT).[416-418] The source of this is not clear though it is again probably produced by astrocytes.[419] The precise role of ACT in plaque formation is not known though, in concert with the apolipoproteins, it may promote the assembly of Aβ peptides into amyloid fibrils.[317]

2.4.6.5. Amyloid P component

Like all systemic amyloids, the Aβ within the diffuse and cored plaques of the cerebral cortex,[420-422] and to a lesser extent that in the diffuse plaques of the cerebellum,[368,399] binds amyloid P component. P component is highly resistant to proteolysis and may be a significant factor in maintaining the persistence of Aβ deposits in brain tissue.[321] However, because P component does not seem to be directly produced in the brain its source is uncertain and is presumed to be hematogenous. If so, it is not clear how P component passes from the circulation into the brain. That it apparently does implies a defect in the blood brain barrier or that it is selectively transported from the circulation into the brain.

2.4.6.6. Other proteins

Aβ deposits have also been noted to contain interleukin-6 (IL-6),[423] basic fibroblast growth factor binding sites,[393,394,424] γ2 macroglobulin (LDL receptor-related protein),[368,423] lactoferrin[368] and complex saccharides.[139,188,425,426] The significance of these latter molecules within the genesis or evolution of plaques remains unknown; some or all may simply be adsorbed onto the amyloid fibrils, either directly or following binding of Aβ to HSPG, ApoE etc, from the extracellular fluid, and are perhaps without biological effect.

While many of these amyloid associated proteins may play a direct role in promoting fibrillogenesis, it is likely that no one single protein is wholly responsible for this; the various proteins present in plaques may act in concert exerting a net effect over amyloid deposition by providing a balance between Aβ solubility and insolubility. Interactions between these molecules are certainly possible; for example between HSPG and ApoE or P component[427] or ACT and ApoE.[317] Differences in distribution of these various proteins may on one hand, determine the region-specific distribution of Aβ deposition or the morphological characteristics of the plaques so formed.

2.4.7. Lysosomal Enzymes

The lysosomal enzymes cathepsin D and β-hexosaminidase A have been demonstrated within the diffuse and cored plaques of the cerebral cortex[428,429] and the diffuse plaques of the cerebellum.[430] These enzymes are normally contained within nerve cells rather than glial cells. In plaque-rich areas of brain, the nerve cells (e.g. pyramidal cells of the cerebral cortex or Purkinje cells of the cerebellum) and their processes become swollen with these hydrolase enzymes, even before any atrophic changes or NFT can be seen. Enzyme alterations like these may represent an abnormality of the endosomal-lysosomal compartment of the nerve cell. Continued enzyme accumulation may lead to metabolic compromise with the excess enzyme being liberated upon cell death from the nerve cell body and its processes into the extracellular space. Given that Aβ can be produced in the endosomal-lysosomal compartment of the nerve cell,[295,306,307] and because this increased lysosomal enzyme activity seems to occur early in the course of plaque evolution[430] it is possible that such an upregulation (or dysregulation) of the acid hydrolase containing compartment of the cell might contribute to the generation of Aβ peptides.

2.4.8. Neurites

The term neurite refers to the altered nerve ending classically present within the neuritic type of plaque of the cerebral cortex. (Fig. 2.1) Neurites however may also occur, though to a much lesser extent, in many of the diffuse plaques of the cerebral cortex.[431] Although they are considered to be mostly axonal in nature it is possible that at least some originate from dendrites. Under the electron microscope neurites contain a multitude of dense and lamellar bodies, these probably representing degenerating mitochondria, and PHF identical to those seen in cell bodies as NFT. (Fig. 2.4) It is the presence of these latter structures that confer the silver staining properties to neurites and because these are composed (see later) of the microtubule-associated protein tau and the 'stress protein' ubiquitin, they can be detected immunohistochemically using antibodies to such molecules. Immunohistochemistry shows neurites also contain much APP[432-438] within lysosomes,[437] and synapse-associated proteins.[438] While the presence of such excess APP within neurites might signal a catabolism of APP and potential formation of Aβ at this site, this might simply reflect a "damming up" of newly formed, but unused, APP continually being delivered by axonal flow[252] to these damaged nerve endings. Some of this APP may be subsequently catabolized and released into the extracellular space as soluble Aβ. However, diffuse plaques are comprised of Aβ but do not display APP containing neurites,[432-438] nor does the presence of these altered nerve terminals predate the appearance of Aβ deposits.[172] Cored plaques also contain much glycosylated material[139,188,425,426] and while the origin of this is uncertain it may be located within neurites and relate either to glycosylated (fragments of) APP or the lamellar/dense bodies, or both. Some neurites also contain the growth associated protein, GAP43[439,440] implying that at least a proportion of neurites may be regenerative sprouts rather than degenerative swellings.

Although probably mostly axonal in origin, it is quite clear that the dystrophic neurites in plaques do not relate to a single neuronal population but consist of the altered nerve endings of a multiplicity of nerve cell types. Hence, neurites have been shown to be immunoreactive to markers of acetylcholine,[441,442] noradrenaline,[443] somatostatin,[444-447] substance P[448] and neuropeptide Y.[447,449] Damage to nerve terminals in plaques may therefore occur "haphazardly", affecting most if not all nerve cell types equally irrespective of their morphological or transmitter characteristics; the balance within any one plaque may depend solely only upon the local mix of nerve endings at the site of plaque formation.

In accordance with the neuritic changes, there is also axonal damage and synapse loss from tissue regions containing neuritic plaques[450-452] though in areas of diffuse amyloid synapse density does not seem to be different from that in non-amyloid containing areas of neuropil.[451] Hence, neuritic plaques can be considered to be malignant plaques whose formation damages the neuropil of the brain, whereas no obvious destructive tissue changes are associated with diffuse amyloid deposits, even those of the cerebral cortex, and such plaques may be thought of as benign.

2.4.9. Microglial Cells

The presence of microglial cells within and surrounding the Aβ deposits of plaques has been repeatedly demonstrated and these are most numerous within the cored and neuritic (classical) plaques of the cerebral cortex while fewer occur in about half the diffuse plaques of this region.[189,223] (Fig. 2.7) Within cored plaques microglial cells are large and plump–"amoeboid" or activated microglia; in diffuse plaques they are mostly small with long thin processes–ramified microglia. Aβ may be directly chemotactic for microglia or may become so following a coating with complement C1[399] or ApoE, particularly the E4 isoform.[363] Electron microscopy of cored plaques[453-455] shows microglial cell processes to be in close contact with Aβ fibrils; indeed in some instances the fibrils appear to be "streaming" into, or away from, the microglial cell. The amyloid of diffuse plaques, on the other hand, is not so closely, if at all, associated with microglial cells or their processes.[456]

In younger persons with DS, diffuse Aβ deposits first appear in the cerebral cortex at around 30 years of age[188-191] but at this stage do not contain microglial cells.[189,195] Between 40 and 50 years of age many microglial cells begin to appear within most of the diffuse plaques and become present by this age in all of the cored plaques.[189] Carboxy-terminal specific Aβ antibodies show[189] that the microglial cells are always (or nearly so) associated with $A\beta_{40}$ and $A\beta_{42}$ containing plaques whereas they are much less common in plaques containing $A\beta_{42}$ alone. Similar observations to these have been made on plaques in mentally able elderly persons[223] and in others with AD.[169,222] Together, these data imply that microglial cells may play a part

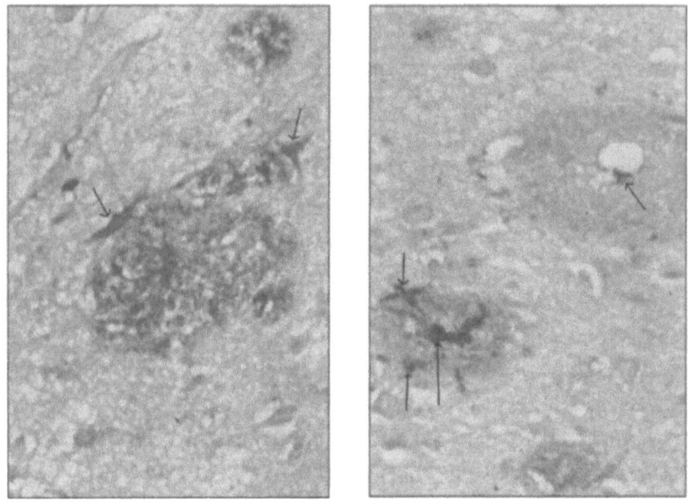

Fig. 2.7. Double labeling for Aβ (gray) and microglial cells (arrows). Cored plaques (a, near right) contain many microglial cells and their processes whereas diffuse plaques (b, far right) contain few or no such cells and processes.

in either the production of Aβ (and $A\beta_{40}$ in particular), its 'remodeling' when deposited, or even its ultimate removal from the brain tissue.

2.4.9.1. Microglial cells and amyloid

Microglial cells are capable of expressing the APP gene, especially the $APP_{751/770}$ isoforms and the equivalent isoforms lacking exon 15 i.e. ($APP_{733/762}$), though mainly only when they become activated or when they come into contact with extracellular matrix proteins;[236,245-247,258] they may also contain low levels of APP protein. However, it is unlikely that they are the primary or initial producers of Aβ within plaques. Many diffuse plaques (the earliest form of plaque), particularly those of the cerebellum, are devoid of (activated) microglial cells in young individuals with DS[189,195] or in elderly persons with or without AD.[189,222,223,457] Nonetheless, the close anatomical proximity of microglial cells to Aβ fibrils[453-455] implies a reactive role. Their preferential association with $A\beta_{40}$ containing plaques suggests they might release exopeptidases that perhaps remove the last two or three carboxy-terminal amino acids from pre-existing $A\beta_{42}$ deposits, thereby generating $A\beta_{40}$. Indeed the reciprocal relationship between $A\beta_{42}$ and $A\beta_{40}$ staining in plaques, with $A\beta_{42}$ immunoreactivity being weak in strongly $A\beta_{40}$ reactive plaques, would accord with this. They may also be responsible for some of the amino-terminal processing of Aβ.[191,220,221] However, some $A\beta_{42}$ containing diffuse plaques with microglial cells display no (detectable) $A\beta_{40}$ though these may still be in very early stages of remodeling where $A\beta_{40}$ levels are too low to be detected by immunohistochemistry. Removal of these two carboxy-terminal amino acids may enhance the efficiency of β-pleating since in this context it is notable that cerebrovascular amyloid is always (or mostly so) β-pleated and contains a much higher proportion of $A\beta_{40}$ (to $A\beta_{42}$) than plaques do, even those of the cored type.[200,210,211] Alternatively, or additionally, $A\beta_{40}$ in plaques may be produced 'de novo' by microglial cells through a processing of their endogenous APP, when activated. This new $A\beta_{40}$ would "seed" upon pre-existing $A\beta_{42}$ in plaques where microglial cells were already present. Furthermore, IL-1 released by microglial cells[457,458] might upregulate APP gene expression and APP catabolism in locally involved nerve cells or in other glial cells[286-288] resulting in the release of further $A\beta_{40}$ (and $A\beta_{42}$) into the extracellular space for incorporation into plaques. The absence of $A\beta_{40}$ in the diffuse, but not the cored, plaques of the cerebellum[167,213,214] (microglial cells are absent in the former but plentiful in the latter), and would support this view. If microglial cells do produce the $A\beta_{40}$ within plaques or cause it to be produced by other cells they may be responsible for the neurotoxicity associated with the cored, neuritic plaques.

On the other hand microglial cells secrete HSPG.[386,387] As already mentioned, this molecule is abundant in cerebral cortical plaques (both diffuse and cored), even from an early stage in their evolution.[376-382] A binding of HSPG to Aβ may induce or facilitate the adoption of the typical β-sheet configuration, but perhaps only after removal of the last two or three amino acids of the Aβ sequence. The lack of HSPG and the absence of $A\beta_{40}$ in diffuse cerebellar plaques,[167,381] where the Aβ is non-congophilic (i.e. not β-pleated), would accord with the lack of microglial cells in this region.[172]

Lastly, microglial cells might assist in the removal of Aβ. The overall "amyloid load" in AD,[146,459] DS[460] or even in elderly non-demented[153] does not increase indefinitely, but eventually plateaus, implying that new Aβ production might slow or even cease in areas where neuronal function becomes compromised, or that a balance is eventually achieved between rates of formation and dissolution. Microglial cells can degrade Aβ (and P3), at least in vitro[461] by the extracellular release of proteases, and may therefore play a key part in achieving this balance.

2.4.9.2. Microglial cells and phagocytosis

Microglial cells are renown for their phagocytotic capabilities though whether they can remove amyloid in this way is uncertain. Aβ deposits in elderly stroke victims seem to be cleared from the tissue by circulating macrophages rather than by microglial cells,[462] though clearly this mechanism can only be invoked when the integrity of the cerebrovasculature has been breached since easy access to tissue deposits of Aβ by macrophages may be denied by an intact blood brain barrier. Microglial cells are capable of "synaptic stripping" following experimental axotomy and may perform such a role in neurological disease following nerve cell injury.[463] By doing so neuronal energy would not be 'wasted' in a luxury conservation of electrochemical gradients leaving the nerve cell free to engage in disease resisting regenerative or restorative changes.

2.4.9.3. Other roles for microglial cells

Apart from participating in Aβ deposition, or remodeling, microglial cells may perform other roles within plaques. They may mediate an 'acute phase response' via secretion of IL-1.[453,454] On one hand, this might drive further production of APP by neurone[286-288] with subsequent channeling down potentially amyloidogenic pathways. Any such enhanced Aβ deposition may in turn attract more microglial cells and with further IL-1 release might thereby drive a vicious cycle of changes. In closed head injury tissue levels of IL-1 increase[464] and there is an upregulation of APP production[283] and a deposition of Aβ as diffuse plaques.[465] Observations like these show that this kind of process can in certain situations result in Aβ deposition. On the other hand microglial secreted IL-1 might stimulate astrocytosis[289,457,458,466] directing the appearance and activity of astrocytes within and around plaques, especially the cored plaques. The lack of astrocytes within many of the diffuse plaques of the cerebral cortex and nearly all of those of the cerebellar cortex[172,467] would be anticipated since in these activated microglial cells are sparse or absent.[172,457]

Microglial cells may also be responsible for secretion of IL-6 and by doing so might contribute to an ongoing acute phase response. However, it is uncertain whether this actually does occur since IL-6 is found mainly in diffuse, rather than in cored, plaques[468] and as already mentioned microglial cells attain maximum numbers in cored plaques with many diffuse plaques being deficient in these.[172,222]

Microglial cells, particularly those in cored and neuritic plaques, strongly express MHC class II antigen,[469,470] the usual purpose of which is to bind and present antigen to T-helper/inducer lymphocytes. They also express, but at much lower levels, MHC class I antigen which normally interacts with T-cytotoxic/suppressor lymphocytes.[470] Whether this increased MHC antigen expression by microglial cells simply reflects their differentiation state or whether it is intended to increase (the potential for) T-lymphocyte reaction is uncertain, though if the latter is the case this would seem unsuccessful since only occasionally do (a few) lymphocytes occur in plaques.[470] Hence microglial changes like these may be indicative of an inflammatory response rather than indicating an autoimmune reaction.

2.4.9.4. Microglial cells and neurotoxicity

Given their phagocytotic capability, microglial cells possess a high capacity for peroxidative activity and could therefore potentially produce excessive amounts of unscavenged free radicals or nitric oxide[336,471,472] which, in the presence of aggregated Aβ, might have the potential to damage nerve cell membranes and perhaps even induce the neurofibrillary and neuritic changes. The latter might also be triggered by $A\beta_{1-42}$ or $A\beta_{1-40}$, released from or produced by microglial cells, again by free radical type reactions.[336] The lack of tau-positive neurites within the microglial free diffuse plaques of the cerebellum[172] would accord with this.

It is therefore possible that microglial cells play an important, and perhaps even a central, role in plaque formation operating at several distinct, though complementary, levels according to their wide range of capabilities. These might include amyloidogenesis, phagocytosis, neurotoxicity, acute phase responses and glial activation. Clarification of the key part(s) microglial cells actually play in plaque formation may lead to therapies that could suppress many of the 'malignant' changes mediated by such cells that take place at this site, leaving perhaps only a relatively "benign" accumulation of diffuse amyloid ($A\beta_{42}$) to bear witness to the presence of disease.

2.4.10. Astrocytes

Astrocytes are frequently associated with neuritic (cored) plaques[172,467,473-476] but are less common in diffuse plaques, especially in those of the cerebellum.[172] In contrast to microglial cells, which are intimately associated with the Aβ (core) of the plaque, astrocytes are typically present around the margins of the plaque but nonetheless direct many of their processes inwards towards the plaque centre. (Fig. 2.8) Factors which stimulate this astrocytic reaction are unknown. As discussed earlier microglial cell released IL-1 may cause activation of local astrocytes[289,457,458] though this might also be triggered directly following damage to nerve endings within plaque confines. What purpose this reactive astrocytosis might achieve is not known though it may, as a physical barrier, limiting the area of tissue damage caused by plaque formation. Alternatively, astrocytes may represent the cellular source of the ACT[419] and ApoE[412-414] present within plaques and through the binding of these molecules to Aβ they may effect an important role in the reorganization of Aβ within evolving plaques. Furthermore, astrocytes can express APP message when activated and may contain APP protein[243,244,477] though whether they contribute to plaque amyloid through a catabolism of their own endogenous APP is unknown. Also, astrocytes produce potentially neurotoxic cytokines and are a major source of nitric oxide; they could contribute to the tissue damage through release of these commodities. Furthermore, S100 protein expression is increased within neuritic plaques.[467,478-480] This may also help to stimulate neuritic changes within such plaques since S100 expression is not increased in the diffuse non-neuritic cerebellar plaques.[467] Lastly, the presence of much glial scar tissue and the secretion of inhibitory factors may limit the regenerative capabilities of damaged neural tissue within the locality of the plaque.

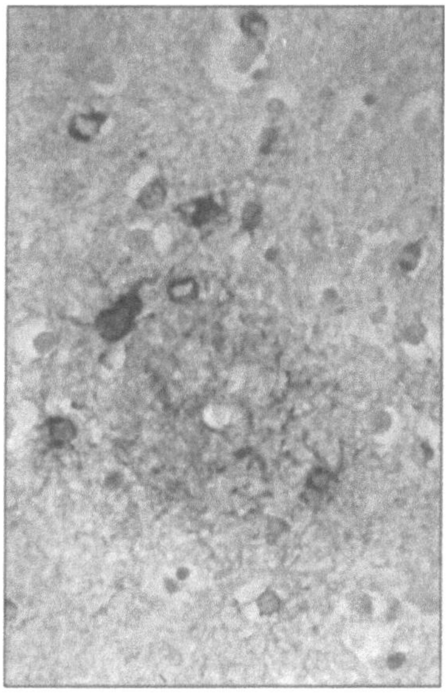

Fig. 2.8. Astrocytic involvement in plaques. Cell bodies are located around the margins of the plaque with their processes being directed inwards towards the amyloid core. From Mann DMA et al (1995). A Colour Atlas and Text of Adult Dementias. Times Mirror International Publishers.

2.5. NEUROFIBRILLARY TANGLES

Neurofibrillary tangle (NFT) was the name given to the abnormal flame-shaped filamentous structures observed in nerve cells in 1907 by Alois Alzheimer in the brain of a 51 year-old woman with dementia.[481] The presence of many NFT in the brain of this patient, along with the many plaques, came to characterize this disorder, which was later to bear the name Alzheimer's disease. However, the first description of these structures was actually made in 1904 by Fragnito,[482] and in 1907 Fuller had also remarked upon such neurofibrillary accumulations.[483] By 1912 many articles had already been published on the occurrence of plaques and tangles in dementia, using the classic Bielschowsky silver staining procedure, among many other variants, to demonstrate this. Even today, silver staining is still routinely used to demonstrate NFT, with the Bielschowsky technique, Bodian's method, Palmgren technique or Gallyas stain being variously favored.

By all these methods, NFT appear as thick, blackened fibrils which tortuously weave around the nucleus of the cell occupying much of the perikaryon and often extending into the axon and proximal dendrites. (Fig. 2.9) Although in most instances the cell outline and nucleus can be seen, sometimes the NFT appears to be devoid of cytoplasmic material, lying freely within the neuropil. (Fig. 2.10) These NFT have been termed 'ghost' tangles or extracellular tangles (skeletons) and presumably represent an end-stage when the nerve cell has died and the cell membranes and other organelles have degenerated and disappeared.

The morphological appearance of the NFT varies greatly according to the cell type in which it is contained. In small neurones of the cerebral cortex the NFT may appear as a single band of material looping around the nucleus, extending from the basal, towards the apical, dendrite. (Fig. 2.11) In larger pyramidal cells the NFT resembles a skein or tangle of wool and in the hippocampus particularly, adopts the archetypical 'torch' or flame' shape. (Fig. 2.9) In subcortical regions such as the nucleus basalis of Meynert, locus caeruleus, dorsal raphe nucleus (Fig. 2.12) and substantia nigra, NFT commonly look like a ball of wool and are termed 'globose'. NFT are not easily seen in routine hematoxylin-eosin (Fig. 2.13a) or Nissl stained (Fig. 2.13b) sections but like plaques can be readily demonstrated using Thioflavin S or Congo red staining.

Major advances in the study of NFT came with the advent of electron

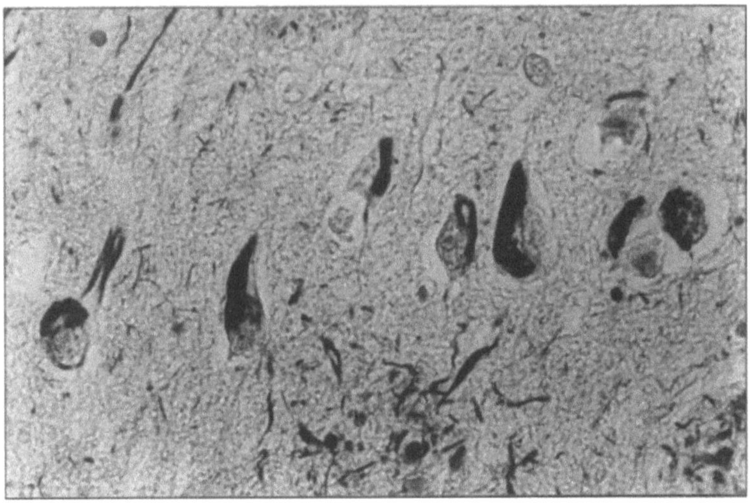

Fig. 2.9. Intracellular 'flame'-shaped NFT in the CA1 region of the hippocampus.

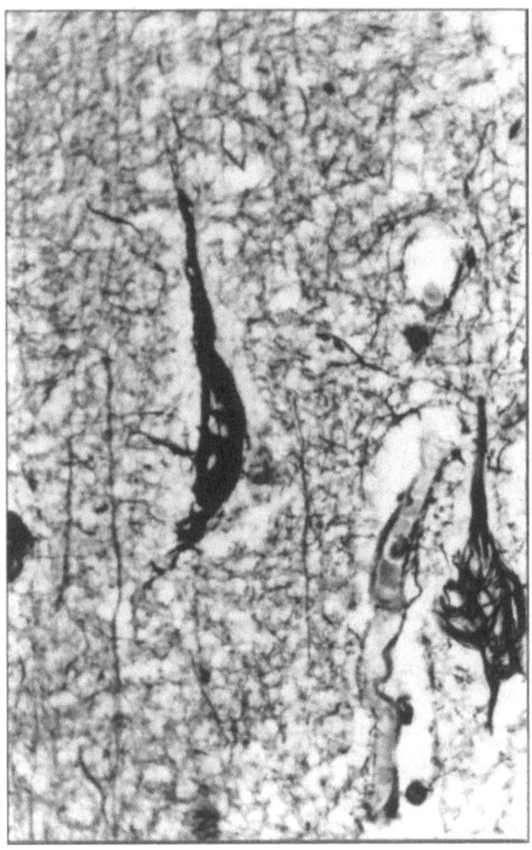

Fig. 2.10. Extracellular (ghost) tangles in layer II of the entorhinal cortex.

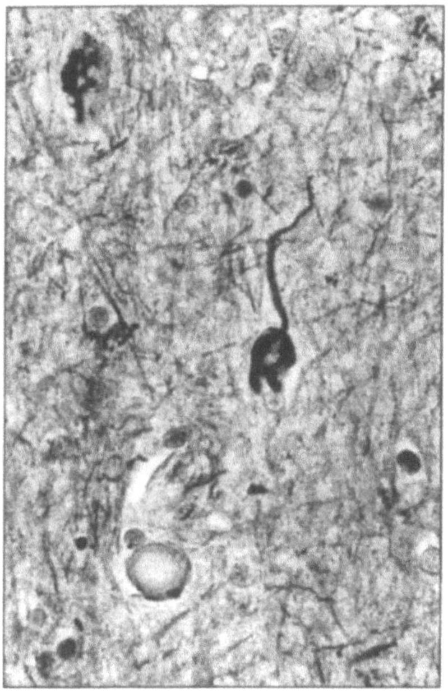

Fig. 2.11. Small intracellular NFT in a pyramidal nerve cell of the cerebral cortex.

microscopy which, for the first time, provided ultrastructural detail[484,485] and led to their delineation as 'paired helical filaments' (PHF). Following their isolation and purification from brain tissue the molecular characteristics of PHF were quickly determined; it was first shown in 1985 by Brion et al[486] that the microtubule associated protein, tau, was a major, if not the principal, constituent of NFT. Since then it has become clear that (at least some of) the tau protein within NFT exists in an abnormally phosphorylated state; many workers believe this is the critical change which results in its altered cellular distribution and its propensity to aggregate into the filamentous structures (PHF) seen microscopically.

2.5.1. Distribution of NFT

A few, widely scattered NFT occur in the brains of many, and perhaps even most, non-demented elderly persons.[32,53,137,140,141,156,487-491] In general, the number of NFT in the brain, and particularly in the hippocampus and inferior temporal cortex, tends to increase with age, but nonagenarians and centenarians seem to suffer NFT formation to a lesser extent than do younger persons (i.e. those aged between 70 and 90 years).[156] When present in such elderly persons, NFT are most likely to occur in the large stellate cells of layer II of the entorhinal cortex (parahippocampal gyrus),[489] in the pyramidal cells of areas CA1 and subiculum of the hippocampus and in cells of the amygdala,[137] particularly those of the cortical and medial nuclei. Outside these regions NFT are rare but can occasionally be seen in the cerebral cortex, especially in the temporal neocortex, or in subcortical regions such as the nucleus basalis of Meynert, substantia nigra, locus caeruleus and the dorsal raphe. In the cerebral cortex NFT are usually present in the pyramidal cells of layers III and V. Some cell types, however, apparently never form NFT, notably the Purkinje cells of the cerebellum and the large motor neurones of the spinal cord, cranial nerves and the precentral gyrus (Betz cells).

Irrespective of their location, NFT are intensely argyrophilic with the same neurofibrillary material extending widely throughout their dendrites, as 'neuropil threads'[186] or into the axon terminals becoming present within plaque confines as

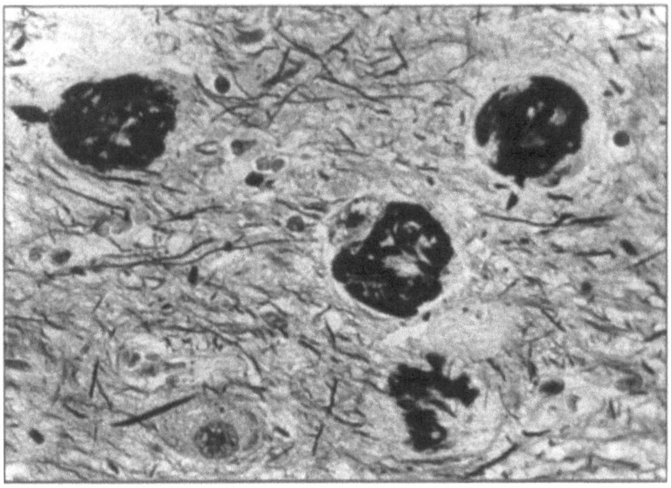

Fig. 2.12. Globose NFT in nerve cells of the dorsal raphe nucleus.

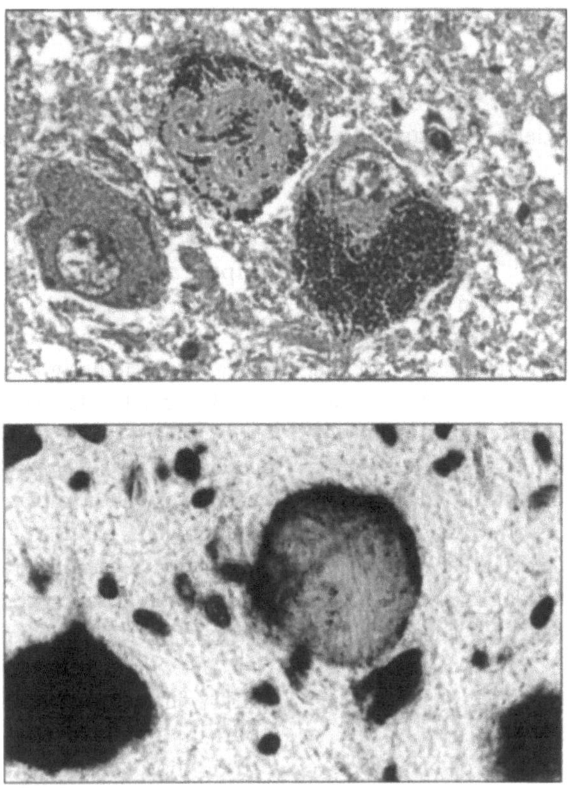

Fig. 2.13. Globose NFT in nerve cells of the locus caeruleus in H & E (a, top) and Nissl (b, bottom) stained sections.

'neurites', to which they give the term 'neuritic plaque'. (Fig. 2.1) NFT can be visualized by Thioflavin S fluorescence or under polarized light as a bright green birefringence following staining with Congo red, this indicating that the neurofibrillary material exists within a β-pleated sheet configuration.

2.5.2. Ultrastructure

NFT were first observed under the electron microscope in 1963 by Kidd[484] and also around that same time by Terry.[485] Ultrastructurally, the bands of fibrillar material comprise numerous long, unbranched but aligned filaments each consisting of a pair of filaments of 15nm diameter wound around each other in left-handed, α-helical fashion. (Fig. 2.14) Such an appearance produced the term 'paired helical filament' (PHF); a term nowadays preferred as a more accurate descriptor of their morphology than NFT. Hence, the dimensions of each PHF are of maximum width about 24nm, minimum width 15nm, and a periodicity of about 150nm. Mixed in with, and sometimes contributing a segment of, the PHF are 'straight filaments', also of 15nm diameter.[492-496] When assembled, each strand of the PHF resembles a long stack of C-shaped subunits, each domain within the subunit relating to a particular region of the tau molecule from which the PHF are assembled[496] a similar configuration is present in the 15nm straight sections (filaments). Crowther has further shown[496] that in PHF the C-shaped subunits are organized "base to base", whereas in straight filaments they are "back to back" thereby producing, instead of a helix, a more rounded cross-section. The straight filament is thus a structural variant of the PHF and not a separate anatomical or molecular entity. The PHF may be cross-linked, this holding them tightly together and accounting for their high insolubility, resistance to proteolysis and retention in the tissue over a period of many years.

2.5.3. Molecular Characteristics of NFT

When first seen in the electron microscope, Terry[485] considered PHF to represent twisted microtubules, though Kidd[484] emphasized their 'filamentous' nature. Little progress was subsequently made in determining their molecular characteristics, largely because of their intense insolubility,[497] until about 10 years ago when it was

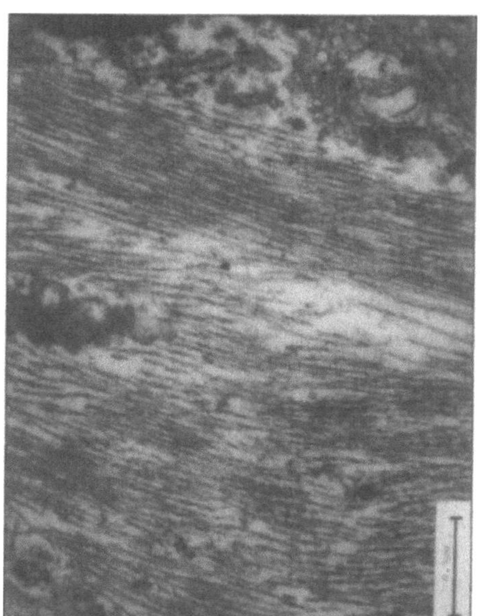

Fig. 2.14. Electronmicroscope appearance of NFT. The NFT is made of bundles of aligned filaments, each consisting of a pair of helically wound structures. From Mann DMA et al (1995). A Colour Atlas and Text of Adult Dementias. Times Mirror International Publishers.

first discovered by Brion and colleagues (1985),[486] and subsequently confirmed by many others, [498-502] that NFT are immunoreactive to antibodies raised against the microtubule associated protein, tau. Previous studies[503-509] had found immunoreactivity for neurofilament protein, but this was subsequently shown[510] to be due in many instances to a cross-reactivity for epitopes shared between anti-neurofilament and anti-tau antibodies; NFT are not now considered to be intrinsically composed of neurofilament protein, though it remains possible that some neurofilament protein may be (loosely) adherent to the PHF structure, as indeed are many other molecules (see later) such as ubiquitin protein,[511-514] MAP 2,[509] vimentin[515] or actin.[516] This mix of cytoskeletal elements within PHF may, at first sight, seem to be at odds with their ordered ultrastructure. This apparent paradox is however resolved by ultrastructural studies[517] of isolated NFT which have shown then to be composed of an inner protease resistant core and an outer protease labile fuzzy coat. After protease digestion a pure core region is left behind that upon sequencing is (seemingly) composed solely of tau protein.[518] Hence, it appears that the PHF may be made up of a core structure of tau, to which a fuzzy coat containing other cytoskeletal and non-cytoskeletal elements (as above), as well as further tau, is adherent.

Subsequently, further protein chemistry has confirmed the presence (by immunohistochemistry) of tau within PHF,[518-521] this seemingly being incorporated into the PHF structure by its carboxyl third which includes the microtubule binding region. How much of the tau molecule is present in PHF is not clear; claims that the entire molecule is present[520] may not be true in all instances since most molecules seem to be partially degraded showing both carboxy- and amino-terminal truncation.[518,522] Furthermore, it is also uncertain as to how much of the overall PHF structure is actually tau; current estimates have shown that only about 10% of the total core mass can be solubilized and definitely shown to be tau. Hence, it remains possible, though perhaps unlikely, that PHF are partly composed of molecules other than tau. Other molecules associated with PHF such as ubiquitin[512] are strongly bound to the PHF but probably within the fuzzy coat region and do not form part of the basic core structure.

2.5.4. The Tau Molecule

Tau is a microtubule associated protein and normally exists in one of six isoforms, these being 352-441 amino acids long.[523] Each isoform is derived by the alternative splicing of a single gene product[523,524] which is encoded on the long arm of chromosome 17.[525] The tau gene is approximately 100KB in length and contains 16 exons.[526] The six isoforms differ according to the number of (tandem) repeat sections (3 or 4) of 31 or 32 amino acids (containing the microtubule binding domain) at the carboxy-terminus, along with the number of 28 amino acid residue inserts (none, 1 or 2) towards the amino-terminus. Each tandem repeat section, although imperfect copies, contains a characteristic Pro-Gly-Gly-Gly motif and being rich therefore in basic amino acids can bind the tau molecule to the acidic domain of tubulin. Binding of tau in this way to tubulin is vital for maintenance of the microtubular structure. Hence microtubular function, and thus intracellular transport, is critically dependent upon this normal tau binding.

2.5.4.1. Tau and NFT

In cells affected by NFT formation there is a major pathoanatomical redistribution of tau from the axonal, to the somatodendritic, compartment together with a change from a soluble form into a form capable of self-assembly into PHF.[527]

On Western blots, tau proteins normally run as a series of bands with Mr between 45-62KDa. Tau proteins extracted from tissues rich in NFT (usually cases of AD) show a shift in mobility such that the proteins instead run as a characteristic triplet at Mr 55, 64 and 69KDa (the so-called

A68 proteins).[528-531] Tau 55 results from hyperphosphorylation of the two tau variants with no amino-terminal inserts, tau 64 from those with one amino-terminal insert, and tau 69 from those with two amino-terminal inserts, each band contains the equivalent tau species with or without the additional carboxy-terminal domain.[529] These three bands appear to be 'diagnostic' for the presence of PHF within brain tissue and are encountered not only in AD[528-531] but in other situations where PHF prevail like in DS[530] and the Parkinsonism dementia complex of Guam,[532] or in the hippocampus and entorhinal cortex of certain non-demented old people.[533]

These bands probably reflect those tau molecules that are located in the fuzzy coat region of the PHF, since being only loosely incorporated into its structure they can be easily extracted from the tissue. This shift in mobility pattern on Western blots appears to be due solely to an overall increase (compared to normal tau) in the phosphorylation state of pathological tau, this incidentally also increasing its resistance to proteolytic degradation. Hence, whereas normal tau has 2.5-3.5 moles of phosphate per mole of protein, tau associated with PHF has 10-12 moles of phosphate for each mole of protein. That tau in PHF is (hyper) phosphorylated can also be deduced from:

i) The abolition of immunostaining of PHF following alkaline phosphatase pretreatment to remove phosphate groups before immunolabeling with phosphorylation dependent tau antibodies.
ii) The fact that certain antibodies against normal tau (e.g. tau-1) will only recognize PHF following alkaline phosphatase treatment.
iii) Protease (pronase) treatment of extracted PHF strips off the fuzzy coat (leaving the core intact) and removes immunolabeling by phosphorylation dependent tau antibodies.

Such observations also imply that these (phosphorylated) epitopes are mostly situated towards the amino-terminus of tau.

How much of the abnormal tau present within the cell exists in a hyperphosphorylated form is not clear though it has been suggested[522] that this may amount to no more than 10%. Furthermore, about half of this abnormally phosphorylated tau seems to occur either loosely bound to the PHF, within the fuzzy coat region, or as soluble cytosolic tau species not (yet) aggregated into the PHF; the rest is present within the PHF core. Thus, perhaps only about 5% of all tau molecules present within the PHF core structure may actually be (over) phosphorylated.[534]

In the process of PHF formation two post-translational modifications of tau seem to occur. Firstly, there is carboxy- and amino-terminal truncation of tau leaving behind, as the main part, the tandem repeat region close to the carboxy terminus; this part of the tau molecule is not phosphorylated.[535] Secondly, other full-length tau molecules become hyperphosphorylated. The former tau derivatives have been termed PHF-tau, the latter, phosphorylated tau.[522] Although often used synonymously these two terms refer to separate tau entities each with distinct molecular locations. PHF-tau is present only within the core structure, phosphorylated tau is mostly (>95%) within the fuzzy coat and the cell cytosol. All 6 tau isoforms seem to be equivalently implicated within the PHF structure with no one form contributing (proportionately) more than the others to the formation of PHF.[523]

It is still not clear whether it is an over-phosphorylation of tau that precedes, and then leads onto, its carboxy- and amino-terminal truncation prior to incorporation into the PHF core, or whether truncation of some tau molecules occurs as a separate event leading to self-formation of a core structure upon which other over-phosphorylated, full length, tau molecules are subsequently deposited within the fuzzy coat component. Both concepts have been argued strongly and the relevance of phosphorylation as a key, or critical, stage in the process of PHF formation is still hotly debated.[522,535-537]

2.5.4.2. Phosphorylation of tau

It is widely agreed that at least some of the tau molecules associated with PHF are indeed over-phosphorylated relative to normal tau molecules. Phosphorylation of tau normally occurs at serine-proline (Ser-Pro) or threonine-proline (Thr-Pro) motifs[528,538-541] of which there are 17 distributed across the carboxy- and amino-terminal domains of full-length tau; 14 of these sites are shared by all six isoforms.

It is important to be able to distinguish whether the over-phosphorylation of PHF associated tau molecules, compared to normal tau, results from an increase in the phosphorylation state across all 17 phosphorylation sites, or whether there is a differential (aberrant) involvement with some sites being affected while others are not. Early autopsy studies showed that, when compared with normal adult CNS tau proteins, the tau associated with PHF (i.e. phosphorylated tau) is phosphorylated principally at Ser^{202},[528,539-543] Ser^{396} and Ser^{404},[528,541,542,544,545] but also at minor sites such as Ser^{235}, Ser^{262}, Thr^{181} and Thr^{231}.[546] This pattern of phosphorylation in tau associated with PHF differs from that seen in normal adult tau and partially recapitulates that pattern present in fetal tau.[539,541] In fetal tau, Ser^{202} and Ser^{396} are the principal sites phosphorylated (fetal tau is composed mostly of the shortest of the six tau isoforms) but neither site is phosphorylated in normal adult tau.

Hence, it initially seemed possible that the over-phosphorylation of tau might be caused by an aberrant reactivation of protein kinases, or an inactivation of phosphatases, usually associated with regulation of the phosphorylation state of tau at critical sites (i.e. Ser^{202}, Ser^{396}) in fetal, rather than adult, tissues. Nonetheless, activation or inhibition of such enzymes in fetal tissue does not result in a neurofibrillary pathology, suggesting that regulation of these phosphorylation sites is not the change that produces, per se, the derivativized forms of tau that are intrinsically capable of assembly into PHF.

However, it has been recently reported[547-549] that normal adult tau, when extracted promptly from biopsy tissues and rapidly analyzed, is phosphorylated, but to a lesser extent, at nearly all the same sites as tau associated with PHF in AD tissues. Such data therefore suggest that certain tau phosphate binding sites may be autopsy labile and that phosphates may be differentially (preferentially) lost from normal tau, compared to pathological (PHF-associated) tau, during the period between death and tissue analysis. Differences in phosphate binding between normal tau and tau associated with PHF thus appear to be solely quantitative rather than qualitative with similar binding sites being phosphorylated in both and only the overall extent of phosphorylation at each site being greater in pathological tau.

However, it is not clear whether hyperphosphorylation at all sites in PHF-tau is of functional importance and in this context (over) phosphorylation of tau at Ser^{262} may be critical since this is the only site within the microtubule binding domain of tau found to be so phosphorylated in PHF preparations from autopsy (AD) tissues.[546,550] Phosphorylation at this particular site will strongly inhibit the binding of tau to microtubules.[551] Moreover, this particular site, Ser^{262}, is also phosphorylated in a proportion of tau molecules in fetal tissue[548,552] yet tau extracted from these tissues can still bind normally to tubulin and promote microtubule assembly. Phosphorylation of tau, particularly at Ser^{262}, may thus form part of the activation/inactivation cycle of tau that regulates microtubule assembly/disassembly.[553] Over-phosphorylation of critical sites, like Ser^{262}, in an excessive proportion of tau molecules may be sufficient for PHF formation to occur.

2.5.4.3. Phosphorylation mechanisms

The net phosphorylation state of normal tau could be elevated either by an increase in protein kinase activity and/or a reduction in protein phosphatase activity. Phosphorylation of tau can occur in vitro at most of the 17 Ser-Pro and Thr-Pro sites by the enzyme, mitogen activated protein kinase (MAP kinase)[529,538,554,555] though glycogen synthetase kinase (GSK) 3α and 3β also can phosphorylate tau at Ser-Pro sites, but mostly at Ser^{235} and Ser^{404}.[556] A proline directed protein kinase[557] and the cyclin-dependent kinases CDK2 and CDK5[558,559] and cyclic AMP-dependent protein kinase (CAMP-PK) and calcium/calmodulin dependent protein kinase (CaMkII)[553] are also capable of in vitro phosphorylation of tau at sites known to be phosphorylated in PHF-tau. Which, if any, of these enzymes is actually responsible for increasing the net phosphorylation state of tau in PHF in vivo is still uncertain though GSK 3β and CAMP-PK and CaMkII are the only candidate enzymes that have so far been shown to be capable of phosphorylating tau in vitro within cells.[553,560] Importantly, CAMP-PK and CaMkII can phosphorylate tau at Ser^{262}– a site critical for the efficient binding of tau to tubulin when non-phosphorylated[553] whereas GSK 3β operates mostly at Ser^{235} and Ser^{404}.[556]

An increased phosphorylation state could also occur through a loss of phosphatase (phosphate-removing) enzyme activity. Of the many such enzymes possible, phosphatase 1, phosphatase 2A and phosphatase 2B (calcineurin) are known to be capable of removing phosphates from Ser-Pro and Thr-Pro motifs[561-564] though only phosphatase 1 and 2A can remove phosphates from Ser^{262} motifs.[552] Moreover, PHF containing neurones do not seem to be deficient in calcineurin immunoreactivity as compared to their non-tangled counterparts[564,565] and although a loss of enzyme activity, without a net loss of protein, could occur in affected cells there is no evidence for this at present.[561]

Transforming growth factor β_2 is thought to play a role in promoting cell survival[566] and its presence in NFT containing cells[567] may represent an attempt to counteract the ongoing pathological process. Transforming growth factor β_1 is known to increase protein phosphatase 1 activity though whether transforming growth factor β_2 has any potential compensatory (stimulatory) affect on protein phosphatase 2A or 2B (calcineurin), enzymes capable of removing phosphates from phosphorylated tau,[561-564] is not known.

Other workers[568,569] have shown a loss in low molecular weight acid phosphatase enzyme in AD, suggesting that the removal of tyrosine-linked phosphate groups may be impaired. Loss of this enzyme might explain the high levels of phosphotyrosine in NFT containing cells and dystrophic neurites,[570] though obviously any increase in tyrosine kinase activity that might occur would also contribute. Although NFT do not themselves contain tyrosine phosphorylation sites, these data may be indicative of a wider disturbance of phosphorylation mechanisms in tangle bearing neurones.

2.5.4.4. The significance of phosphorylation

Although the significance of phosphorylation of tau in terms of PHF formation is still contentious[522,535-537] the effect of this change with respect to microtubule function is clear. In vitro studies show that phosphorylated tau is less effective at binding to tubulin than non-phosphorylated tau and suggest that (over) phosphorylation in vivo may reduce the degree of tau binding to tubulin thereby destabilizing microtubules and militating against fast axonal transport. The redistribution of tau into the somato-dendritic component will also contribute. Hence, with disruption to the microtubule network it will become progressively harder for the nerve cell to dispatch macromolecules from the sites in the cell body where they are produced to the points in the cell periphery where they are needed. Derangement of cell metabolism and function will occur; continued damage will undoubtedly cause the nerve cell to fail, as a functional unit, and eventually to die.

However, as mentioned earlier, it is less clear whether this change in phosphorylation state of tau is critical or even

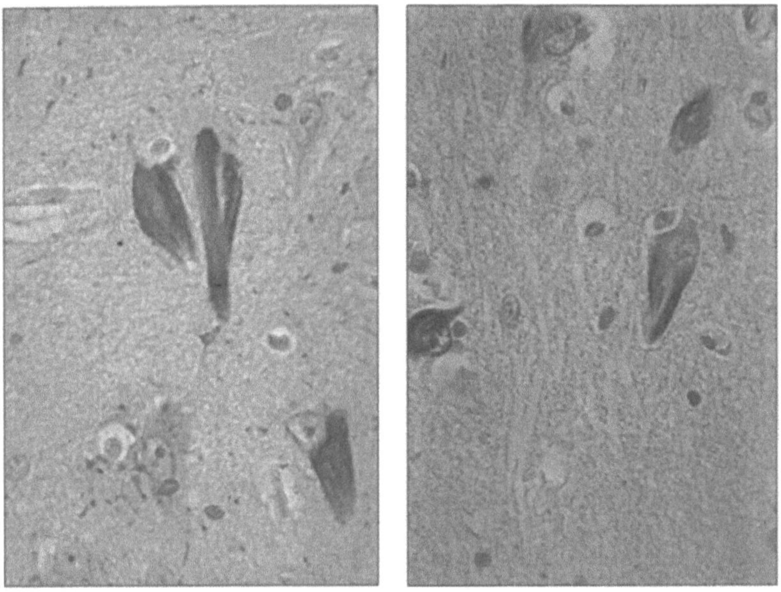

Fig. 2.15. Tau (a, near right) and ubiquitin (b, far right) immunostained NFT in hippocampal pyramidal cells. From Mann et al (1995). A Color Atlas and Text of Adult Dementias. Times Mirror International Publishers.

necessary for the production of PHF. Experimental studies[571,572] have shown that non-phosphorylated truncated fragments of tau containing the microtubule binding domain (tandem repeat section) can self-assemble into fibrils similar to PHF. Furthermore, this self-assembly does not vary according to phosphorylation state, with phosphorylated and non-phosphorylated tau fragments assembling equally well providing the tandem repeat section is present; even full-length tau can form PHF-like structures, irrespective of its phosphorylation state.

Therefore, it is presently uncertain as to how critical a part phosphorylation of tau plays in the generation of PHF. The phosphorylation of tau molecules may occur as part of a wider derangement of cellular metabolism, or even a more extensive disturbance of protein phosphorylation, that carries along in its wake a certain proportion of tau molecules that would in any case already have been predestined towards PHF formation even had phosphorylation of these not occurred. Nonetheless, it is clear that this combined derangement of tau structure and function (i.e. PHF formation and phosphorylation) will result in a sick neurone incapable of directing the products of its metabolism to the correct sites of action and that such a cellular derangement will have only one outcome–the death of that cell.

From even a cursory microscopic inspection of NFT within nerve cells it is not hard to imagine that cells affected in this way are likely to be incapable of normal function, at least during the later stages of PHF accumulation when the cell may become densely packed, and that this change might ultimately lead to the death of these cells. Indeed, in the subiculum in AD, the number of surviving healthy neurones plus those containing NFT, when added to the number of extracellular NFT present, equals the total number of normal cells in control cases. [491] Such data imply that, in this brain region at least, all cell death in AD proceeds through NFT formation and accumulation.

Indices of protein synthetic capacity such as nucleolar size[573,574] or cytoplasmic RNA[128] show these to be lowered in tangle bearing cells in AD compared to their non-tangle bearing counterparts. At electron-microscope level[575] a progressive accumulation of PHF is associated with decreases in the number of ribosomes, the extent of endoplasmic reticulum and the density of mitochondria. (Fig. 2.16) However, Salehi and colleagues[576] have shown that protein synthesis is equally impaired in AD in tangle and non-tangle bearing neurones.

Likewise, Mann reported in AD reductions in the RNA content of neurones even in areas where NFT did not occur.[129] This, together with the derangement of tau function within NFT bearing cells, leads naturally to the conclusion that in cells affected by NFT the capacity to form proteins is much reduced and the ability to transport those that are formed to the parts of the cell where they are required is impaired. These metabolic disturbances may be sufficient in themselves to cause cell death once a threshold level of tolerance has been crossed. However, because non-tangled cells also show (lesser) reductions in protein synthesis, these particular changes in protein production could simply be indicative of a wider derangement of cell function possibly involving oxidative stress[332,577-580] or an elevation of intracellular calcium[331] leading to necrosis[331] or apoptosis.[327-330,581] Such oxidative stress might be exacerbated by ongoing stimulation from excitatory amino acid neurotransmitters causing continued electrophysiological activity in these ailing cells since NFT containing cells are seemingly still capable of function, at least during the early stages of NFT formation.[576]

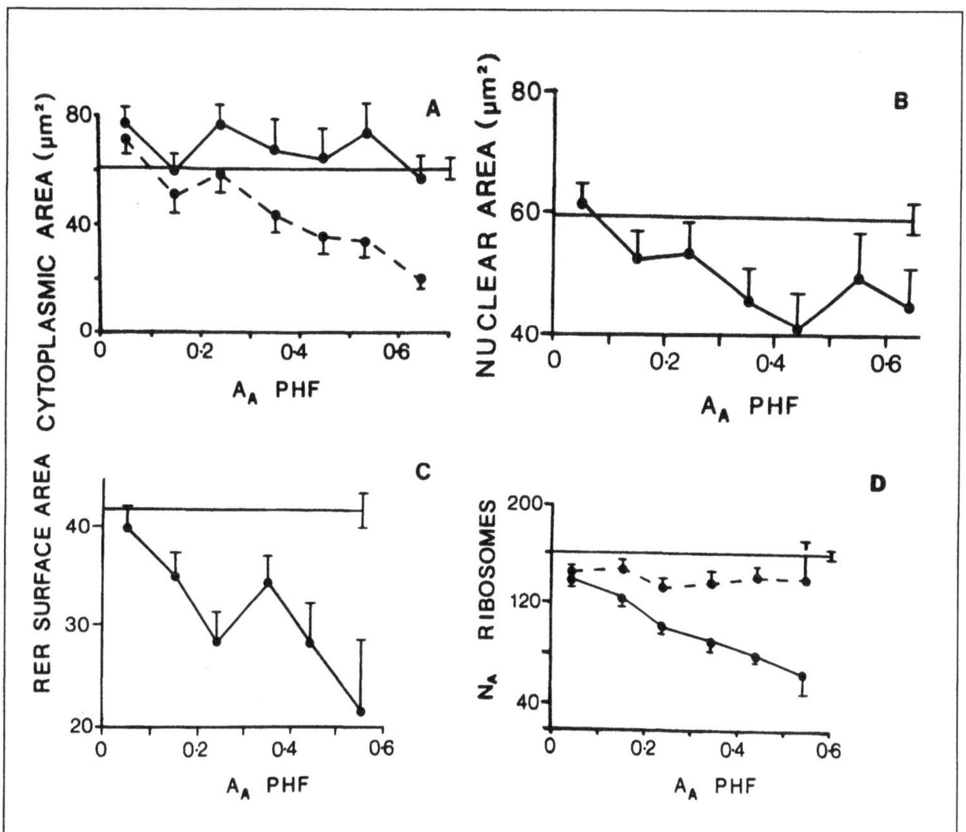

Fig. 2.16. The effects of intracellular accumulation of paired helical filaments (PHF) in cortical pyramidal cells. As area proportion A_A of PHF increases, cell size remains constant (solid line), close to the normal value (horizontal line) (A). However, the amount of "useful" cytoplasm (ie total cell area - area occupied by tangle) decreases with increasing A_A PHF (A), as does the cross-sectional area of the nucleus (B), the surface area of rough endoplasmic reticulum (C) and the overall numerical density (Na) of ribosomes (solid line) (D). In (D) the broken line indicates that the numerical density of ribosomes in the remaining "useful" cytoplasm is close to normal, suggesting a maintenance of protein synthesis in unaffected (by PHF) regions of the cell.

2.5.5. Histochemical Studies on PHF formation

Several studies[513,582,583] have shown that phosphorylated tau can be detected, as a granular material, in cell types at risk of NFT formation, even before any definite fibrillary material can be seen. (Fig. 2.17) These early tau deposits appear to be amino-terminally intact and are phosphorylated at $Ser^{198/202}$, as well as at other sites.[529] At this time the deposits are amorphous and lack a clear PHF structure[584] but are nonetheless carboxy-terminally truncated.[535,536,584] Later amino-terminal truncation and PHF assembly take place and fully-formed NFT appear in the nerve cell; these are composed of a core of carboxy- and amino-terminally truncated tau with a fuzzy coat of full-length tau, some of which is abnormally phosphorylated.[535,536] Nerve cells burdened with PHF will eventually die due to disruption of their intracellular transport network and perhaps also through the perikaryal crowding out of endoplasmic reticulum, ribosomes and mitochondria,[575] organelles vital for the maintenance of cell physiology and function. (Fig. 2.16) When this occurs the outer membrane is lost and the perikaryal and cell terminal remains are liberated into the extracellular space. Such extracellular tangles (and neurites) progressively lose their tau immunoreactivity[498,513,585-591] (Fig. 2.15a) due to proteolytic removal of amino-terminal (phosphorylated) fragments within the fuzzy coat, though conformational changes affecting epitope accessibility may also occur. Proteolysis of PHF may result from microglial action since these cells are common in the vicinity of extracellular tangles[513,592,593] though phagocytosis of PHF by microglia is unlikely.[456] Eventually when much, or all, of the amino-terminal tau has been removed the tangle becomes immunoreactive to carboxy-terminal tau (PHF) antibodies[522,535,588] as the core structure, resistant to further degradation due to an extensive cross-linking of fibrils,[577-590] is left behind. This PHF skeleton may eventually become infiltrated by astrocytes and their processes (and show GFAP reactivity) and these cells may further degrade or phagocytose the residuum.[594-596] Extracellular NFT also exhibit

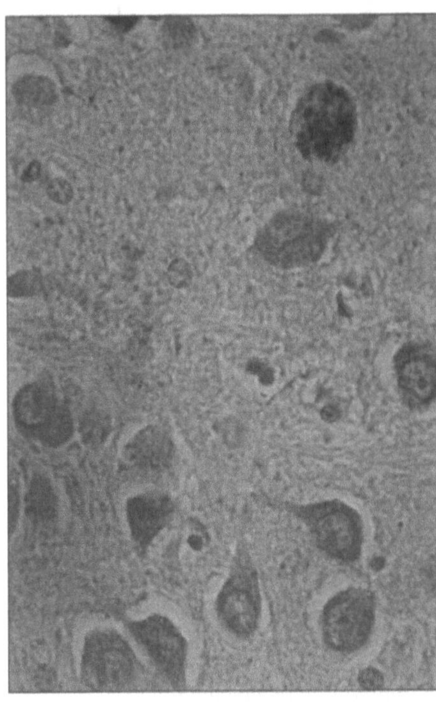

Fig. 2.17. Early involvement of hippocampal neurones by neurofibrillary degeneration shows the abnormal tau immunoreactivity to be fine and granular, contrasting with the "clumped" appearance of that in cells with overt NFT.

basic fibroblast growth factor,[424] this probably binding to NFT via the HSPG present in the fuzzy coat.

NFT formation and degradation appear to proceed along the same routes in AD and DS, as in the non-demented elderly, whether in terms of morphology, biochemistry or immunology; few, if any, differences of importance in these respects exist between the various clinical settings in which NFT are present.

2.5.6. NFT and Apolipoprotein E

Given the genetic association between the possession of an ApoE E4 allele and the incidence of AD it has been suggested[597] that the possession of an ApoE E3 (or ApoE E2) allele might be protective against NFT formation, since ApoE E2 and ApoE E3 protein isoforms can bind, in vitro, to tau whereas the E4 isoform cannot to any great extent. Hence, possession of the E4 protein may fail to protect against PHF assembly. This scenario, however, seems unlikely for several reasons. Firstly, no differences in the levels of either PHF-bound tau or phosphorylated tau have been found in patients with AD with or without an E4 allele[362] nor does the actual density of NFT so vary[341,344,364,366,367] (but see refs. 358, 359). Secondly, in cortical Lewy body disease, where Aβ plaques are numerous but NFT are usually sparse or absent, the ApoE E4 allele frequency is elevated similar to that in AD.[362,364,598] Third, about half of all AD patients do not possess an E4 allele, yet they still develop a typical (neurofibrillary) pathology and patients with DS, in whom a normal ApoE allele frequency distribution is seen,[353,599,600] invariably develop a robust pathology. Fourth, ApoE E4 allele frequency is also normal in other disorders, apart from DS, as in the amyotrophic lateral sclerosis/Parkinsonism dementia complex of Guam[601] or progressive supranuclear palsy,[602] despite the widespread frequency of NFT in both such conditions. Fifthly, even if the E4 allele is implicated it is unclear how tau and ApoE proteins might come into contact in the cell. Tau is a neuronal cytosolic protein whereas ApoE is produced by glial cells[412-414] but can be taken up by neurones[339] through LRP and LRP-related protein receptors on their cell surface. In this way extracellular ApoE can be internalized in neurones but would presumably be restricted to the endosomal/lysosomal compartment and so separated from cytosolic tau. Lastly, no evidence has been produced showing that tau requires ApoE in order to perform its normal function as a microtubule binding protein; indeed the interactions between tau and ApoE reported in vitro[597] take place at relatively high concentrations and it is uncertain whether this same interaction might occur in vitro at physiological levels or in vivo within the constraints of a cell system.

Hence even though ApoE protein will bind to most extracellular and some intracellular NFT,[341,342,345,601] there is no compelling evidence that any particular isoform(s) can directly influence the extent of formation of NFT. Hence, a role for ApoE in NFT formation is unclear, though it may help to induce its tertiary (i.e. β-pleated) structure.

2.5.7. NFT and Ubiquitin

It is well known from both immunohistochemistry[511-514] (Fig. 2.15b) and protein chemistry[512] that NFT contain ubiquitin, probably within the fuzzy coat region. Ubiquitin is a polypeptide of 76 amino acids whose function it is to label effete or redundant proteins for proteolysis via non-lysosomal pathways. Because such a labeling of NFT takes place it is believed that the affected cell 'recognizes' PHF (or phosphorylated) tau as abnormal and attempts to degrade it along established routes. Presumed failure to achieve this (successfully) leads to PHF accumulation in the cell. Following cell death, ubiquitin immunoreactivity, like that of tau, is lost[513,586-590] as the fuzzy coat of the NFT is degraded and removed by microglial cells or astrocytes.[513,592-594] Other "stress" proteins such as HSP 27, [603] HSP 70 [604] and heme oxygenase-1 (HSP 32)[605-607] appear in neurites and NFT and heme oxygenase-mRNA expression is increased[607] in tissue regions where NFT are plentiful. Such changes may be further indications of an ongoing oxidative stress in tangle bearing cells. Nonetheless, other proteins produced by neurones when under "stress", like αB crystallin, are not increased in AD; when present in neurones, αB crystallin does not necessarily relate only to NFT containing cells and is just as likely to occur in non-tangle bearing cells.[608]

2.5.8. NFT and Other Proteins

2.5.8.1. NFT and proteoglycans

Nerve cells containing NFT stain strongly for HSPG,[376-380] dermatan sulphate proteoglycan,[388] chondroitin sulphate proteoglycan,[389] and keratin sulphate proteoglycan[390] these molecules again being bound to the PHF within the fuzzy coat region. What purpose these might serve is currently unknown though, given that another β-pleated protein, Aβ, likewise binds to HSPG,[376-380] it may be presumed that HSPG in NFT also plays a role in favoring the β-pleated configuration which may permit the binding of individual PHF into the broad bands of material that are accumulated in the cell body as NFT.

2.5.8.2. NFT and Aβ

Extracellular tangles may act as secondary sites of deposition for Aβ protein; several studies using various Aβ antibodies have detected an immunoreactivity for this in extracellular NFT.[587,590,609-612] Extracellular NFT can also be stained by methenamine silver, one of the more modern silver staining techniques widely used to detect Aβ deposits. Such observations are indicative of a secondary deposition of Aβ peptides on PHF when these are present in the extracellular space and should not be taken to infer that NFT are actually composed of Aβ. However, some studies have shown that intracellular NFT can also be labeled by anti-Aβ antibodies[609] and these may also contain APP.[613,614] APP fragments containing Aβ could perhaps act as a "seed" for the growth of PHF polymers.

2.5.8.3. NFT and glycoproteins

Because NFT persist within the tissue for many years, changes may take place in them over time that confer such stability. It is well known that extracellular matrix proteins become increasingly insoluble and resistant to proteolysis during aging following their participation in advanced Maillard reactions brought on by the binding of such proteins, in Schiff-type condensation reactions, to reducing sugars such as glucose. These sugar bound molecules then undergo a series of non-enzymatic glycation and oxidation reactions to generate end-products such as pyrroline and pentosidine; lysine rich proteins such as tau are particularly reactive.[615] Immunohistochemistry[577-580] has shown these sorts of glycated molecules to be present in NFT and neurites and that tau in NFT is itself glycated.[579,580,615] Aβ deposits in plaques and blood vessel walls is similarly glycated. Such Maillard-type reactions might occur because of a delayed turnover of tau, following its phosphorylation, in the presence of such reducing sugars. Reactions of this kind include much cross-linking of proteins and this may partially explain why the proteinaceous material in lesions like NFT or Aβ deposits is difficult to remove from the cell or tissue by either proteolysis or through microglial action, or by experimental solubilization.

Lectin histochemistry[231,349,616] has likewise demonstrated complex glycoproteins in NFT, though whether these relate to the same advanced glycation end-products of the Maillard reaction detected immunohistochemically is not clear.[577-580] Accumulated glycation end-products can impart oxidative stress upon cells in which they are contained[579,580] and this can be manifest in various ways, including the induction of heme oxygenase-1 (this is present in NFT and neurites[605-607]) and the formation of malonaldehyde epitopes of lipid peroxidation products.[589,590]

Hence, changes of this kind may, over a long term, become superimposed upon pre-existing alterations (i.e. phosphorylation) in tau which lead to its reduced turnover and PHF formation. Glycation of PHF tau may promote its stability within the cell and, as demonstrated by the induction of heme oxygenase,[605-607] allow participation in the generation of reactive oxygen species and free radicals capable of inducing, or contributing to, oxidative stress within affected cells and in doing so helping to bring about neuronal death. Recent studies[580] imply that these glycated end-products of tau can induce the expression of interleukins within the cell which might lead to an elevation of APP formation with breakdown into Aβ. Hence, mechanisms leading to the formation of glycated end-products may, over time, potentiate Aβ deposition within plaques.

2.5.8.4. Changes in other cytoskeletal proteins

While some earlier observations of neurofilament protein immunoreactivity within NFT[503-509] were explained in terms of epitope cross-reactivity with tau,[510] other workers[617] have recently shown that some neurofilament protein does indeed exist within NFT and that this too has undergone post-translational modifications rendering it, like modified tau, resistant to proteolysis and decreasing its solubility. Such modification of neurofilament protein may contribute, along with the changes involving tau, and perhaps also other susceptible proteins such as tropomyosin[618] to a widespread or generalized disruption and collapse of the cells cytoskeleton.

2.5.9. NFT and Aluminum

In some studies[619,620] the amount of aluminum within NFT bearing nerve cells, relative to non-tangled neighbors, has been claimed to be increased, this being located within the nucleus of affected cells. However, others have not confirmed this finding[621,622] and as with aluminum and Aβ deposits, the issue remains unresolved. Nonetheless, exposure of neuroblastoma cells to aluminum induces a neurofilamentous accumulation that, while structurally distinct from the PHF of NFT, does share some antigenic similarities to PHF.[623] Aluminum can also potentiate iron-induced lipid peroxidation and an increase in iron has also been claimed to occur in NFT containing cells.[620] Hence, while evidence supporting a causative role for aluminum in NFT formation is lacking its presence, or selective accumulation, within affected cells might potentiate existing physiological difficulties and contribute towards cellular demise.

2.6. LEWY BODIES

2.6.1. The Prevalence and Morphology of Lewy Bodies

These intracytoplasmic inclusions were first described in 1912 by Lewy[624] and so-named in 1919 by Tretiakoff.[625] Under the light microscope these characteristically appear as rounded profiles, 5-30μm in diameter though, on tracing of consecutive sections, Lewy himself perceived them as more elongated or serpentine structures which only appeared to be circular when cut in cross-section. They can be easily detected in routine hematoxylin-eosin stained sections (Fig. 2.18) but are especially well demonstrated using the phloxine-tartrazine method[626] and when stained in this way typically possess a dense core, a body with or without concentric lamellae, and a clear surrounding halo.[627] They may occur singly or multiply in affected cells at any given level of section.

Lewy bodies are most commonly seen in the pigmented neurones of the substantia nigra, locus caeruleus and dorsal motor vagus[80,626-631] though occur widely in many other aminergic neurones throughout the mid brain and brain stem.[630] Lewy bodies can occur in other nerve cell types with the cholinergic cells of the substantia innominata (nucleus basalis of Meynert) being frequently affected; indeed involvement of this region was stressed in Lewy's original study. Nerve cells in the sympathetic[632] and parasympathetic ganglia[633] can be affected as can those in the nerve plexi of the gastrointestinal tract.[634] Although their widespread presence in these regions is considered to be pathognomic for the idiopathic form of Parkinson's disease (PD) they can also commonly occur in these same brain regions in cases of AD.[627,628,635-637] Their presence in the cerebral cortex, in association with an Alzheimer-type pathology,[638-641] may be indicative of the condition now known as cortical Lewy body disease (CLBD) (see later). Nonetheless, about 3-10% of elderly persons in the general population who apparently suffer neither from clinical AD nor PD also occasionally show Lewy bodies within cells of the substantia nigra[627,629,635] and locus caeruleus,[80,631] with the proportion of cells so affected increasing with each decade. Such cases have been termed[627] "incidental Lewy body disease" and might in fact represent ones of preclinical PD (see later). Lewy bodies may sometimes be seen lying extracellularly in the neuropil and following the death of the affected neuron they may even be removed from the tissue, though how this is achieved is not known.

It has recently been recognized[642] that, along with well-formed Lewy bodies, affected cells of the substantia nigra, locus caeruleus and nucleus basalis can also contain "pale-bodies". These are larger, rounded, homogeneously eosinophilic structures whose relevance is not known and relationship to Lewy bodies uncertain though they may represent the morphological forerunners of these.

Although Lewy bodies are common in brainstem structures, in CLBD they are additionally and widely present in neocortical regions such as the temporal, cingu-

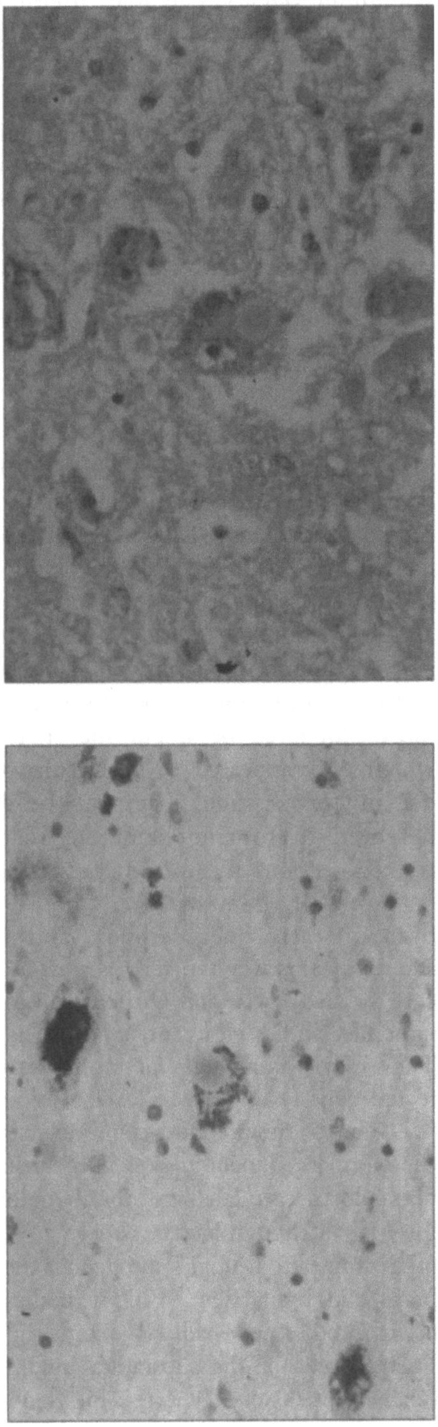

Fig. 2.18. Brain stem Lewy bodies in H & E (a, top) and anti-ubiquitin immunostaining (b, bottom). (b) from Mann DMA et al (1995). A Colour Atlas and Text of Adult Dementias. Times Mirror International Publishers.

late, insular and parahippocampal cortex[638-641] especially within the large pyramidal cells of layers V and VI. These cortical Lewy bodies only occasionally closely resemble their brainstem counterparts. They are usually less stereotypic appearing as ill-defined pale, homogeneous, slightly hyaline structures that tend to displace the nucleus peripherally and as such are morphologically similar to the pale bodies in brainstem neurones.[642] (Fig. 2.19a) They can be easily missed on routine hematoxylin-eosin stained sections but are, however, conspicuous on anti-ubiquitin immunostaining; (Fig. 2.19b) indeed this is the preferred method for their demonstration.[643] Cortical Lewy bodies like these are not seen, or are so only very occasionally, in the cerebral cortex of the non-demented, non-Parkinsonian elderly.

2.6.2. The Composition of Lewy Bodies

Electron microscopy [644,645] shows Lewy bodies to consist of a dense central mass of filaments and granular material with a periphery of looser radiating fibrils, 7-20nm in diameter. However, given their scarcity within the tissue, it has not as yet proved possible to purify Lewy bodies sufficiently for direct biochemical analysis though immunohistochemistry has provided some insight into their composition. Lewy bodies immunostain strongly, especially the core regions, with antibodies against ubiquitin protein[514,640,643,646] (Figs. 2.18b, 2.19b) and enzymes associated with ubiquitin mediated proteolysis.[647] They also contain all three members of the neurofilament protein triplet[646,648,649] and MAP2[646] but seemingly do not contain tau protein or actin.[646] Some are also immunoreactive to αB crystallin.[608,646,649] These data strongly imply that Lewy bodies are composed of neurofilament protein and like NFT and Hirano bodies may represent a disordering of the cytoskeleton. In common with other intracellular inclusions, like NFT, Lewy bodies, both in the brainstem and cortex contain HSPG[379] and chondroitin sulphate proteoglycan.[650] The role of these is unclear but, again as with NFT, the presence of proteoglycan within this inclusion may facilitate the interlinking of filaments allowing these to resist proteolysis and to persist in the cell for a long period.

Brainstem Lewy bodies in CLBD may differ slightly from their counterparts in PD. For example in CLBD, some brainstem and cortical Lewy bodies may contain tau,[651] and tropomyosin,[652] in contrast to those in PD. Brainstem Lewy bodies in PD contain HSPG[379] whereas those in CLBD do not, even though chondroitin sulphate proteoglycan is present in Lewy bodies of both conditions.[650] The neurofilaments in Lewy bodies in PD are insoluble[653] while those in CLBD can be solubilized.[654] Whether these differences are due to variations in the basic mechanism underlying the formation of Lewy bodies in PD and CLBD is not known; it is possible that they simply reflect local differences in the internal micro-environment of neurones related to the particular disturbance in function ongoing in each disorder.

2.6.3. The Effects of Lewy Bodies

It is not known whether the processes leading to the formation of Lewy bodies, or indeed the structures themselves, exert any direct and detrimental effects on cell function, though any disruption to the neurofilament axis they might cause, or result from, could precipitate a "dying back" of the axon disconnecting the nerve cell from its afferent target. The presence of ubiquitin[514,640,643,646] and ubiquitin-associated enzymes[647] in Lewy bodies suggests a cellular attempt to get rid of abnormal or defunct cytoskeletal proteins and they may therefore be formed as part of a protective cell stress response. The presence of iron within Lewy bodies[655,656] suggests they might participate in free radical mediated nerve cell damage though the Lewy body may simply represent an inert 'sink' for this

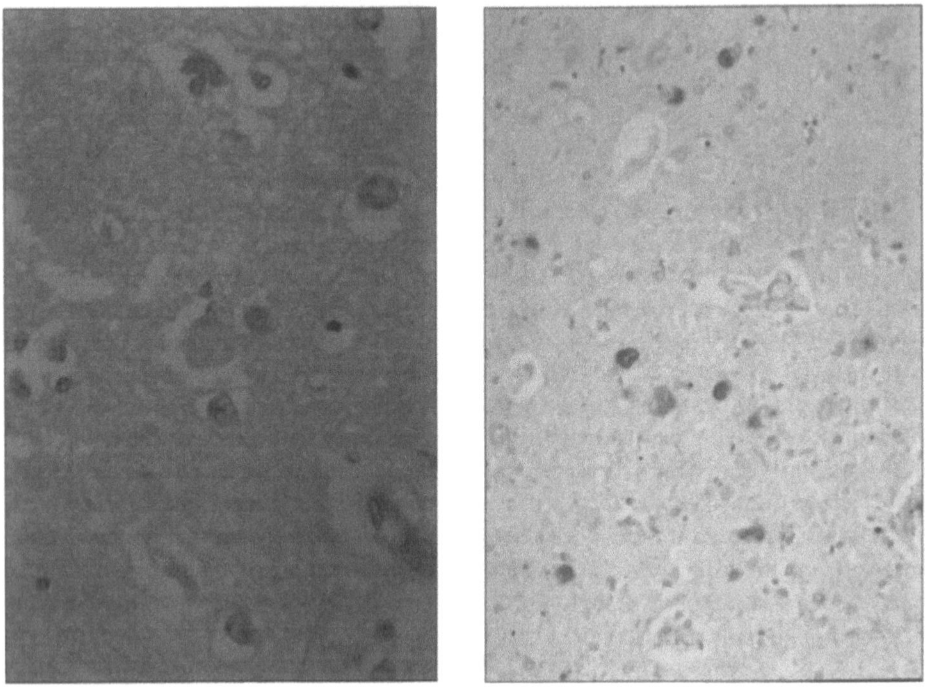

Fig. 2.19. Cortical Lewy bodies in H & E (a, top) and anti-ubiquitin immunostaining. (b, bottom) from Mann DMA et al (1995). A Colour Atlas and Text of Adult Dementias. Times Mirror International Publishers.

Fig. 2.20. Hirano bodies in CA1 region of the hippocampus. From Mann DMA et al (1995). A Colour Atlas and Text of Adult Dementias. Times Mirror International Publishers.

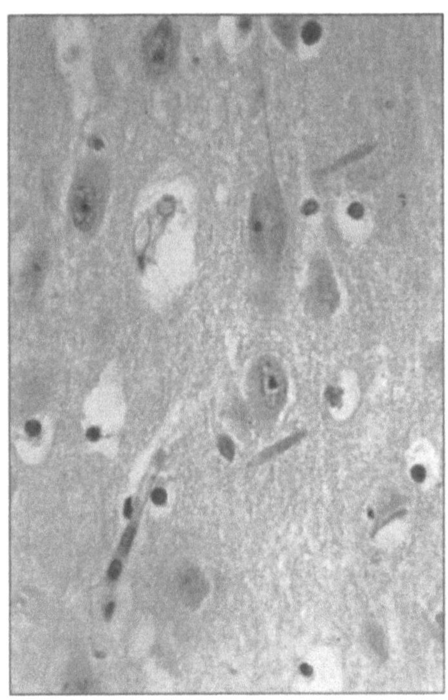

potentially neurotoxic element. On the other hand Lewy body formation may represent just one facet of a wider spectrum of degenerative changes going on in such diseased cells and they may occur simply as a "pathological onlooker".

2.7. HIRANO BODIES

2.7.1. THE PREVALENCE OF HIRANO BODIES

These structures were first described by Hirano in 1961, as present in the pyramidal cell layer, and especially in area CA1, of the hippocampus of patients with the Guam Parkinsonism-dementia amyotrophic lateral sclerosis complex.[657] The name, Hirano body (HB) was given in 1968 by Schochet et al[658] while describing the fine structure of these in a case of Pick's disease. Since then it has become recognized that HB, like NFT and GVD commonly occur in this region, albeit at low frequency, in many non-demented old people[46,659,660] but are also present in much greater numbers in persons with AD[46,659,660] and other dementias[658,661] and in chronic alcoholism.[662] In humans HB are sometimes seen in brain regions other than the hippocampus, especially the cerebellum[663,664] and can even occur in tissues other than the CNS.[665-668]

2.7.2. THE MORPHOLOGY OF HIRANO BODIES

HB are easily seen in hematoxylin-eosin stained sections appearing as bright eosinophilic structures, ovoid, circular or rod-like in shape. (Fig. 2.20) Usually, they appear to be within the neuropil (probably in dendrites) and sometimes seem to lie next to pyramidal cells.[669] In other studies,[670,671] however, a glial location has been emphasized. HB do not stain with PAS, Congo red, thioflavins or silver methods. They have a hyaline appearance and sometimes distinct striations can be seen. HB measure up to 15mm in short axis and sometimes exceed 30mm in long axis. Electron microscopy[658,669,670,672-675] indicates a paracrystalline configuration composed of a lattice of crossing filaments, each measuring 6-10nm in diameter and 60-100nm in length (Fig. 2.21).

2.7.3. THE COMPOSITION OF HIRANO BODIES

Immunohistochemistry suggests the protein lattices to be immunoreactive to actin antibodies, especially F-actin,[674-676] though other studies have indicated the presence of actin-associated proteins such as α-actinin, vinculin and tropomyosin,[674] tau protein,[677] MAP2 [678] or neurofilament protein.[679] They are not, however, reactive to ubiquitin antibodies[680] nor in most instances to antibodies against Aβ or PHF.

2.7.4. THE SIGNIFICANCE OF HIRANO BODIES

The precise molecular source and mode of formation of HB is still unclear and the preferential involvement of the CA1 region of the hippocampus in the elderly remains unexplained. Nonetheless, as NFT represent a disordering of the neuronal microtubular network the formation of HB might reflect an alteration within the microfilament compartment of the cytoskeleton, this involving mostly actin fibers. Whether these structures contribute directly to the dementia of AD, or the amnesia of chronic alcoholics, is not known though because they seem to be located within the dendrites of CA1 pyramidal cells, which make contact with branches of the perforant pathway and the Schaffer collaterals of areas CA3/4, they may cause an impairment, or reflect a breakdown in the functional efficiency, of such pathways along with other lesions common in this brain region such as NFT and GVD.

2.8. GRANULOVACUOLAR DEGENERATION

2.8.1. THE PREVALENCE OF GRANULOVACUOLAR DEGENERATION

The term granulovacuolar degeneration (GVD) refers to a change, first described

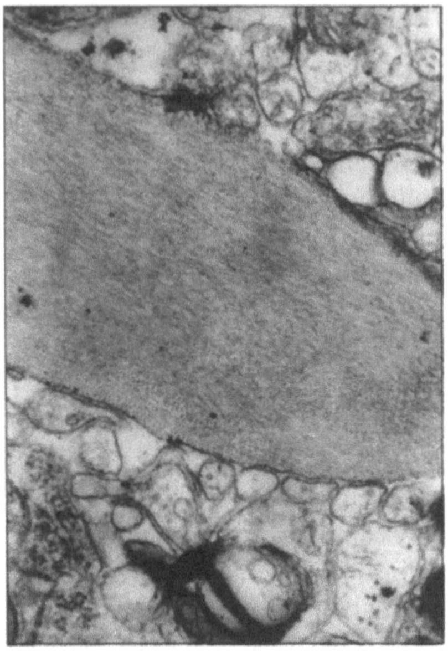

Fig. 2.21. Electronmicrograph of a Hirano body. The structure is composed of arrays of aligned filaments.

in 1911 by Simchowicz,[681] that occurs within nerve cells and is characterized by the presence of one or more (sometimes as many as 20) clear, rounded vacuoles each measuring 3-5mm in diameter and containing a single haematoxylinophilic, argyrophilic granule 0.5-1.5mm in width.[32,682-685] (Fig. 2.22) When present in large numbers, the vacuoles distend the cell outline and displace the nucleus towards the margin of the cell. Nerve cells affected in this way are nearly always the pyramidal cells of the hippocampus, particularly those of the CA1 region, though nerve cells in other regions of the hippocampus may also be involved, but less severely so, as occasionally are other nerve cell types such as those in the amygdala, nucleus basalis of Meynert, the frontal, temporal and cingulate cortices, the olfactory bulb and certain brainstem nuclei. The number of hippocampal cells that may be affected by this pathological change seems to increase with age but in non-demented persons remains low with less than 10% maximally of cells being involved.[32,682-685] In AD, however, the proportion of cells affected and the severity of individual cellular involvement both increase greatly.[133,682-685] Involvement is not restricted to AD; GVD is severe and commonplace in other neurodegenerative diseases such as Pick's disease or in the Parkinsonism-dementia, amyotrophic lateral sclerosis complex of Guam.[672]

2.8.2. The Composition of the Granulovacuoles

Preservation of GVD in autopsy specimens is usually poor and electron microscopy has added little to our understanding. In this, GVD is seen simply as an electron-dense amorphous granular core within an electron-lucent membrane bound sphere.[672] Histochemistry is also largely uninformative, the granules being negative to stains for carbohydrates, nucleic acids or glycosaminoglycans. Immunohistochemistry has shown tubulin-like immunoreactivity[686] neurofilament-like immunoreactivity[687] and tau-immunoreactivity[688-690] within the granular component. The anti-tau immunoreaction seems to relate to phosphorylated epitopes located towards the amino-terminus[689] implying that the sequestered tau within GVD is either partially degraded or exists in a modified conformational state in which the carboxy-terminal regions are rendered inaccessible to antibodies specific to such parts of the molecule. Although, as with NFT, the anti-neurofilament reactivity could represent a "cross-reactivity" with antigenic sites in common with tau protein, such staining nonetheless suggests that, again like NFT, formation of GVD may involve a widespread abnormality of the neuronal cytoskeleton. Indeed, on occasions GVD does co-exist with NFT in the same cell though this is rather exceptional. Although some studies have shown GVD to be immunoreactive to ubiquitin protein,[514,689,691] others[680] have suggested this not to be so.

2.8.3. The Significance of Granulovacuolar Degeneration

GVD might represent an autophagic abnormality in which (ubiquitinated) cytoskeletal elements are degraded. Clearly the presence of GVD within affected neurones is detrimental to function, or is at least indicative of an unhealthy cell, since affected cells are much deficient in ribonucleic acid.[692] However, the precise manner by which GVD might cause a derangement of function, or stem from the same, is uncertain. Further, it is not known why hippocampal and not other nerve cell types should be so vulnerable. Because GVD is limited in distribution throughout the brain this suggests that it may play only a minor part in producing the disturbance of brain function that represents AD. It may, however, act more significantly at a local level by disrupting connections within the hippocampus involving the CA1 region and subiculum–areas that form the principal output pathways from this region thereby contributing to the generation of memory disturbances. Whatever the true significance of this change might be, it does appear to represent yet another marker of cellular aging which, like plaques, NFT and Hirano bodies, differs only in quantitative terms between the brains of non-demented old people and those of the demented elderly.

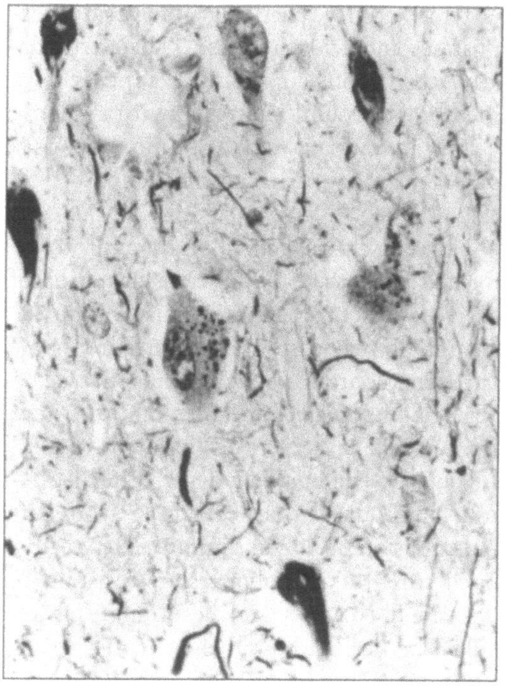

Fig. 2.22. Granulovacuolar degeneration of hippocampal pyramidal cells.

2.9. NEUROPIGMENTS

Many nerve cells of diverse types come to contain and progressively accumulate pigment granules, as they get older. In most nerve cell types the pigment is known as lipofuscin though in the pigmented cells of the substantia nigra, locus caeruleus, ventral tegmentum and dorsal motor vagus nucleus this is termed neuromelanin. These two neuropigments share many structural and chemical properties and it is now widely accepted that neuromelanin is actually a 'melanized' form of lipofuscin, being present in these particular cell types because of their use, and catabolism, of catecholamines as neurotransmitter.

2.9.1. Lipofuscin

Lipofuscin granules have a native yellow-brown color and are strongly autofluorescent yellow when subjected to illumination by near ultraviolet light (at a wavelength of 430nm). Although unstained by haematoxylin-eosin, the granules are metachromatic green with Nissl stain. Lipofuscin granules are perhaps best demonstrated with Periodic-acid Schiff, Nile Blue A or S stain, or by Sudan Black B methods. Lipofuscin appears to be a heterogeneous mix of carbohydrate, lipid and protein and contains much hydrolytic enzyme activity, particularly for acid phosphatases and acid esterases.[693] Ultrastructurally, the lipofuscin granule appears as a membrane bound material with a dense granular matrix and a homogeneously pale globular lipid component[644,645,694] (Fig. 2.23).

Lipofuscin is generally thought to be formed from the gradual transformation of active lysosomes into residual bodies,[695] though it may also arise from autophagic vacuoles containing acid phosphatase.[696] Various biological molecules can form into lipofuscin-like structures following cross-linkage by malonaldehyde or other carbonyl compounds that originate from the peroxidation of polyunsaturated fatty acids. Such cross-linked polymers are resistant to lysosomal enzymes and accumulate as residual bodies. Although the chemical composition of lipofuscin does not suggest a functional role, it could act as a 'trap' for other substances that might damage the cytoplasm. Most workers, however, believe that it simply represents an accumulation of metabolically inactive residues.

There is a characteristic topography to the distribution of lipofuscin in the brain. Although widespread through neurones of the CNS, it is especially prominent in cells of the inferior olives (where it is even present at infancy) and in the dentate nucleus of the cerebellum,[697-699] in anterior horn cells[94] and other motor neurone types and dorsal root ganglion cells, and in cells of the lateral geniculate bodies.[700] By contrast, Purkinje cells of the cerebellum are almost free from pigment, even at extreme old age.[698] The pyramidal cells of the cerebral cortex and hippocampus and neurones of the thalamus and globus pallidus are usually intermediate in terms of lipofuscin accumulation though occasional cells can become densely packed.[698,699]

Lipofuscin appears in some cells, such as those of the inferior olives, as early as the perinatal period[697] and then accumulates with age in almost a linear fashion.[697-701] Nevertheless, any effect its accumulation might have on cellular metabolism or function are not known. Accumulation of pigment has been linked to a gradual loss of ribosomal RNA and a decrease in the size of the nucleolus[697,698] suggesting an associated decline in protein synthetic capability. Yet, even in those cell types showing in later life the greatest accumulation of pigment and where RNA losses are maximal, i.e. the inferior olives[697,698] such changes do not apparently lead to any decline in neuronal number.[75,76] (but see ref. 74). Nonetheless, Barden[702] did note a reduction in the Golgi apparatus enzyme within cells of the aged inferior olives. It

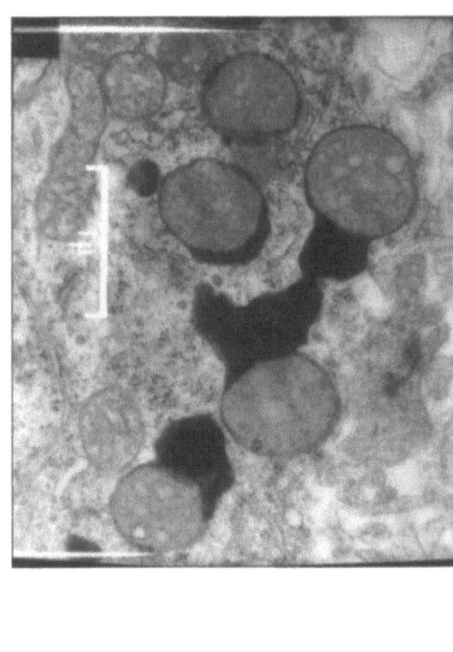

Fig. 2.23. Electronmicroscopy shows lipofuscin granules to be made of a globular, electron lucent lipid component and a coarse, electron dense protein matrix.

therefore remains possible that cellular metabolism or its efficiency could be mechanically disrupted in these cells by an expanding pigment bulk, though not to such an extent as to lead to overt cell death. Clearly other age-related processes which damage cells could occur in parallel and be responsible for these latter perikaryal changes.

In those cell types where pigment accumulation is low (e.g. Purkinje cells and pyramidal cells of the hippocampus) protein synthetic capacity is reduced by less than 10% in old age[698] and the Golgi apparatus is unaltered.[702] Nevertheless, the number of Purkinje cells is reduced in old age by 25%[63] and those in the hippocampus by about 20%[53-57] (but see refs. 59,60). Cerebral cortical neurones also accumulate only modest amounts of lipofuscin, yet their numbers also may be much reduced in later life[39,46] (but see refs. 27, 35, 36, 49-51). In all these latter cell types factors other than lipofuscin accumulation alone must be responsible for their downfall in later life. It still remains uncertain, therefore, whether lipofuscin is accumulated in cells as an innocent bystander, simply marking the passage of time, or whether in some insidious way it contributes to, or even hastens, a deterioration in function in later life in certain vulnerable cell types. Lipofuscin is also often present in glial cells and extracellular deposits are frequently seen around blood vessels suggesting that glial cells may facilitate its removal from neurones following the death of such cells; such removal could perhaps even occur during their lifetime with gradual failure to do this being responsible for its accumulation in nerve cells in later life.

2.9.2. NEUROMELANIN

The neuromelanin containing cells of the brain form a continuous column extending along the brainstem from a position central to the motor nucleus of the oculomotor nerve to the caudal end of the medulla, terminating in direct continuity with the intermediolateral grey column of the spinal cord.[703] Of all the cell groups present within this column, the substantia nigra, (Fig. 2.24a) locus caeruleus and the dorsal motor vagus nucleus are the most pigmented and probably the best known. Neuromelanin, like lipofuscin, stains for acid phosphatases and acid esterases [693]and likewise at ultrastructural level has the same granular matrix and lipid components, though additional coarse electron dense particles are present on the matrix.[644,645,694,704] (Fig. 2.24b) These similarities of staining and structure imply that neuromelanin and lipofuscin are related, neuromelanin being a 'melanized' form of lipofuscin. Their slightly differing structure probably results from interactions between the lipofuscin granule and the products (quinones) of catecholamine auto-oxidation.[705] The granules also appear to be rich in iron[706] and thus might play a role in free radical mediated death of such cells in later life.

Neuromelanin first appears in cells of the locus caeruleus around the time of birth,[707,708] but cells of the substantia nigra are not regularly pigmented until around 18 months of postnatal life.[708] The beginnings of pigmentation may reflect the onset of sustained physiological function by such cells. Nonetheless, in both cell types pigment is then accumulated steadily up to about 60 years of age[708,709] after which the average cell levels of pigment decrease, (Fig. 2.25) this perhaps being due to a preferential fallout, from this age onwards, of the more heavily pigmented cells[710] leaving the lesser pigmented cells to survive further into old age. Such observations suggest neuromelanin may carry with it a potential for cytotoxicity, or that its excessive presence reflects a 'weakening' of cells bearing such high amounts.

Indeed, as with lipofuscin, age-related increases in neuromelanin in the substantia nigra are associated with reductions in nerve cell cytoplasmic RNA content and nucleolar volume[90] and decreases in the content of dopamine synthetic enzymes[89,711] and dopamine production[712-715] in the corpus striatum. In old age therefore there is

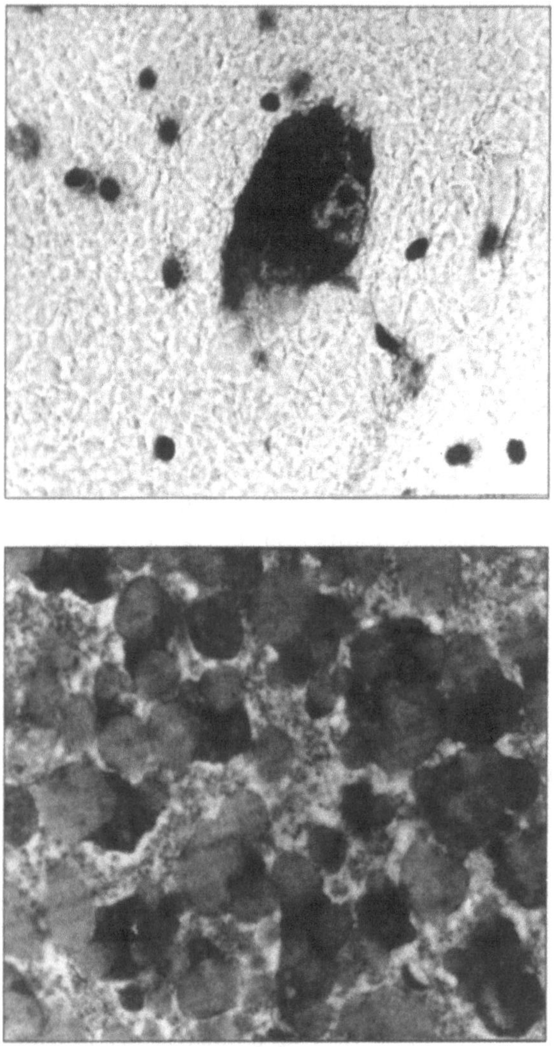

Fig. 2.24. Neuromelanin containing cells of the substantia nigra (a, top). The pigment, like lipofuscin, contains a globular, electron lucent component, on a coarse electron dense protein matrix, though additional granular material is present.

a 25% decrease in protein synthetic capacity of cells of the substantia nigra,[90] though the actual loss of nerve cells far exceeds this, averaging some 50%.[88-92] However, in the other major neuromelanin containing cell group, the locus caeruleus, there is less cell loss[77-85] and better maintenance of protein synthetic capacity;[77,78,80] levels of the noradrenaline synthesizing enzyme, dopamine β-hydroxylase[716] as well as noradrenaline itself[713,715] within projection areas of the cerebral cortex are not reduced. This lessening of severity of change in cells of the locus caeruleus, compared to the substantia nigra, may partly reflect their lower pigment concentration[90] though the degradative products of noradrenaline are less cytotoxic than those of dopamine.[705] Furthermore, cells of the locus caeruleus have highly extensive axonal connections throughout the cerebral cortex and may thus have substantially higher 'reserves' of metabolic potential for functional compensation than those of the substantia nigra when faced with a loss of neighboring neurones.

Consequently, as with lipofuscin, it is uncertain whether the accumulation of neuromelanin within these cell types carries with it any metabolic or cytotoxic (free radical) repercussions; it may likewise simply accumulate as a marker of the passage of time upon such cells.

2.10. OTHER NEURONAL CHANGES

2.10.1. Intranuclear Inclusions

Inclusions such as paracrystalline rodlets[717] (Fig. 2.26a) or spherical bodies[718] (Fig. 2.26b) are sometimes seen under the electron microscope within the nucleus of nerve cells, and the frequency of these may increase with aging.[719,720] The composition of these structures remains uncertain but they may consist of microfilaments.[720] Any relevance they might have to cell function is unknown, though they could be important in terms of changes in DNA transcription and protein synthesis since their appearance has been said to increase in situations of metabolic disturbance associated with, for example, viral infections such as SSPE[721-723] or in Creutzfeldt-Jakob disease.[724]

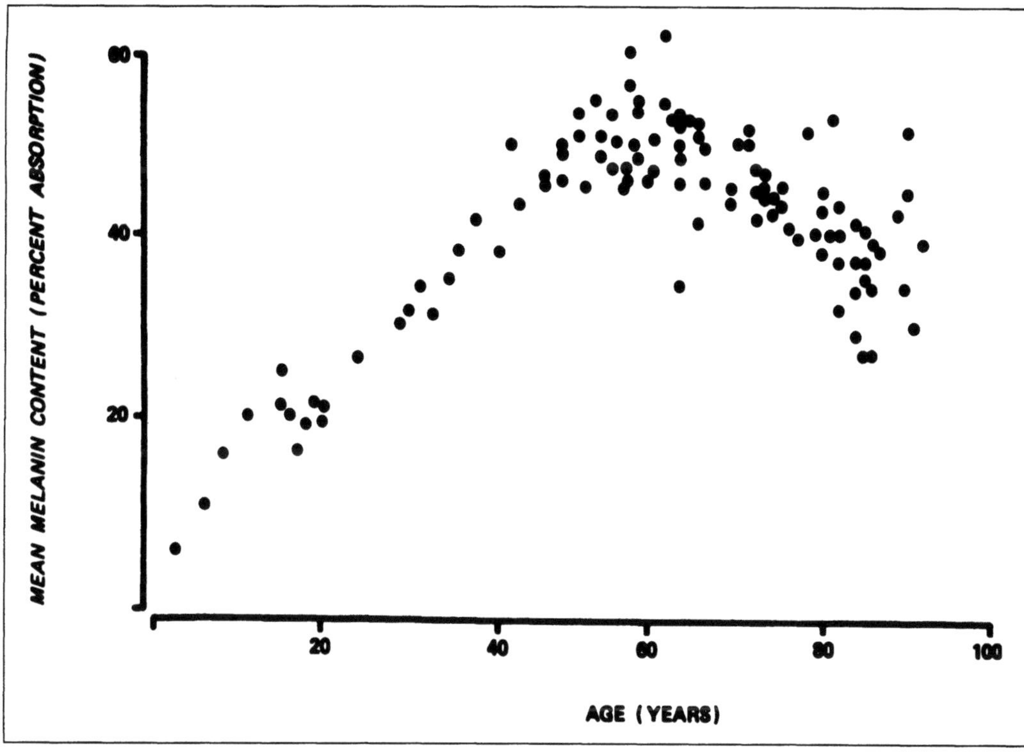

Fig. 2.25. The average amount of neuromelanin pigment in cells of the substantia nigra increases with age until about 60 years, then declines thereafter.

Eosinophilic inclusions have been reported in the nucleus of pigmented neurones of the substantia nigra and locus caeruleus.[725-727] These structures, called Marinesco bodies,[727] are round or oval in shape and stain pink with eosin or Masson Trichrome but are PAS negative; up to four inclusions, each measuring 2-10μm in diameter may be seen in a single nucleus. They consist of non-membrane bound, aggregated, osmiophilic granular material occasionally associated with filamentous material.[728] Their origin and significance is unknown and while they may be age-related[727] their presence seems to be increased in encephalopathies associated with liver[725] or lung[729] disease.

2.10.2. Intracytoplasmic Inclusions

In most normal people, and especially in the elderly, clusters of eosinophilic granules can be seen in the cytoplasm, lying among the pigment granules, of cells of the substantia nigra and locus caeruleus. Their ultrastructure is described[730] as groups of parallel filaments about 8nm in diameter, connected by finer filaments, which are present within the distended cisternae of the rough endoplasmic reticulum. These sort of granules have also been seen within altered mitochondria of the same cell types.[731] Similar structures have been reported in neurones of the dorsal root ganglia[732] and thalamus.[733,734] Although their origin is uncertain and their significance unknown, they are considered to represent an age-related process occurring within affected cells.

2.10.3. Polyglucosan Bodies

These consist mostly of glycogen-like material (glucose polymers) mixed with a small amount of protein and glycosaminoglycan.[735-741] The best known form of these are the corpora amylacea, these being

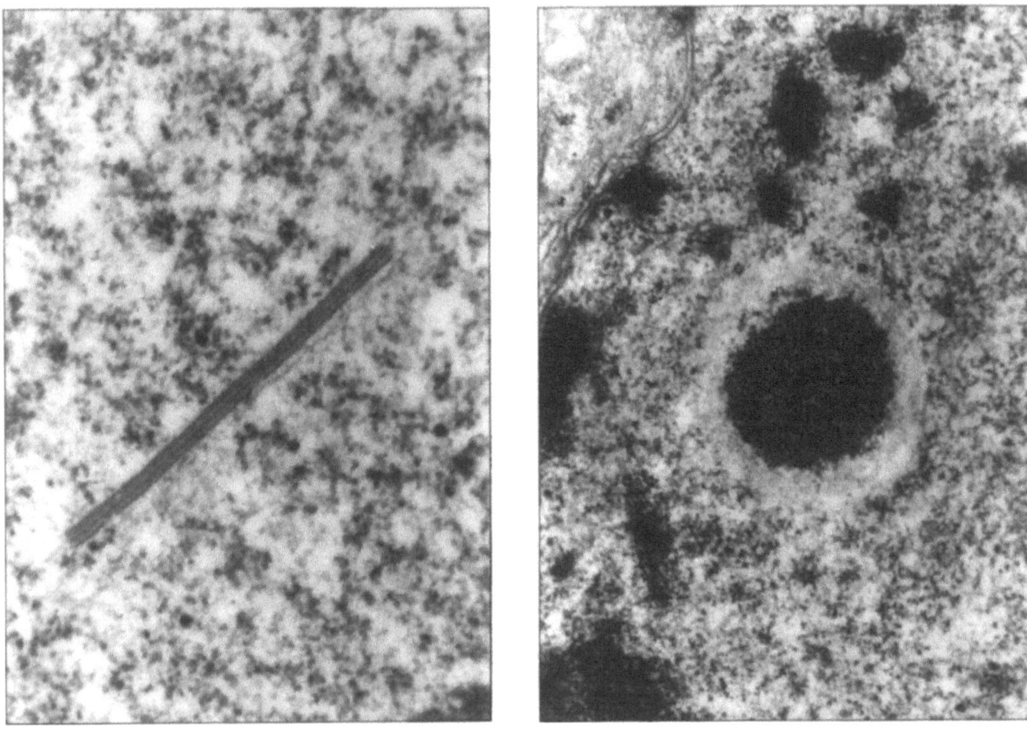

Fig. 2.26. Intranuclear rodlets (a, top) and spherical bodies (b, bottom) in nerve cells.

rounded (5-20µm in diameter) structures having a bluish or grey color in haematoxylin-eosin stained sections. They are strongly PAS positive and frequently have a denser core.

Corpora amylacea are mostly seen in subpial, perivascular or subependymal areas of grey and white (especially) matter and increase in number with age or in the presence of neurodegenerative disease.[742,743] Ultrastructurally, they are usually present in the processes of astrocytes,[744] but sometimes in neurones too,[743,745-747] and consist of randomly arranged 6.5nm filaments which form whorls and are not bounded by a limiting membrane. The filaments often become more compact within the center and may contain electron dense floccular material.

The low amounts of protein within corpora amylacea are, however, strongly immunoreactive for markers of oligodendroglial cells or axonal (myelin) elements.[748] They are also ubiquitinated[735,736,748,749] and contain low molecular weight heat shock proteins[749,750] and complement factors.[751] The intracellular source of corpora amylacea remains unknown though the presence within them of mitochondrial proteins and DNA suggests[742] they might represent the autophagic remains of such organelles which have undergone an oxidative stress mediated degeneration leading to their accumulation within astrocytic, or other cellular, processes. Death of affected cells results in their extracellular deposition and build up in white matter (especially periventricular) regions. X-ray microprobe analysis[743,752] has shown them to contain much sodium, phosphorus, sulphur and chloride with minor amounts of aluminum, silicon, copper, iron, potassium and calcium. They may thus contain inorganic ions or compounds such as sodium chloride or calcium phosphate. Although widely regarded as being of no pathological significance, it has been suggested[751] that corpora amylacea could serve a useful purpose, acting perhaps as 'traps' for the end-products of myelin damage or neuronal degeneration brought about by complement activation. In this latter context they might prevent recognition of such immunogenic proteins by lymphocytes or microglial cells and by doing so restrain further tissue damage. Others[752] consider corpora amylacea may firstly adsorb, then absorb, inorganic material extravasated from blood and cerebrospinal fluid and taken up by astrocytes following failures in blood brain or blood cerebrospinal fluid barriers with age or neurodegeneration.

2.10.4. Tissue Mineralization

2.10.4.1. Basal ganglia calcification

Severe calcification of the basal ganglia, thalamus and dentate nucleus is common in persons suffering from parathyroid hormone deficiency.[753] However, a much milder calcification is frequently seen in the brains of apparently mentally normal old people and persons of all ages with a wide variety of neurological illnesses,[754-760] but particularly so in AD and DS[758-760] though not in cerebrovascular disease.[757,760] The calcification either involves the walls of the large arteries of the globus pallidus, which lose their muscle and elastic coats and become thickly fibrosed and calcified, (Fig. 2.27) or is seen as "strings" of calcospherites lying alongside capillaries or occurring freely in the neuropil.[760] Occasionally, the dentate nucleus of the cerebellum or the fascia dentata of the hippocampus can be affected. The significance of this pathology is not known though it may represent some kind of aging process in blood vessels of the globus pallidus that is exacerbated in AD and DS,[760] but one that is not obviously related to co-existing cerebrovascular disease. Chemical analysis shows the calcospherites to contain variable amounts of polysaccharide and protein enriched with calcium, iron, manganese and aluminum.[753] The increased presence of these particular elements is probably not however indicative of any homeostatic disorder of metals.

2.10.4.2. Iron deposition

In, or close to, areas of old infarction surviving nerve cells may become coated with iron and calcium, this conferring a positive Prussian Blue reaction for iron and a positive von Kossa reaction for calcium. This is known as ferrugination or mineralization of nerve cells. Hemosiderin deposition following subarachnoid haemorrhage gives an orange-yellow discoloration to the meninges. A similar pigmentation over the surface of, or within, the brain tissue marks the site of a prior contusion. While both of these changes may be commonly seen in the brains of elderly persons, they are incidental findings not obviously related to any distinct aging process.

2.10.5. Axonal Changes

Axonal torpedoes are argyrophilic swellings typically seen on the proximal part of the axon of Purkinje cells of the cerebellum before the origin of collateral branches. They are most frequently seen in old age and in neurodegenerative disorders, especially those involving the cerebellum such as multisystem atrophy and Creutzfeldt-Jakob disease.[761,762] The axon torpedo is characterized by a central accumulation

Fig. 2.27. Basal ganglia calcification. The walls of the large arteries in the globus pallidus show calcification in H & E (a, top) and von Kossa (b, bottom) stained sections.

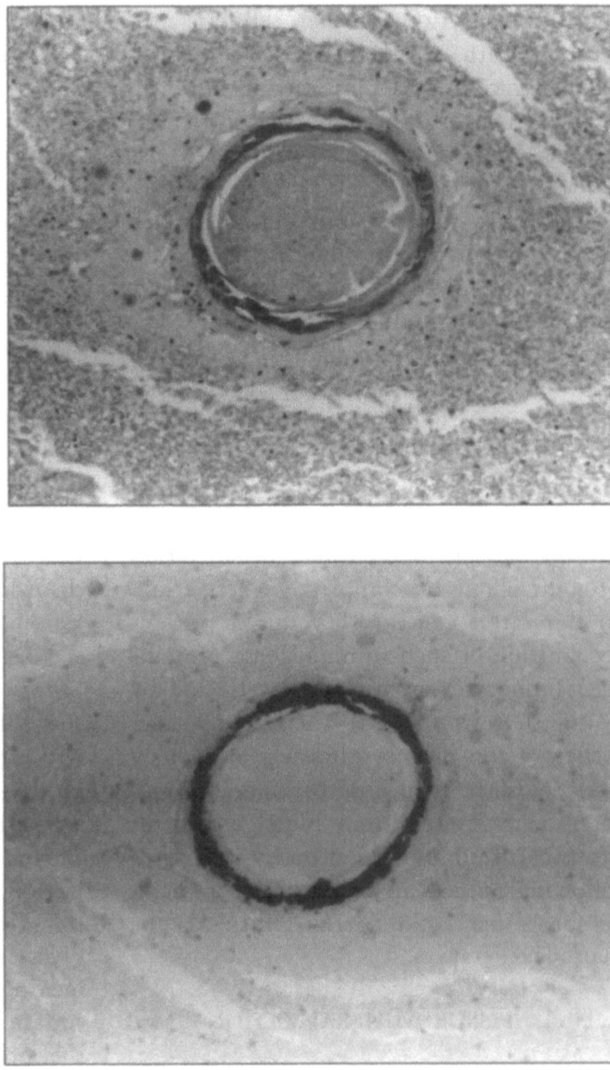

of haphazardly arranged 10nm neurofilaments which displace the usual mitochondria and endoplasmic reticulum to the periphery.[763] Axon torpedoes are distinct from the retraction bulbs that typically form at the ends of damaged or severed axons following traumatic head injury and which contain many lamellar, multivesicular and dense bodies. As these latter changes are not seen in axon torpedoes they may represent a regenerating, rather than a degenerating, state within the nerve cell. Similar, but more rounded, structures measuring over 20μm in diameter have been seen in the axons of motor neurones, particularly in amyotrophic lateral sclerosis.[764]

Granular glycogen bodies, consisting of densely packed accumulations of α or β (but not both) glycogen granules can sometimes be seen in both myelinated and unmyelinated axons.[765,766] (Fig. 2.28) These may measure from 5-50μm in diameter, and in AD are often seen lying in close proximity to the dystrophic neurites of the senile plaque. Although so far only reported in AD, [765,766] such bodies may be more common with failure to comment more widely on these reflecting the difficulty in obtaining suitable tissues for their demonstration at cerebral biopsy. Their presence may relate to disturbances in the transport of carbohydrate within nerve processes perhaps because of interruptions in axonal flow.

2.11. CHANGES IN GLIAL CELLS

Glial cells, and particularly astrocytes, show changes in response to nearly all forms of central nervous system injury or disease. Not surprisingly therefore changes in all three major glial cell types–astrocytes, oligodendrocytes and microglial cells–are commonly seen in the brains of elderly persons. Such changes are usually of a reactive nature, involving mainly an increase in their size and morphology with alterations in their pattern of protein expression though an overall increase in number can occur through local immigration of cells or by mitotic division of the preexisting population. Only in situations where their viability is threatened (e.g. in ischaemic hypoxia) is the number of glial cells actually reduced and this will occur only as part of a wider tissue damage. In this particular respect oligodendrocytes are more vulnerable than other glial cell types. Hence, hypoxia-ischemia may deplete the tissue of oligodendrocytes and myelin while astrocytes survive and react.

2.11.1. Astrocytes

Astrocytes are normally, but sparsely, located throughout all grey and white matter regions, but are principally found in tissue areas in close juxtaposition to the cerebrospinal fluid such as the outer layer of the cerebral and cerebellar cortex where their processes form a glia limitans–a boundary of cell processes that serves as a barrier between the cerebrospinal fluid and the brain parenchyma. They also abut cerebral microvessels via their end-feet–processes that, along with the vascular basement membrane and the endothelium, form a major component of the blood brain barrier.

Although in the past, silver impregnation or haematoxylin-based methods (e.g. phosphotungstic-acid haematoxylin) have been used to demonstrate the location and morphology of astrocytes this is nowadays best shown by the immunohistochemical detection of glial fibrillary acidic protein (GFAP), the intermediate filament protein exclusively produced by these cells. In young persons GFAP immunohistochemistry shows only the thinnest of laminar staining at the glia limitans. However, this staining increases in old age, and particularly so when neurodegenerative disease is present, leading to a considerable thickening, often maximally over the gyral crests, in the glia limitans. Indeed, in elderly brains there is an 3-4 fold increase in the expression of GFAP m-RNA,[767] consistent with this increase in GFAP immunostaining, though the actual number of astrocytes remains constant. [768,769] Similarly, the staining of vascular end-feet is inconspicuous in young persons, though in the elderly, and especially when arteriosclerosis is present, astrocytes and their processes become prominent these often concentrating around the affected vessel. (Fig. 2.29) In addition, patchy foci of astrocytic reaction can frequently be seen in the cerebral cortex or the hippocampus of many old people, these relating to the cored or neuritic plaques of which they are an important component.[172,473-477] These astrocytic changes may reflect responses to the secretion of cytokines, such as IL-1[457,458] or TGFβ-1,[770] by activated microglial cells, whose numbers also tend to increase with age.

When activated by damage to the blood brain barrier in cerebral edema or infarction or by the presence of extraneous molecules within the extracellular fluid, as for example in arteriosclerosis, astrocytes become plump and rounded and contain much of the extravasated fluid and protein. (Fig. 2.29) These are called gemistocytic (swollen bodied) astrocytes. Such cells have a densely eosinophilic homogeneous cytoplasm, due to the absorbed protein, and contain a slightly enlarged and often eccentric nucleus (occasionally paired nuclei may be present); they are strongly immunoreactive for GFAP. Resolution of acute

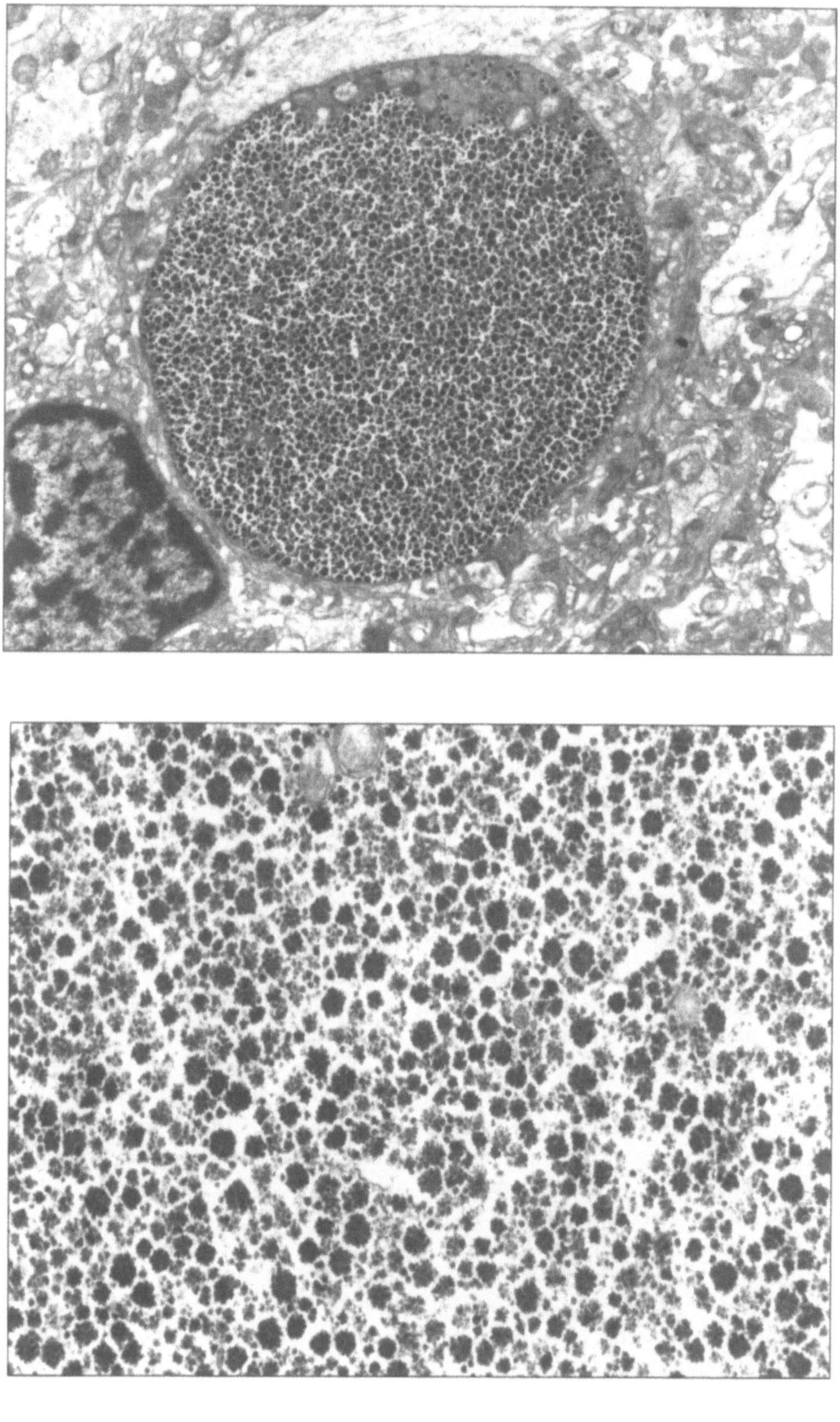

Fig. 2.28. Granular glycogen body in an unmyelinated axon at low (a, top), high (b, bottom) magnification.

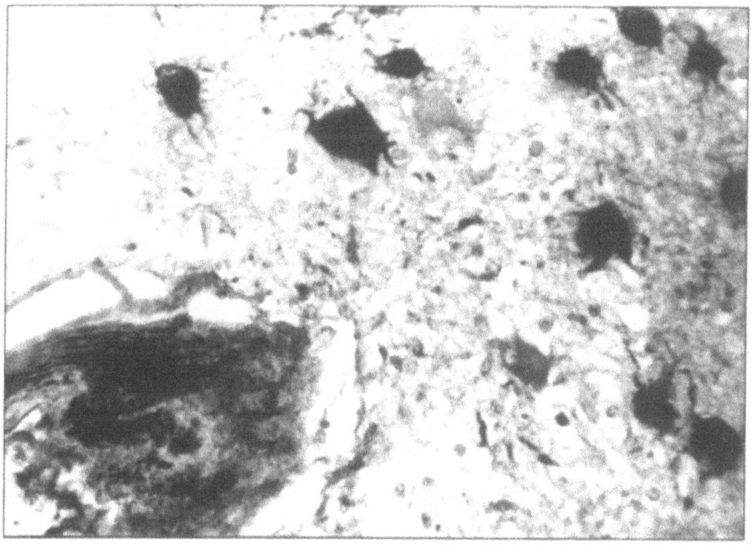

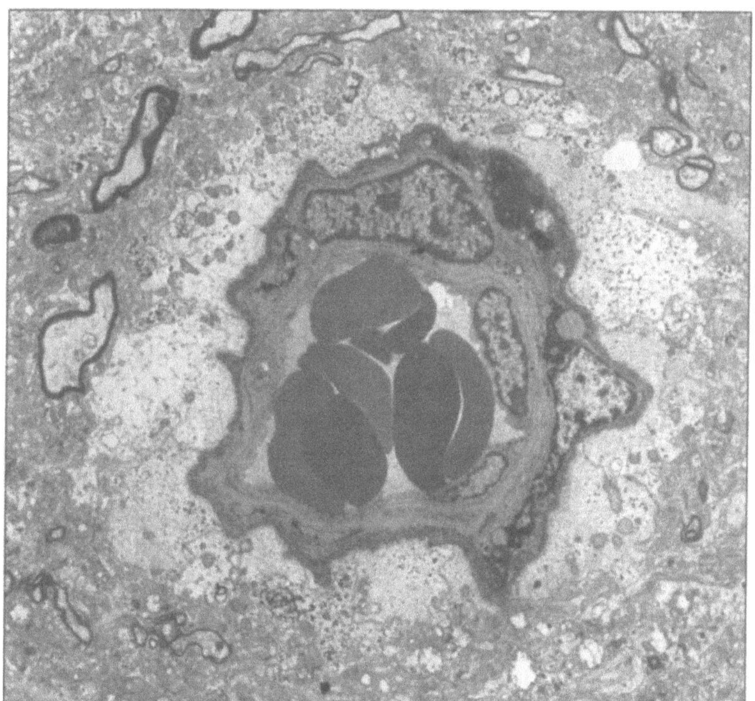

Fig. 2.29. Hypertrophied astrocytes surround a damaged blood vessel and contain much serum protein, as shown by their intense anti-albumin immunoreactivity (a, top). Similar immunoreaction within the blood vessel wall indicates a transudation of protein. Hypertrophied foot processes of astrocytes surround a blood vessel whose wall shows a thickening and reduplication of the basement membrane.

lesions like these leads to shrinkage of these cells and the extensive formation of glial fibers. In the presence of neurodegeneration astrocytes greatly extend their processes, produce much GFAP, and are often termed fibrous astrocytes. These astrocytic reactions are part of a response to injury or damage, presumably 'sealing off' areas of tissue destruction from healthy tissue by the formation of a glial 'scar'. Nonetheless, a phagocytotic role has also been ascribed and in this context astrocytes seem capable of ingesting the PHF of extracellular NFT,[594,595] the lipofuscin from dead neurones or the hemosiderin pigment following haemorrhage.

Other specific pathologies involving astrocytes may lie in the formation and accumulation of corpora amylacea (see earlier) or grumose (foamy spheroid) bodies.[771] This latter change appears to preferentially involve astrocytes in the substantia nigra pars reticulata and the globus pallidus, but also those in the caudate nucleus and putamen to a lesser extent.[771] In this, affected cells are filled with colorless ill-defined granules. Grumose bodies are immunoreactive to ubiquitin and around their margin to GFAP.[771] They are occasionally also reactive for synaptophysin, chromogranin A, neurofilament (68KDa) protein or calbindin, but immunostain inconsistently for tau and MAP5 and do not contain neurofilament (160KDa and 200KDa) protein, phosphorylated neurofilament protein, PHF, Aβ, vimentin, desmin or actin.[771] Ultrastructurally, grumose bodies show dense bodies, ranging from 0.1-1.0μm in diameter, surrounded and infiltrated by glial fibers. Thus, although they may contain elements of the neuronal cytoskeleton, they are probably astrocytic in origin. While present in some aged brains, grumose bodies are frequent in various Parkinsonian-type neurodegenerative disorders like progressive supranuclear palsy and corticobasal degeneration but are not seen in PD itself.[771] It is uncertain whether they are of any clinical relevance.

2.11.2. Oligodendrocytes

These are the myelin forming cells of the central nervous system. In haematoxylin-eosin stained section, they seem to have scant cytoplasm and possess a small, dark, round nucleus closely resembling a lymphocyte in size and appearance. The perikaryon is small and shows only a few protoplasmic processes in silver impregnation. In the white matter they often occur as longitudinal clusters lying between the nerve fibers, whereas in the grey matter they are present around nerve cell bodies as "satellite" cells. Their reaction to injury usually involves their degeneration and loss. Changes in blood supply (ischaemia-hypoxia) in old age, associated for example with arteriosclerosis, will deplete the tissue of oligodendrocytes and result in a net reduction in the amount of myelin within the white matter.

2.11.3. Microglial Cells

Microglial cells have strong phagocytotic capabilities and are generally considered to be the macrophages of the brain. Normally, they are present as 'ramified' (also known as 'resting') microglia whose sparse but long processes project extensively through the brain tissue. They are often seen in close contact with blood vessels or neuronal perikarya. Upon activation the cell body enlarges and rounds up and the processes become shorter and thicker. Such cells then become known as 'activated' or 'amoeboid' microglia. Because they are derived from the macrophage/monocyte cell line microglial cells express, particularly upon activation, many of the cell surface markers associated with white cells such as major histocompatibility class I antigen (HLA-DR) or class II antigen,[772] leukocyte common antigen,[773] CD45 etc. When activated they contain much ferritin protein,[774] this presumably being related to their high myeloperoxidative capabilities when in this state. Microglial cells can also be detected using various highly toxic lectins like ricin (*Ricinus communis*)

(RCA_{120}),[775] abrin (*Abrus precatorius*) (APA) or mistletoe (*Viscum album*) (VVA)[776] lectin but can also be demonstrated using the less toxic elder bark lectin, *Sambucus nigra* (SNA).[172] However, for demonstration in routinely prepared, formalin fixed, paraffin embedded tissue, these latter techniques are best used; cell surface markers such as HLA-DR become lost, or inaccessible, during formaldehyde fixation and tissue processing,[777] though antigen retrieval through microwave treatment may (partially) circumvent this problem. Immunohistochemical techniques with white cell markers are used to best advantage with cryostat, or minimally fixed vibratome sectioned, tissues.

Microglial cells are 'wandering' scavengers moving towards, and concentrating within, areas of tissue damage (due for example to ischaemic necrosis or amyloid plaque formation) where they may break down and ingest dead tissue and nerve cells frequently leaving a glial "star" as witness to their previous action. Although phagocytosis has been assumed to be their principal role more specific duties in terms of amyloidosis,[453-455] IL-1 mediated astrocytosis,[457,458] or the production of free radicals[336,471,472] have been claimed. Because they are capable of expressing APPmRNA and producing their own endogenous APP[245-247] they may also contribute to the deposition of Aβ within plaques.

2.12. CEREBROVASCULAR CHANGES

The energy requirements of the brain are such that about 40% of all the oxygen taken up into the circulation, and much of the glucose synthesized by the body, is used up in the brain. Most of the energy so formed is required for the maintenance of ionic gradients across cell membranes for the propagation of electrical impulses along nerve fibers. As neither of these basic commodities is stored in the brain, the brain is highly dependent upon an adequate and sustained supply of blood for delivery of these vital substances to all parts of the organ. Clinical studies suggest that neurones in certain vulnerable areas of the brain (e.g. hippocampus) may not survive for long if their supplies of oxygen and glucose are lowered below a critical threshold, even for as little as a few minutes. Consequently, the brain is well supplied with blood via a dual arterial circulation–the basilar-vertebral system and the internal carotid system–these being united through a series of 'communicating arteries', at the base of the brain, forming the so-called circle of Willis. The major lobes of the cerebral cortex are served by the anterior and middle cerebral arteries from the carotid circulation, and by the posterior cerebral arteries from the vertebral and basilar circulation. Subcortical structures such as the basal ganglia, brainstem and cerebellum are served by penetrating arteries from the middle cerebral artery and basilar artery respectively.

2.12.1. Ischemia

Failure to supply the brain with adequate oxygen and glucose may lead to irreversible brain damage the scale of which will depend upon the extent of the vascular deficiency and the time period over which this might extend. The proximal cause for such a lack of vital factors may result either from a failure of perfusion (ischaemia) or a reduction in blood oxygenation (hypoxia) even when an adequate flow is maintained; usually a combination of both these damaging events is present. Ischaemia can be focal, due to a failure within a local artery (e.g. middle cerebral artery) or can be more generalized due to an overall poor cardiac output or a major reduction within one or more of the principal supplying vessels (e.g. internal carotid).

While blood flow through the distributive vessels of the brain is appropriately regulated by sympathetic nervous control under normal cardiac output, should cardiac output fall dramatically and compensatory sympathetic action become inadequate certain so-called watershed areas (border zones) of brain, at the boundaries between anterior and middle cerebral artery territories, between middle and posterior cerebral artery territory, or between superior and inferior cerebellar arteries and at the end points of the penetrating arteries, fall at risk of becoming compromised through inadequate perfusion. Hence, infarctions, caused by failures of perfusion, occur particularly in the parieto-occipital cortex or in pre- and post-central gyri, or in the cerebellum and basal ganglia. The extent of damage depends upon the degree of oxygenation of the blood at the time of ischaemia and the period taken for the (successful) restoration of normal blood flow to the brain. Moreover, chronic inflammatory or degenerative damage within neck vertebrae (as in osteoarthritis), causing a collapse of the vertebral column, may result in the vertebral arteries becoming spiral or tortuous, increasing the risk of thrombosis and so putting watershed zones at risk.

The initial effect of ischaemia/hypoxia on brain tissue is to cause microvacuolation of the nerve cell perikaryon leading on to a shrinkage of the cell body, which may then become strongly eosinophilic, with the nucleus becoming misshapen and darkly staining. Later, dark 'encrustations' appear on the nerve cell surface and necrosis ensues. Glial cells are more resistant than neurones to ischaemic damage and may respond by astrocytic proliferation. Phagocytosis of neuronal debris by microglial cells may also take place.

In the cerebral cortex the depths of the sulci rather than the crowns of the gyri tend to be affected most by ischaemia, with the pyramidal cells of layers III and V being more vulnerable than cells in other layers. In the hippocampus, areas CA1, CA4 and CA5 are vulnerable with CA2 and the dentate gyrus most resilient. The head of the caudate nucleus is most frequently at risk in the basal ganglia and in the cerebellum it is the Purkinje and basket cells that are lost while the granule and Golgi cells and the cells of the dentate nucleus are more resistant.

2.12.2. Atherosclerosis

Some degree of atherosclerosis is present in the walls of the major extracerebral distributive arteries in the great majority of middle-aged and elderly people, this being most prominent in the vertebral/basilar/posterior cerebral arteries or in the internal carotid/middle cerebral artery system.[32] In most instances the luminal reduction imposed by the atheroma is insufficient to cause a significant fall in blood flow through the vessel and overall or local cerebral perfusion is usually maintained (particularly if collateral circulation via the circle of Willis is good). A reduction in luminal area of more than 75% is generally necessary to affect flow. However, if plaque ulceration or superadded thrombosis occurs in carotid or vertebral arteries occlusive infarctions, particularly in middle cerebral artery territory (basal ganglia, temporal lobe) or posterior cerebral artery territory (occipital lobe) and cerebellum, may result. Moreover, the losses of elasticity and musculature that accompany the deposition of fatty and fibrous tissue may render the vessels unresponsive to sympathetic regulation, limiting their flexibility of response in terms of potential augmentation of blood flow. Such thickened and rigid vessels, which often become calcified or distended (ectactic) and tortuous, transform into passive conductors of blood through which, the rate of blood flow is "determined" more by overall cardiac output than by local regulatory factors.

2.12.3. Hypertension and the Brain

It has long been known that the value for cerebral blood flow in patients with elevated systemic blood pressure is similar to that in individuals whose systemic blood pressure is normal. Hence, hypertensive patients are more susceptible to brain ischaemia than normals because they are less able to compensate for sudden falls in blood pressure. Brain damage is liable during periods of systemic hypotension which may reduce the overall rate of flow through the brain rendering the boundary zones especially susceptible to ischaemic damage.

In addition to such cardiovascular mediated physiological changes, hypertension produces overt pathological changes in the brain. Routine autopsy discloses such changes in the brains of many elderly subjects, even in the absence of observable clinical symptoms. Firstly, it may aggravate (pre-existing) atheroma within the large extracerebral arteries, causing an extension of damage into regions of vessels, or actual vessels, not normally affected in normotensives, for example the penetrating arteries of the brain stem or basal ganglia. Secondly, in large and medium-sized arteries, following a period of muscle cell hypertrophy and thickening and reduplication of the internal elastic laminae, there is a replacement of muscle by fibrous tissue and fragmentation and absorption of the elastic elements. The artery becomes thickened and rigid (arteriosclerotic), its lumen dilated and overall becomes longer and tortuous (ectactic). In small arteries (i.e. those <1mm in diameter) internal thickening also occurs, producing a concentric increase in fibrous tissue (onion skinning), with substantial narrowing of the lumen. Hyaline change through the vessel wall may be seen. Thirdly, arteriosclerosis within intraparenchymal arteries produces a spiraling and elongation of the vessel under the effects of a sustained elevated blood pressure, forming "lacunae", measuring 3-20µm in diameter. Here the artery has lost contact with the surrounding brain parenchyma and runs in a 'tunnel' of expanded perivascular space. The term for this is 'etat lacunaire' when the cavity is present in the grey matter; 'etat crible' describes similar white matter cavities. The cavities are lined by a narrow band of gliosis. Occlusion of such vessels may lead to (micro) infarction within the cerebral cortex, basal ganglia or brainstem. Lacunae are common in normal elderly subjects, occurring in at least 10% of all routine autopsies. When numerous, the cumulative damage caused by such lesions may be sufficient to cause mental deterioration. Such changes are readily detected on CT[1-4] or NMR[5-9] scanning and appear as regions of translucency (leukoaraiosis) or hyperintensity, respectively, in the white matter, especially in a periventricular location (see earlier). Lastly, microhaemorrhages, associated with microaneurysms or fibrinoid necrosis, may occur, particularly in those cases where a rapid rise in blood pressure has evolved over a short period of time (i.e. in malignant hypertension).

2.12.4. Microvascular Changes

Microvascular changes take place in aged persons otherwise free from overt disease of the large or medium sized intra- and extracellular blood vessels. In these there is a net reduction in old age in the density of microvessels.[778-780] Pathological changes involving, a looping or coiling of vessels leading to a tortuosity,[780,781] atrophy[782] or a thickening of basement membrane[783] are seen in many remaining vessels. Whether these vascular changes represent primary disease, or age-related, damage is uncertain. They might simply reflect a remodeling of the microcirculation to meet the falling energy demands of a declining nerve cell population.[781] Nevertheless, the basement membrane thickening might indeed represent actual age-changes in the microvasculature, these perhaps reflecting the progressive increase in blood pressure towards later life that occurs in many elderly subjects.

2.12.5. Amyloidosis

In many elderly persons, and particularly in those where a cerebral atrophy is apparent, the leptomeningeal blood vessels (especially) (Fig. 2.30) and, most times also, the intraparenchymal arteries and arterioles show a thickening and intense eosinophilia of their walls. This is most prominently seen in those arteries supplying the occipital lobes and the cerebellar cortex. The proteinaceous material within the walls of these arteries stains with periodic acid-Schiff and displays a typical (of amyloid) fluorescence when stained with Thioflavin S or T and subjected to ultraviolet illumination, or a characteristic birefringence when stained with Congo red and viewed under polarized light. The media and adventitia are affected with disruption and loss of the elastic lamina. The distribution of amyloid within the wall is often discontinuous and patchy. Severely affected vessels are said to display cerebral 'congophilic angiopathy' and the brain, 'cerebral amyloid angiopathy'.[784,785] Sometimes, the walls of capillaries are affected, these showing a 'dyshoric angiopathy' in which the amyloid appears to "stream" away from the vessel wall as fine radiating fibrils into the surrounding parenchyma. Severely affected vessels may cause ischaemic damage to surrounding tissue. They may also become thrombosed producing microinfarctions or even rupture producing microhaemorrhages. Occasionally, but more commonly in certain inherited forms of disease present in some Dutch and Icelandic families, fatal intracerebral or subarachnoid haemorrhage may occur.

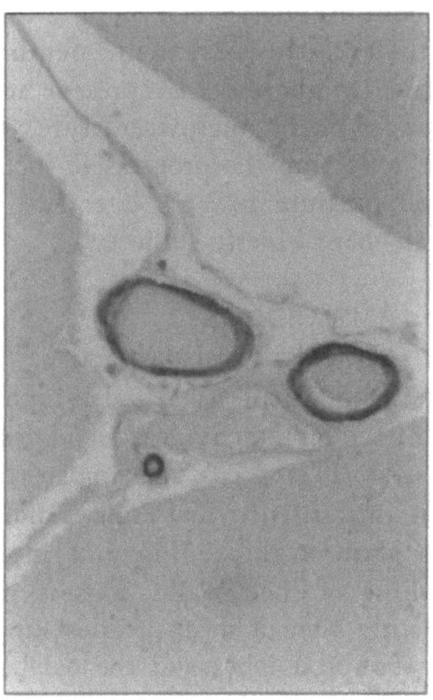

Fig. 2.30. Amyloid angiopathy within a leptomeningeal artery in the cerebellum as seen in anti-Aβ immunostaining.

Mostly, cerebral cortical vessels are affected; those in the white matter and basal ganglia are rarely involved.

2.12.5.1. Composition of the amyloid

It has long been recognized that this amyloid angiopathy forms a basic part of the histopathology of AD;[784-786] indeed most, and nearly all, cases of AD and DS show some amyloidosis within cerebellar and cerebral (occipital) cortical blood vessels.[166,207-211,787,788] Hence, it has been suspected that the amyloid within the blood vessel wall in AD might bear some biochemical relationship to that within plaques in the brain parenchyma. Certainly, at ultrastructural level the two proteins seem similar with both being made of apparently the same 10nm filament.[789] However, the observation of rare cases showing severe plaque formation without amyloid angiopathy, or the converse, imply that while these two pathological processes leading to a deposition of amyloid frequently co-exist they are not necessarily interdependent, each can exist in the absence of the other.

Initial studies isolating the amyloid protein from blood vessels in AD[207] and DS[790] showed this to contain a peptide 40 amino acids long. This structure was subsequently confirmed[196,200,208-211] and a clear homology with the Aβ extracted from plaques, at least over the initial 40 amino acid sequence of the latter ($A\beta_{1-40}$), demonstrated.[148,196,197,200] Later work has now shown that the Aβ extracted from vessels also contains lesser though variable amounts of species like $A\beta_{1-39}$ and $A\beta_{1-42}$.[210,211] Immunohistochemical studies[190,210-213,216] confirm that $A\beta_{40}$ is a major component of vascular Aβ, but again show that other Aβ species, terminating at ala-42 ($A\beta_{42}$) and thr-43 ($A\beta_{43}$) are present but to a lesser extent. As with plaque amyloid the amino-terminal properties of cerebrovascular amyloid are variable with major Aβ peptide species commencing at asp-1, with or without isomerization and racemization, and pyroglutamate-3 though, interestingly, more Aβ species commencing at pyroglutamate-11 and leucine-17 (P3 peptide) are seen in vascular amyloid deposits than in plaque amyloid.[191,220,221]

Hence, it appears that the Aβ peptides deposited both within plaques and blood vessel walls are essentially the same. Both are derived from the same amyloid precursor protein; variations in the sequence extent of the peptides present, and the relative amounts of each, may reflect either local differences in cellular source or patterns of catabolism.

2.12.5.2. Source of Aβ in amyloid angiopathy

For many years argument has raged as to whether vascular Aβ is derived from initial parenchymal deposits that are subsequently deposited in blood vessel walls, or vice versa, or even that both have separate but haematogenous origins. It is now, however, accepted that both proteins probably have separate sources with each being produced from local tissue elements. Hence, plaque Aβ is thought to be derived essentially and primarily from neuronally produced APP (probably APP_{695}) upon which microglial cell produced Aβ may be secondarily incorporated. Smooth muscle cells express and contain much APP[791-795] and these are the most likely source of the vascular amyloid,[796-798] though here again a contribution by perivascular cells or microglial cells[796] might take place.

2.12.5.3. Amyloid angiopathy and apolipoprotein E

As with plaque amyloid, Aβ within blood vessel walls strongly binds ApoE.[191,193,341,345] Whether possession of the ApoE E4 allele has any influence over the amount of Aβ deposited within affected vessels or the overall extent of amyloid angiopathy through the vasculature of the brain is uncertain. Some groups[344,799] have suggested that the severity of amyloid angiopathy, like extent of plaque amyloid deposition, in AD may be increased in the presence of the ApoE E4 allele, though other workers have not found this to be so.[366,367] Hence, the effect of ApoE E4 on extent of amyloid angiopathy still remains to be resolved.

2.12.5.4. Amyloid angiopathy and other proteins

Like plaque amyloid, the amyloid within blood vessel walls contains the amyloid-associated proteins, HSPG,[376,377,380] ACT[416-418] and amyloid P component.[420-422] The purpose of these is not clear though they probably likewise play a role in conferring the β-pleated nature to the amyloid fibrils and increasing their persistence in the vessel wall.

Other proteins, not normally present in blood vessel walls, such as heme oxygenase 1 occur in vessels affected by amyloid angiopathy.[608] The elevated expression of this molecule in smooth muscle cells may again occur as a response to oxidative stress, this possibly being induced by Aβ deposition.

2.13. BRAIN AGING: COMPENSATION VERSUS REDUNDANCY

Faced with this wealth of information pointing to the conclusion that the majority of, if not all, the brains of elderly subjects show some evidence of damage over time whether this be in terms of atrophy and loss of nerve cells, the presence of diverse degenerative changes within, or outside of, surviving nerve cells, or problems in sustaining the required supply of oxygen and glucose, it is tempting to accept that such changes reflect a system that has entered into a stage of passive decline in later life; functional compromise would seem an inevitable consequence of such accumulated pathology. Yet clinical studies (Chapter 1) show that this is not so. Not everyone entering the later decades of life experience a significant decline in their faculties, mental or physical. This means that there is either a tremendous 'slack' or redundancy in the system, such that seemingly substantial amounts of cell or tissue damage can be sustained without undue impairment of brain function, or that loss of, or damage to, critical elements might spark off a recovery or a compensatory action in anatomically neighboring or functionally related regions that increase their capacity for function accordingly. Dysfunction might occur only when the cumulative degree of damage exceeds the capacity for compensation or when such compensatory mechanisms fail, perhaps because of an age-related attrition that involves this capability along with other aspects of brain function.

This notion of 'slack in the system' or 'capacity for compensation' has led to the concept of 'functional reserve'. The chance of becoming impaired in later life might thus depend partly upon the extent and frequency of damaging events which the system may sustain and partly upon the powers of 'reserve' one might have as an individual. The greater the reserve, the more the amount of damage that can be tolerated.

What this concept of functional reserve might represent in anatomical terms is uncertain. Clearly, a high degree of nerve cell loss can on occasions be tolerated without apparent functional deficit. For example, in PD it is well recognized that over 50% of the original complement of nerve cells in the substantia nigra is required to be lost before the clinical symptomatology of the disease becomes apparent.[91] This implies that in normal circumstances much of the available nerve pathway between the substantia nigra and the corpus striatum is underused and that even after a substantial decimation of the original complement of neurones and fibers (as in PD) sensory information can still be effectively channeled down the remaining elements of that pathway and function maintained. This kind of redundancy in the system might be established at an early time of life, perhaps during brain development when neurones are produced and then "trimmed" according to required levels by a combination of growth factors and apoptotic mechanisms. In such a situation it is possible that certain individuals are left as adults with a "happy excess" of nerve cells and synapses whereas in others the merest of surpluses is available. Obviously, if this were to be the case, individuals with low reserve would fall at greater risk of functional decline following cumulative small amounts of damage accrued over their lifetime or as a result of a single event in later life when the damage this might produce may be more widespread and severe.

Alternatively, we might all start out as adults on a fairly common basis, in crude anatomical terms of cell number etc, but with wide variation in the capacity to respond to, or repair, damage being the critical factor. In this context, synaptic or dendritic remodeling and factors which might influence these could be instrumental in promoting a healthy and sustainable reaction. That plasticity responses like these normally occur in later life is unarguable (see section 2.3). Even in the devastating neurodegenerative diseases of old age compensatory changes can take place. For example, in PD, following loss of their connections with nigral neurones, adaptive changes take place in the synapses on striatal neurons made by incoming fibers from cortical pyramidal cells.[800] Of course, the dopaminergic system with its extensive striatal ramifications might represent a 'special case' with other cellular systems being unable to match this potential, though this does not seem so. In AD, even at late stages of the disease, the granule cells of the dentate gyrus of the hippocampus seem able to extend their dendritic fields to compensate for loss of incoming fibers from the entohinal cortex via the perforant pathway.[801,802]

Diseases such as AD and PD may only appear clinically when the powers of compensation have been finally overwhelmed, even though in pathological terms the disease may have been ongoing for some time prior. Therefore, instead of being thought of as an inevitable and 'all downhill' process which the brain is powerless to influence, aging in the brain may be an active process which in many people successfully meets the challenges of stress and damage in later life; only in a minority of individuals do these responses fail with the damage so inflicted culminating in the familiar disorders of later life.

In practice, it is likely that 'functional reserve' relates to a combined effect of both these properties (i.e. redundancy and compensation) though the balance between them, as well as the extent of each, may vary much between individuals.

References

1. LeMay M. Radiologic changes of the aging brain and skull. Am J Radiol 1984; 143: 383-389.
2. Schwartz M, Creasey H, Grady CL et al. Computed tomographic analysis of brain morphometrics in 30 healthy men, aged 21 to 81 years. Ann Neurol 1985; 17: 146-157.
3. Pfefferbaum A, Zatz LM, Jernigan TL. Computer-interactive method for quantifying cerebrospinal fluid and tissue in brain CT scans: Effects of aging. J Comput Assist Tomogr 1986; 10: 571-578.
4. Drayer BP. Imaging of the aging brain. Radiology 1988; 166: 785-796.
5. Creasey H, Rapoport SI. The aging human brain. Ann Neurol 1985; 17: 2-10.
6. Bradley WG, Waluch V, Brant-Zawadzki M et al. Patchy, periventricular white matter lesions in the elderly: A common observation during NMR imaging. Noninvas Med Imaging 1984; 1: 35-41.
7. Brant-Zawadski M, Fein G, Van Dyke C et al. MR imaging of the aging brain: patchy white-matter lesions and dementia. Am J Neuroradiol 1985; 6: 675-682.
8. George AE, de Leon MJ, Kalnin A et al. Leukoencephalopathy in normal and pathologic aging: II MRI of brain lucencies. Am J Neuroradiol 1986; 7: 561-570.
9. Fazekas F, Chawluk JB, Alavi A et al. MR signal abnormalities at 1.5 T in Alzheimer's dementia and normal aging. Am J Neuroradiol 1987; 8: 421-426.
10. Sullivan EV, Marsh L, Mathalon DH et al A. Age-related decline in MRI volumes of temporal lobe gray matter but not hippocampus. Neurobiol Aging 1995; 16: 591-606.
11. Coffey CE, Wilkinson WE, Parashos IA et al. Quantitative cerebral anatomy of the aging human brain–A cross-sectional study using magnetic resonance imaging. Neurology 1992; 42: 527-536.
12. Bhatia S, Bookheimer SY, Gaillard WD et al. Measurement of whole temporal lobe and hippocampus for MR volumetry: Normative data. Neurology 1993; 43: 2006-2010.
13. Jack CR, Petersen RC, O'Brien PC et al. MR-based hippocampal volumetry in the diagnosis of Alzheimer's disease. Neurology 1992; 42: 183-188.
14. Golomb J, De Leon MJ, Kluger A et al. Hippocampal atrophy in normal aging: An association with recent memory impairment. Arch Neurol 1993; 50: 967-973.
15. Kuhl DE, Metter EJ, Riege WH et al. Effects of human aging on patterns of local cerebral glucose utilization determined by the (18 F) flurodeoxyglucose method. J Cereb Blood Flow Metab 1982; 2: 163-171.
16. Stoessl AJ, Tuokko H, Martin WRW et al. Cerebral glucose metabolism in normal aging. Neurology 1986; 36: 104.
17. Chawluk JB, Alavi A, Dann R et al. Positron emission tomography in aging and dementia: Effect of cerebral atrophy. J Nucl Med 1987; 28: 431-437.
18. Yoshii F, Barker WW, Chang JY et al. Sensitivity of cerebral glucose metabolism to age, gender, brain volume, brain atrophy and cerebrovascular risk factors. J Cereb Blood Flow Metab 1988; 8: 654-661.
19. DeCarli C, Atack JR, Ball MJ et al. Post-mortem regional neurofibrillary tangle densities but not senile plaque densities are related to regional cerebral metabolic rates for glucose during life in Alzheimer's disease patients. Neurodegeneration 1992; 1: 113-121.
20. De Leon MJ, Ferris SH, George AE et al. Positron emission tomographic studies of aging and Alzheimer disease. Am J Neuroradiol 1983; 4: 568-571.
21. Duara R, Margolin RA, Robertson-Tschabo EA et al. Cerebral glucose utilization as measured with positron emission tomography in 21 resting healthy men between the ages of 21 and 83 years. Brain 1983; 106: 761-775.
22. Martin WRW, Palmer MR, Patlak CS et al. Nigrostriatal function in humans studied with positron emission topography. Ann Neurol 1989; 26: 535-542.

23. Sawle GV, Colebatch JG, Shah A et al. Striatal function in normal aging: implications for Parkinson's disease. Ann Neurol 1990; 28: 799-804.
24. Blinkov SM, Glezer II. The human brain in figures and tables. In: Basic Books Inc, Plenum Press, New York 1968.
25. Dekaban AS, Sadowsky D. Changes in brain weights during the span of human life: Relation of brain weights to body heights and body weights. Ann Neurol 1978; 4: 345-356.
26. Davis PJM, Wright EA. A new method for measuring cranial cavity volume and its application to the assessment of cerebral atrophy at autopsy. Neuropath Appl Neurobiol 1977; 3: 341-358.
27. Terry RD, De Teresa R, Hansen LA. Neocortical cell counts in normal human adult aging. Ann Neurol 1987; 21: 530-539.
28. Hatazawa J, Ito M, Yamaura H et al. Sex difference in brain atrophy during aging: a quantitative study with computed tomography. J Amer Geriat Soc 1982; 30: 235-239.
29. Hubbard BM, Anderson JM. Sex difference in age-related brain atrophy. Lancet 1983; I: 1447-1448.
30. Last RJ, Tompsett DH. Casts of the cerebral ventricles. Br J Surg 1953; 40: 525-543.
31. Barron SA, Jacobs L, Kinkel WR. Changes in size of normal lateral ventricles during aging determined by computerized tomography. Neurology 1976; 26: 1011-1013.
32. Tomlinson BE, Blessed G, Roth M. Observations on the brains of non-demented elderly people. J Neurol Sci 1968; 7: 331-356.
33. Hubbard BM, Anderson JM. Age, senile dementia and ventricular enlargement. J Neurol Neurosurg Psychiatry 1981; 44: 631-635.
34. Miller AKH, Alston RL, Corsellis JAN. Variation with age in the volumes of grey and white matter in the cerebral hemispheres of man: measurements with an image analyser. Neuropath Appl Neurobiol 1980; 6: 119-132.
35. Haug H. Macroscopic and microscopic morphometry of the human brain and cortex. A survey in the light of new results. Brain Pathol 1984; 1: 123-149.
36. Haug H. Are neurones of the human cerebral cortex really lost during aging? A morphometric examination. In: Traber J, Gispern WH eds. Senile Dementia of the Alzheimer Type. Early Diagnosis, Neuropathology and Animal Models. Springer-Verlag, Berlin 1985; 150-163.
37. Harris GJ, Schlaepfer TE, Peng LW et al. Magnetic resonance imaging evaluation of the effects of aging on grey-white ratio in the human brain. Neuropath Appl Neurobiol 1994; 20: 290-293.
38. Hodge CF. Changes in ganglion cells from birth to senile death. Observations on man and honey bee. J Physiol 1894; 17: 129-134.
39. Brody H. Organization of the cerebral cortex III. A study of aging in the human cerebral cortex. J Comp Neurol 1955; 102: 511-556.
40. Brody H. Structural changes in the aging nervous system. Interdiscip. Top. Gerontol 1970; 7: 9-21.
41. Hanley T. 'Neuronal fallout' in the aging brain: A critical review of the quantitative data. Age Aging 1974; 3: 133-151.
42. Colon EJ. The elderly brain. A quantitative analysis in the cerebral cortex of two cases. Psychiat Neurol Neurochir 1972; 75: 261-270.
43. Shefer VF. Absolute numbers of neurones and thickness of the cerebral cortex during aging, senile and vascular dementia, and Pick's and Alzheimer's diseases. Neurosci Behav Physiol 1973; 6: 319-324.
44. Henderson G, Tomlinson BE, Gibson P. Cell counts in human cerebral cortex in normal adults throughout life using an image analysing computer. J. Neurol Sci 1980; 46: 113-136.
45. Anderson JM, Hubbard BM, Coghill GR et al. The effect of advanced old age on the neurone content of the cerebral cortex. Observations with an automatic image analyser point counting method. J Neurol Sci 1983; 58: 233-244.
46. Mann DMA, Yates PO, Marcyniuk B. Some morphometric observations on the cerebral cortex and hippocampus in presenile Alzheimer's disease, senile dementia of Alzheimer type and Down's syndrome in middle age. J Neurol Sci 1985; 69: 139-159.

47. Devaney KO, Johnson HA. Changes in cell density within the human hippocampal formation as a function of age. Gerontology 1984; 30: 100-108.
48. Hubbard BM, Anderson JM. A quantitative study of cerebral atrophy in old age and senile dementia. J Neurol Sci 1981; 50: 135-145.
49. Haug H, Kuhl S, Mecke E et al. The significance of morphometric procedures in the investigation of age changes in cytoarchitectonic structures of human brain. J Hirnforsch 1984; 25: 353-374.
50. Haug H, Eggers R. Morphometry of the human cortex cerebri and corpus striatum during aging. Neurobiol Aging 1991; 12: 336-338.
51. Eggers R, Haug H, Fischer D. Preliminary report on macroscopic age changes in the human prosencephalon. A stereologic investigation. J Hirnforsch 1989; 25: 129-139.
52. Braak H, Braak E. Ratio of pyramidal cells versus non-pyramidal cells in the human frontal isocortex and changes in ratio with aging and Alzheimer's disease. In: Swaab DF, Fliers E, Mirmiran M, Van Gool WA, Van Haaren F eds. Progress in Brain Research. Elsevier Science Publishers BV (Biomedical Division) 1986; 70: 185-212.
53. Ball MJ. Neuronal loss, neurofibrillary tangles and granulovacuolar degeneration in the hippocampus with aging and dementia. A quantitative study. Acta Neuropathol 1977; 37: 111-118.
54. Shefer VF. Hippocampal pathology as a possible factor in the pathogenesis of senile dementias. Neurosci Behav Physiol 1977; 8: 236-239.
55. Mouritzen Dam A. The density of neurones in the human hippocampus. Neuropath Appl Neurobiol 1979; 5: 249-263.
56. Miller AKH, Alston RL, Mountjoy CQ et al. Automatic differential cell counting on a sector of the normal human hippocampus. The influence of age. Neuropath Appl Neurobiol 1984; 10: 123-141.
57. West MJ. Regionally specific loss of neurons in the aging human hippocampus. Neurobiol Aging 1993; 14: 287-293.
58. Brown MW, Cassell MD. Estimates of the number of neurones in the human hippocampus. Journal of Physiology 1980; 301: 58-59P.
59. Mani RB, Lohr JB, Jeste DV. Hippocampal pyramidal cells and aging in the human; a quantitative study of neuronal loss in sectors CA1 to CA4. Exp Neurol 1986; 94: 29-40.
60. Davies DC, Horwood N, Isaacs SL et al. The effect of age and Alzheimer's disease in pyramidal neurone density in the individual fields of the hippocampal formation. Acta Neuropathol 1992; 83: 510-517.
61. Devaney KO, Johnson HA. Neuron loss in the aging visual cortex of man. J Gerontol 1980; 35: 836-841.
62. Herzog AG, Kemper TL. Amygdaloid changes in aging and dementia. Arch Neurol 1980; 37: 625-629.
63. Hall TC, Miller AKH, Corsellis JAN. Variations in the human Purkinje cell population according to age and sex. Neuropath Appl Neurobiol 1975; 1: 267-292.
64. Ellis RS. A preliminary quantitative study of Purkinje cells in normal, subnormal and senescent human cerebella with some notes on functional organization. J Comp Neurol 1919; 30: 229-252.
65. Ellis RS. Norms for some structural changes in the human cerebellum from both to old age. J Comp Neurol 1920; 32: 1-33.
66. De Lorenzi D. Constanza numerica delle cellule de Purkinje in individui di varia eta. Boll Soc Ital Biol Sper 1931; 6: 80.
67. Torvik A, Torp S, Lindboe CF. Atrophy of the cerebellar vermis in aging. A morphometric and histologic study. J Neurol Sci 1986; 76: 283-294.
68. Hopker von W. Das altern des nucleus dentate. Z Altersforsch 1951; 5: 256-261.
69. Konigsmark BW, Murphy EA. Neuronal populations in the human brain. Nature 1970; 228: 1335-1336.
70. Konigsmark BW, Murphy EA. Volume of the ventral cochlear nucleus in man–its relationship to neuronal populations and age. Neuropathol Exp Neurol 1972; 31: 304-316.
71. Vijayashankar N, Brody H. Aging in the human brain stem. A study of the nucleus of the trochlear nerve. Acta Anat 1977; 99: 169-172.

72. Vijayashankar N, Brody H. A study of aging in the human abducens nucleus. J Comp Neurol 1977; 173: 433-438.
73. Van Buskirk C. The seventh nerve complex. J Comp Neurol 1945; 171: 501-516.
74. Sandoz P, Meier-Ruge W. Age related loss of nerve cells from the human inferior olives and unchanged volume of its grey matter. IRCS Med Sci 1977; 5: 376.
75. Monagle RD, Brody H. Effects of age upon the main nucleus of the inferior olive in the human. J Comp Neurol 1974; 155: 61-66.
76. Moatamed F. Cell frequencies in the human inferior olivary nuclear complex. J Comp Neurol 1966; 128: 109-116.
77. Mann DMA, Yates PO, Marcyniuk B. Monoaminergic neurotransmitter systems in presenile Alzheimer's disease and in senile dementia of Alzheimer type. Clin Neuropathol 1984; 3: 199-205.
78. Mann DMA, Yates PO, Marcyniuk B. Alzheimer's presenile dementia, senile dementia of Alzheimer type and Down's syndrome in middle age form an age-related continuum of pathological changes. Neuropath Appl Neurobiol 1984; 10: 185-207.
79. Vijayashankar N, Brody H. A quantitative study of the pigmented neurones in the nuclei locus caeruleus and subcaeruleus in man as related to aging. J Neuropathol Exp Neurol 1979; 38: 490-497.
80. Mann DMA, Yates PO, Hawkes J. The pathology of the human locus caeruleus. Clin Neuropathol 1983; 2: 1-7.
81. Tomlinson BE, Irving D, Blessed G. Cell loss in the locus caeruleus in senile dementia of Alzheimer type. J Neurol Sci 1981; 49: 419-428.
82. Marcyniuk B, Mann DMA. The topography of nerve cell loss from the locus caeruleus in elderly persons. Neurobiol Aging 1989; 10: 5-9.
83. German DC, Manaye KF, White CL et al. Disease specific patterns of locus caeruleus cell loss: Parkinson disease, Alzheimer's disease and Down's syndrome. Ann Neurol 1992; 32: 667-676.
84. Chan-Palay V, Asan E. Quantitation of catecholamine neurons in the locus coeruleus in humn brains of normal young and older adults and in depression. J Comp Neurol 1989; 287: 357-372.
85. Manaye KF, McIntire DD, Mann DMA et al. Locus coeruleus cell loss in the aging human brain: a non-random process. J Comp Neurol 1995; 358: 79-87.
86. Wree A, Braak H, Schleicher A, Zilles K. Biomathematical analysis of the neuronal loss in the aging human brain of both sexes, demonstrated in pigment preparations of the pars cerebellaris loci caerulei. Anat Embryol 1980; 160: 105-119.
87. Mann DMA, Yates PO, Hawkes J. The noradrenergic system in Alzheimer and multi-infarct dementias. J Neurol Neurosurg Psychiatry 1982; 45: 113-119.
88. Hirai S. Aging of the substantia nigra. Adv Neurol Sci 1968; 12: 845-849.
89. McGeer PL, McGeer EG, Suzuki JS. Aging and extrapyramidal function. Arch Neurol 1977; 34: 33-35.
90. Mann DMA, Yates PO. The effects of aging on the pigmented nerve cells of the human locus caeruleus and substantia nigra. Acta Neuropathol Berl 1979; 47: 93-97.
91. Fearnley JM, Lees AJ. Aging and Parkinson's disease: Substantia nigra regional selectivity. Brain 1991; 114: 2283-2301.
92. German DC, Manaye KF, Smith WK et al. Mid brain dopaminergic cell loss in Parkinson's disease: computer visualization. Ann Neurol 1989; 26: 607-614.
93. Halliday GM, McRitchie DA, Cartwright H et al. Midbrain neuropathology in idiopathic Parkinson's disease and diffuse Lewy body disease. J Clin Neurosci 1996; 3: 1-9.
94. Tomlinson BE, Irving D. The number of limb motor neurones in the human lumbosacral cord throughout life. J Neurol Sci 1977; 34: 213-219.
95. Wilkinson A, Davies I. The influence of age and dementia on the neurone population of the mammillary bodies. Age Aging 1978; 7: 151-160.
96. Goudsmit E, Hopman MA, Fliers E et al. The supraoptic and paraventricular nuclei of the human hypothalamus in relation to sex, age and Alzheimer's disease. Neurobiol Aging 1990; 11: 529-536.
97. Fliers E, Swaab DF, Pool CW et al. The vasopressin and oxytocin neurons in the

human supraoptic and paraventricular nucleus; changes with aging and in senile dementia. Brain Res 1985; 342: 45-53.
98. Swaab DF, Fliers E, Partiman TS. The suprachiasmatic nucleus of the human brain in relation to sex, age and senile dementia. Brain Res 1985; 342: 37-44.
99. Zhou J-N, Hofman MA, Swaab DF. VIP neurons in the human SCN in relation to sex, age, and Alzheimer's disease. Neurobiol Aging 1995; 16: 571-576.
100. Bugiani O, Salvarani S, Perdelli F et al. Nerve cell loss with aging in the putamen. Eur Neurol 1980; 17: 286-291.
101. Bottcher J. Morphology of the basal ganglia in Parkinson's disease. Acta Neurol Scand 1975; 52(Suppl62): 1-87.
102. McGeer PL, McGeer EG, Suzuki JS et al. Aging, Alzheimer's disease and the cholinergic system of the basal forebrain. Neurology 1984; 34: 741-745.
103. Mann DMA, Yates PO, Marcyniuk B. Changes in nerve cells of the nucleus basalis of Meynert in Alzheimer's disease and their relationship to aging and the accumulation of lipofuscin pigment. Mech Aging Dev 1984; 25: 189-204.
104. De Lacalle S, Iraizoz I, Ma Gonzalo L. Differential changes in cell size and number in topographic subdivisions of human basal nucleus in normal aging. Neuroscience 1991; 43: 445-456.
105. Whitehouse PJ, Parhad IM, Hedreen JC et al. Integrity of the nucleus basalis of Meynert in normal aging. Neurology 1983; 33 suppl 2: 159.
106. Clark AW, Parhad IM, Folstein SE et al. The nucleus basalis in Huntington's disease. Neurology 1983; 33: 1262-1267.
107. Chui HC, Bondareff W, Zarow C, Slager U. Stability of neuronal number in the human nucleus basalis of Meynert with age. Neurobiol Aging 1984; 5: 83-88.
108. Arendt T, Bigl V, Arendt A et al. Loss of neurones in the nucleus basalis of Meynert in Alzheimer's disease, paralysis agitans and Korsakoff's disease. Acta Neuropathol 1983; 61: 101-108.
109. Cullen KM, Halliday GM. Neurofibrillary tangles in chronic alcoholics. Neuropath Appl Neurobiol 1995; 21: 312-318.
110. Schade JP, Baxter CF. Changes during growth in the volume and surface area of cortical neurones in the rabbit. Exp Neurol 1960; 2: 158-178.
111. Scheibel ME, Lindsay RD, Tomiyasu U et al. Progressive dendritic changes in the aging human cortex. Exp Neurol 1975; 47: 392-403.
112. Scheibel ME, Lindsay RD, Tomiyasu U et al. Progressive dendritic changes in the aging human limbic system. Exp Neurol 1976; 53: 420-430.
113. Scheibel ME, Tomiyasu U, Scheibel AB. The aging human Betz cell. Exp Neurol 1977; 56: 598-609.
114. Buell SJ, Coleman PD. Dendritic growth in the aged human brain and failure of growth in senile dementia. Science 1979; 206: 854-856.
115. Buell SJ, Coleman PD. Quantitative evidence for selective dendritic growth in normal human aging but not in senile dementia. Brain Res 1981; 214: 23-41.
116. Buell SJ. Golgi-cox and rapid Golgi methods as applied to autopsied human brain tissue: widely disparate results. J Neuropathol Exp Neurol 1982; 41: 500-507.
117. Flood DG, Buell SJ, De Fiore CH et al. Age related dendritic growth in the dentate gyrus of human brain is followed by regression in the "oldest' old". Brain Res 1985; 345: 366-368.
118. Flood DG, Buell SJ, Horwitz GJ et al. Dendritic extent in human dentate gyrus granule cells in normal aging and senile dementia. Brain Res 1987; 402: 205-216.
119. Flood DG, Guarnaccia M, Coleman PD. Dendritic extent in human CA2/3 hippocampal pyramidal neurones in normal aging and senile dementia. Brain Res 1987; 409: 88-96.
120. Nakamura S, Akiguchi M, Kameyama M et al. Age-related changes of pyramidal cell basal dendrites in layers III and V of human motor cortex: A quantitative Golgi study. Acta Neuropathol 1985; 65: 281-284.
121. Coleman PD, Flood DG. Neurone numbers and dendritic extent in normal aging and Alzheimer's disease. Neurobiol Aging 1987; 8: 521-545.

122. Cragg BG. The density of synapses in neurons in normal, mentally defective and aging human brains. Brain 1975; 98: 81-90.
123. Adams I. Plasticity of the synaptic contact zone following loss of synapses in the cerebral cortex of aging humans. Brain Res 1987; 424: 343-351.
124. Huttenlocher PR. Synaptic density in human frontal cortex–developmental changes and effects of aging. Brain Res 1979; 163: 195-205.
125. Gibson PH. EM Study of the numbers of cortical synapses in the brains of aging people and people with Alzheimer-type dementia. Acta Neuropathol 1983; 62: 127-133.
126. Masliah E, Mallory BS, Hansen L et al. Quantitative synaptic alterations in the human neocortex during normal aging. Neurology 1993; 43: 192-197.
127. Scheff SW, Price DA. Synapse loss in the temporal lobe in Alzheimer's disease. Ann Neurol 1993; 33: 190-199.
128. Doebler JA, Markesbery WR, Anthony A et al. Neuronal RNA in relation to neuronal loss and neurofibrillary pathology in the hippocampus in Alzheimer's disease. J Neuropathol Exp Neurol 1987; 46: 28-39.
129. Mann DMA, Sinclair KGA. The quantitative assessment of lipofuscin pigment, cytoplasmic RNA and nucleolar volume in senile dementia. Neuropath Appl Neurobiol 1978; 4: 129-135.
130. Blocq P, Marinesco G. Sur les lesions et la pathogenie de l'epilepsie dite essentielle. Semaine Medicale 1892; 12: 445-456.
131. Redlich E. Uber miliare sklerose der hirnrinde bei seniler atrophie. Jahrbucher fur Psychologie und Neurologie 1898; 17: 208-216.
132. Dayan AD. Quantitative histological studies on the aged human brain. II Senile plaques and neurofibrillary tangles in senile dementia. Acta Neuropathol Berl 1970; 16: 95-102.
133. Tomlinson BE, Blessed G, Roth M. Observations on the brains of demented old people. J Neurol Sci 1970; 11: 205-242.
134. Mirra SS, Heyman A, McKeel D et al. The consortium to establish a registry for Alzheimer's disease (CERAD). Part II. Standardization of the neuropathologic assessment of Alzheimer's disease. Neurology 1991; 41: 479-486.
135. Khachaturian ZS. Diagnosis of Alzheimer's disease. Arch Neurol 1985; 42: 1097-1105.
136. Dayan AD. Quantitative histological studies on the aged human brain. I Senile plaques and neurofibrillary tangles in "normal" patients. Acta Neuropathol 1970; 16: 85-94.
137. Mann DMA, Tucker CM, Yates PO. The topographic distribution of senile plaques and neurofibrillary tangles in the brains of non-demented persons of different ages. Neuropath Appl Neurobiol 1987; 13: 123-139.
138. Ulrich J. Alzheimer changes in nondemented patients younger than sixty-five: possible early stages of Alzheimer's disease and senile dementia of Alzheimer type. Ann Neurol 1984; 17: 273-277.
139. Mann DMA, Brown AMT, Prinja D et al. A morphological analysis of senile plaques in the brains of non-demented persons of different ages using silver, immunocytochemical and lectin histochemical staining techniques. Neuropath Appl Neurobiol 1990; 16: 17-25.
140. Price JL, Davis PB, Morris JC et al. The distribution of plaques, tangles and related immunohistochemical markers in healthy aging and Alzheimer's disease. Neurobiol Aging 1991; 12: 295-312.
141. Arriagada PV, Marzloff K, Hyman BT. Distribution of Alzheimer-type pathologic changes in non-demented elderly individuals matches the pattern in Alzheimer's disease. Neurology 1992; 42: 1681-1688.
142. McKee AC, Kosik KS, Kowall NW. Neuritic pathology and dementia in Alzheimer's disease. Ann Neurol 1991; 30: 156-165.
143. Wisniewski HM, Terry RD. Re-examination of the pathogenesis of the senile plaque. Prog Neuropathol 1973; 2: 1-26.
144. Hyman BT, West HL, Rebeck GW et al. Quantitative analysis of senile plaques in Alzheimer disease: Observation of log-normal size distribution and molecular epidemiology of differences associated with apolipoprotein E genotype and trisomy 21 (Down syndrome). Proc Natl Acad Sci (USA) 1995; 92: 3586-3590.

145. Armstrong RA, Myers D, Smith CUM. What determines the size frequency distribution of β-amyloid (Aβ) deposits in Alzheimer's disease patients? Neurosci Lett 1995; 187: 13-16.
146. Hyman BT, Marzloff K, Arriagada PV. The lack of accumulation of senile plaques or amyloid burden in Alzheimer's disease suggests a dynamic balance between amyloid deposition and resolution. J Neuropathol Exp Neurol 1993; 52: 594-600.
147. Armstrong RA. Factors determining the morphology of β-amyloid (Aβ) deposits in Down's syndrome. Neurodegeneration 1995; 4: 179-186.
148. Masters CL, Simms G, Weinman NA et al. Amyloid plaque core protein in Alzheimer disease and Down syndrome. Proc Natl Acad Sci (USA) 1985; 82: 4245-4249.
149. Davies L, Wolska B, Hilbich C et al. A4 amyloid protein deposition and the diagnosis of Alzheimer's disease : prevalence in aged brains determined by immunocytochemistry compared with conventional neuropathologic techniques. Neurology 1988; 38: 1688-1693.
150. Rumble B, Retallack R, Hilbich C et al. Amyloid (A4) protein and its precursor in Down's syndrome and Alzheimer's disease. N Engl J Med 1989; 320: 1446-1452.
151. Ogomori K, Kitamoto T, Tateishi J et al. β amyloid protein is widely distributed in the central nervous system of patients with Alzheimer's disease. Am J Pathol 1989; 134: 243-251.
152. Ikeda S-I, Allsop D, Glenner GG. The morphology and distribution of plaque and related deposits in the brains of Alzheimer's disease and control cases: an immunohistochemical study using amyloid β protein antibody. Lab Invest 1989; 60: 113-122.
153. Mackenzie IRA. Senile plaques do not progressively accumulate with normal aging. Acta Neuropathol 1994; 87: 520-525.
154. Delaere P, He Y, Fayet G et al. βA4 deposits are constant in the brain of the oldest old: an immunocytochemical study of 20 French centenarians. Neurobiol Aging 1993; 14: 191-194.
155. Coria F, Moreno A, Rubio I et al. The cellular pathology associated with Alzheimer β-amyloid deposits in non-demented aged individuals. Neuropath Appl Neurobiol 1993; 19: 261-268.
156. Giannakopoulos P, Hof PR, Mottier S et al. Neuropathological changes in the cerebral cortex of 1258 cases from a geriatric hospital: retrospective clinicopathological evaluation of a 10-year autopsy population. Acta Neuropathol 1994; 87: 456-468.
157. Morris JC, McKeel JDW, Storandt M et al. Very mild Alzheimer's disease: informant-based clinical, psychometric, and pathologic distinction from normal aging. Neurology 1991; 41: 469-478.
158. Joachim CL, Morris JH, Selkoe DJ. Diffuse senile plaques occur commonly in the cerebellum in Alzheimer's disease. Am J Pathol 1989; 135: 309-319.
159. Wisniewski HM, Bancher C, Barcikowska M et al. Spectrum of morphological appearance of amyloid deposits in Alzheimer's disease. Acta Neuropathol 1989; 78: 337-347.
160. Yamaguchi H, Hirai S, Morimatsu M et al. A variety of cerebral amyloid deposits with brains of Alzheimer-type dementia demonstrated by β-protein immunostaining. Acta Neuropathol 1988; 76: 541-549.
161. Yamaguchi H, Hirai S, Morimatsu M et al. Diffuse type of senile plaques in the cerebellum of Alzheimer-type dementia detected by β-protein immunostaining. Acta Neuropathol 1989; 77: 314-319.
162. Suenaga T, Hirano A, Llena JF et al. Modified Bielschowsky and immunocytochemical studies on cerebellar plaques in Alzheimer's disease. J Neuropathol Exp Neurol 1990; 49: 31-40.
163. Suenaga T, Hirano A, Llena JF et al. Modified Bielschowsky staining and immunohistochemical studies on striatal plaques in Alzheimer's disease. Acta Neuropathol 1990; 80: 280-286.
164. Cole G, Neal JW, Singrao SK et al. The distribution of amyloid plaques in the cerebellum and brain stem in Down's syndrome and Alzheimer's disease: a light microscopical analysis. Acta Neuropathol 1993; 85: 542-552.
165. MacKenzie IRA, McKelvie PA, Beyreuther K et al. βA4 amyloid protein deposition in

the cerebellum in Alzheimer's disease and Down's syndrome. Dementia 1991; 2: 237-242.
166. Mann DMA, Jones D, Prinja D et al. The prevalence of amyloid (A4) protein deposits within the cerebral and cerebellar cortex in Down's syndrome and Alzheimer's disease. Acta Neuropathol 1990; 80: 318-327.
167. Mann DMA, Iwatsubo T. Diffuse plaques in the cerebellum and corpus striatum in Down's syndrome contain amyloid β protein (Aβ) only in the form of Aβ42(43). Neurodegeneration 1996; 5: 115-120.
168. Allsop D, Haga S-I, Haga C et al. Early senile plaques in Down's syndrome brains show a clear relationship with cell bodies of neurones. Neuropath Appl Neurobiol 1989; 15: 531-542.
169. Cras P, Kawai M, Siedlak S et al. Neuronal and microglial involvement in β-amyloid protein deposition in Alzheimer's disease. Am J Pathol 1990; 137: 241-246.
170. Pappolla MA, Omar RA, Sambamurti K et al. The genesis of the senile plaque. Further evidence in support of its neuronal origin. Am J Pathol 1992; 141: 1151-1159.
171. Probst A, Langui D, Ipsen S et al. Deposition of β/A4 protein along neuronal plasma membranes in diffuse senile plaques. Acta Neuropathol 1991; 83: 21-29.
172. Mann DMA, Younis N, Jones D et al. The time course of pathological events concerned with plaque formation in Down's syndrome with particular reference to the involvement of microglial cells. Neurodegeneration 1992; 1: 201-215.
173. Luthert PJ, Williams JA. A quantitative study of the coincidence of blood vessels and A4 protein deposits in Alzheimer's disease. Neurosci Lett 1991; 126: 110-112.
174. Kawai M, Kalaria RN, Harik SI et al. The relationship of amyloid plaques to cerebral capillaries in Alzheimer's disease. Am J Pathol 1990; 137: 1435-1446.
175. Candy JM, Oakley AE, Klinowski J et al. Aluminosilicates and senile plaque formation in Alzheimer's disease. Lancet 1986; i: 354-357.
176. Duckett S, Galle P. Mise en evidence de l'aluminum dans les plaques seniles de la maladie d'Alzheimer, etude la microsonde de lastaing. CR Acad Sci 1976; 282: 393-395.
177. Nikaido T, Austin JH, Trueb L et al. Studies in aging of the brain. II Microchemical analyses of the nervous system in Alzheimer patients. Arch Neurol 1972; 27: 549-554.
178. Chafi AH, Hauw J-J, Rancurel G et al. Absence of aluminum in Alzheimer's disease brain tissue: electron microprobe and ion microprobe studies. Neurosci Lett 1991; 123: 61-64.
179. Landsberg JP, McDonald B, Watt JF. Absence of aluminum in neuritic plaque cores in Alzheimer's disease. Nature 1992; 360: 65-68.
180. Tagliavini F, Giaccone G, Frangione B et al. Preamyloid deposits in the cerebral cortex of patients with Alzheimer's disease and non demented individuals. Neurosci Lett 1988; 93: 191-196.
181. Verga L, Frangione B, Tagliavini F et al. Alzheimer patients and Down patients: cerebral preamyloid deposits differ ultrastructurally and histochemically from the amyloid of senile plaques. Neurosci Lett 1989; 105: 294-299.
182. Yamaguchi H, Nakazato Y, Hirai S et al. Electron micrograph of diffuse plaques. Am J Pathol 1989; 135: 593-597.
183. Hachimi KH, Verga G, Giaccone G et al. Relationship between non-fibrillary amyloid precursors and cell processes in the cortical neuropil of Alzheimer patients. Neurosci Lett 1991; 129: 119-122.
184. Davies CA, Mann DMA. Is the "preamyloid" of diffuse plaques in Alzheimer's disease really non-fibrillar? Am J Pathol 1993; 143: 1594-1605.
185. Yamaguchi H, Nakazato Y, Shoji M et al. Ultrastructure of diffuse plaques in senile dementia of the Alzheimer-type: comparison with primitive plaques. Acta Neuropathol 1991; 82: 13-20.
186. Braak H, Braak E, Grundke-Iqbal I et al. Occurrence of neuropil threads in the senile human brain and in Alzheimer's disease: a third location of paired helical filaments outside of neurofibrillary tangles and neuritic plaques. Neurosci Lett 1986; 65: 351-355.

187. Mann DMA. The pathological association between Down syndrome and Alzheimer disease. Mech Aging Dev 1988; 43: 99-136.
188. Mann DMA, Brown AMT, Prinja D et al. An analysis of the morphology of senile plaques in Down's syndrome patients of different ages using immunocytochemical and lectin histochemical methods. Neuropath Appl Neurobiol 1989; 15: 317-329.
189. Mann DMA, Iwatsubo T, Fukumoto H et al. Microglial cells and amyloid β protein (Aβ) deposition; association with Aβ40 containing plaques. Acta Neuropathol 1995; 90: 472-477.
190. Iwatsubo T, Mann DMA, Odaka A et al. Amyloid β protein (Aβ) deposition: Aβ42(43) precedes Aβ40 in Down syndrome. Ann Neurol 1995; 37: 294-299.
191. Lemere CA, Blusztajn JK, Yamaguchi H et al. Sequence of deposition of heterogeneous amyloid β-peptides and APO E in Down syndrome: Implications for initial events in amyloid plaque formation. Neurobiol Dis 1996; 3: 16-32.
192. Wisniewski HM, Wegiel J, Kotula L. Some neuropathological aspects of Alzheimer's disease and its relevance to other disciplines. Neuropath Appl Neurobiol 1996; 22: 3-11.
193. Wisniewski T, Lalowski M, Bobik M et al. Amyloid β1-42 deposits do not lead to Alzheimer's neuritic plaques in aged dogs. Biochem J 1996; 313: 575-580.
194. Wegiel J, Wisniewski HM, Dziewiatkowski J et al. Fibrillar and non-fibrillary amyloid in the brain of aged dogs. In: Iqbal K, Mortimer JA, Winblad B, Wisniewski HM, eds. Research Advances in Alzheimer's Disease and Related Disorders. John Wiley & Sons Ltd 1995; 703-707.
195. Cork LC, Masters C, Beyreuther K et al. Development of senile plaques. Relationship of neuronal abnormalities and amyloid deposits. Am J Pathol 1990; 137: 1383-1392.
196. Miller DL, Papayannopoulos IA, Styles J et al. Peptide compositions of the cerebrovascular and senile plaque core amyloid deposits of Alzheimer's disease. Arch Biochem Biophys 1993; 301: 41-52.
197. Selkoe DJ, Abraham CR, Podlisny MB et al. Isolation of low-molecular-weight proteins from amyloid plaque fibres in Alzheimer's disease. J Neurochem 1986; 46: 1820-1834.
198. Gorevic P, Goni F, Pons-Estel B et al. Isolation and partial characterization of neurofibrillary tangles and amyloid plaque core in Alzheimer's disease: Immunohistological studies. J Neuropathol Exp Neurol 1986; 45: 647-664.
199. Roher A, Wolfe D, Palutke M et al. Purification, ultrastructure, and chemical analysis of Alzheimer disease amyloid plaque core protein. Proc Natl Acad Sci (USA)1986; 83: 2662-2666.
200. Roher AE, Lowenson JD, Clarke S et al. β amyloid (1-42) is a major component of cerebrovascular amyloid deposits; implications for the pathology of Alzheimer's disease. Proc Natl Acad Sci (USA) 1993; 90: 10836-10840.
201. Roher AE, Lowenson JD, Clarke S et al. Structural alterations in the peptide backbone of β-amyloid core protein may account for its deposition and stability in Alzheimer's disease. J Biol Chem 1993; 268: 3072-3083.
202. Mori H, Takio K, Ogawara M et al. Mass spectrometry of purified amyloid β protein in Alzheimer's disease. J Biol Chem 1992; 267: 17082-17086.
203. Gowing E, Roher AE, Woods AS et al. Chemical characterization of Aβ17-42 peptide, a component of diffuse amyloid deposits of Alzheimer disease. J Biol Chem 1994; 269: 10987-10990.
204. Naslund J, Schierhorn A, Hellman U et al. Relative abundance of Alzheimer Aβ amyloid peptide variants in Alzheimer disease and normal aging. Proc Natl Acad Sci (USA) 1994; 91: 8378-8382.
205. Tamaoka A, Odaka A, Ishibashi Y et al. APP717 mis-sense mutation affects the ratio of amyloid β protein species (Aβ1-42/43 and Aβ1-40) in familial Alzheimer's disease brain. J Biol Chem 1994; 269: 32721-32724.
206. Gravina SA, Ho L, Eckman CB et al. Amyloid β protein (Aβ) in Alzheimer's disease brain. J Biol Chem 1995; 270: 7013-7016.

207. Glenner GG, Wong CW. Alzheimer's disease: initial report of the purification and characterization of a novel cerebrovascular amyloid protein. Biochem Biophys Res Commun 1984; 120: 885-890.
208. Joachim CL, Duffy LK, Morris JH et al. Protein chemical and immunocytochemical studies of meningovascular beta-amyloid protein in Alzheimer's disease and normal aging. Brain Res 1988; 474: 100-111.
209. Prelli F, Castano E, Glenner GG et al. Differences between vascular and plaque core amyloid in Alzheimer's disease. J Neurochem 1988; 51: 648-651.
210. Shinkai Y, Yoshimura M, Ito Y et al. Amyloid β-proteins 1-40 and 1-42(43) in the soluble fraction of extra- and intracranial blood vessels. Ann Neurol 1995; 38: 421-428.
211. Suzuki N, Iwatsubo T, Odaka A et al. High tissue content of soluble β1-40 is linked to cerebral amyloid angiopathy. Am J Pathol 1994; 145: 452-460.
212. Iwatsubo T, Odaka N, Suzuki N et al. Visualization of Aβ42(43)-positive and Aβ40-positive senile plaques with end-specific Aβ monclonal antibodies: Evidence that an initially deposited species is Aβ1-42(43). Neuron 1994; 13: 45-53.
213. Mann DMA, Iwatsubo T, Ihara Y et al. Predominant deposition of Aβ42(43) in plaques in cases of Alzheimer's disease and hereditary cerebral haemorrhage associated with mutations in the amyloid precursor protein gene. Am J Pathol 1996; 148: 1257-1266.
214. Mann DMA, Iwatsubo T, Cairns NJ et al. Amyloid (Aβ) deposition in chromosome 14-linked Alzheimer's disease: predominance of Aβ42(43). Ann Neurol 1996; 40: 149-156.
215. Murphy GM, Forno LS, Higgins L et al. Development of a monoclonal antibody specific for the COOH-terminal of β-amyloid 1-42 and its immunohistochemical reactivity in Alzheimer's disease and related disorders. Am J Pathol 1994; 144: 1082-1088.
216. Mak K, Yang F, Vinters HV et al. Polyclonals to β-amyloid (1-42) identify most plaque and vascular deposits in Alzheimer cortex, but not striatum. Brain Res 1994; 667: 138-142.
217. Fukumoto H, Asami-Odaka A, Suzuki N et al. Amyloid β protein (Aβ) deposition in normal aging has the same characteristics as that in Alzheimer's disease: predominance of Aβ42(43) and association of Aβ40 with cored plaques. Am J Pathol 1996; 148: 259-265.
218. Yamaguchi H, Sugiahra S, Ishiguro K et al. Immunohistochemical analysis of COOH-termini of amyloid beta protein (Aβ) using end-specific antisera for Aβ40 and Aβ42 in Alzheimer's disease and normal aging. Amyloid: Int J Clin Invest 1995; 2: 7-16.
219. Kida E, Wisniewski KE, Wisniewski HM. Early amyloid-β deposits show different immunoreactivity to the amino- and carboxy-terminal regions of β-peptide in both Alzheimer's disease and Down's syndrome brain. Neurosci Lett 1995; 193: 1-4.
220. Iwatsubo T, Saido TC, Mann DMA et al. Full-length Aβ(1-42(43)) as well as amino-terminally modified and truncated Aβ42(43) deposit in diffuse plaques. Am J Pathol 1996; (In press):.
221. Saido TC, Iwatsubo T, Ihara Y et al. Dominant and differential deposition of distinct β-amyloid peptide species, AβN3(PE) in senile plaques. Neuron 1995; 14: 457-466.
222. Ohgami T, Kitamoto T, Shin R-W et al. Increased senile plaques without microglia in Alzheimer's disease. Acta Neuropathol 1991; 81: 242-247.
223. Fukumoto H, Asami-Odaka A, Suzuki N et al. Association of Aβ40 positive senile plaques with microglia cells in the brains of patients with Alzheimer's disease and non-demented aged individuals. Neurodegeneration 1996; 5: 13-17.
224. Clinton J, Roberts GW, Gentleman SM et al. Differential pattern of β-amyloid protein deposition within cortical sulci and gyri in Alzheimer's disease. Neuropath Appl Neurobiol 1993; 19: 277-281.
225. Gentleman SM, Allsop D, Bruton CJ et al. Quantitative differences in the deposition of βA4 protein in the sulci and gyri of frontal temporal isocortex in Alzheimer's disease. Neurosci Lett 1992; 136: 27-30.
226. Praprotnik D, Smith MA, Richey PL et al.

Plasma membrane fragility in dystrophic neurites in senile plaques of Alzheimer's disease: an index of oxidative stress. Acta Neuropathol 1996; 91: 1-5.
227. Kang J, Lemaire H-G, Unterbeck A et al. The precursor of Alzheimer's disease amyloid A4 protein resembles a cell surface receptor. Nature 1987; 325: 733-736.
228. Goldgaber D, Lerman MI, MacBride OW et al. Characterization and chromosomal localization of a cDNA encoding brain amyloid of Alzheimer's disease. Science 1987; 235: 877-880.
229. Tanzi RE, St. George-Hyslop PH, Haines JH et al. The genetic defect in familial Alzheimer's disease is not tightly linked to the amyloid precursor protein gene. Nature 1987; 329: 156-157.
230. Lemaire HG, Salbaum JM, Multhaup G et al. The $PreA4_{695}$ precursor protein of Alzheimer's disease A4 amyloid is encoded by 16 exons. Nucl Acids Res 1989; 17: 517-522.
231. Yoshikai S-I, Sasaki H, Doh-ura K et al. Genomic organization of the human amyloid beta-protein precursor gene. Gene 1990; 87: 257-263.
232. Rooke K, Talbot C, James L et al. A physical map of the human APP gene in YACs. Mammalian Genome 1993; 4: 662-669.
233. Ponte P, Gonzalez DeWhitt P, Schilling J et al. A new A4 amyloid mRNA contains a domain homologous to serine proteinase inhibitors. Nature 1988; 331: 331-525.
234. Tanzi RE, McClatchey AI, Lamperti ED et al. Protease inhibitor domain encoded by an amyloid protein precursor mRNA associated with Alzheimer's disease. Nature 1988; 331: 528-530.
235. Kitaguchi N, Takahashi Y, Tokushima Y et al. Novel precursor of Alzheimer's disease amyloid protein shows protease inhibitory activity. Nature 1988; 331: 530-532.
236. Konig G, Monning U, Czech C et al. Identification and differential expression of a novel alternative splice form of the βA4 amyloid precursor protein (APP) mRNA in leucocytes and brain microglial cells. J Biol Chem 1992; 267: 10804-10809.
237. de Sauvage F, Octave J-N. A novel mRNA of the A4 amyloid precursor gene coding for a possibly secreted protein. Science 1989; 245: 651-653.
238. Golde TE, Estus S, Usiak M et al. Expression of β amyloid protein precursor mRNAs: recognition of a novel alternatively spliced form and quantitation in Alzheimer's disease using PCR. Neuron 1990; 4: 253-267.
239. Jacobsen JS, Muenkel HA, Blume AJ et al. A novel species-specific RNA related to alternatively spliced amyloid precursor protein mRNAs. Neurobiol Aging 1991; 12: 575-583.
240. Goedert M. Neuronal localization of amyloid β protein precursor mRNA in normal human brains and in Alzheimer's disease. EMBO J 1987; 6: 3627-3632.
241. Bahmanyar S, Higgins GA, Goldgaber D et al. Localization of amyloid β protein messenger RNA in brains from patients with Alzheimer's disease. Science 1988; 237: 77-80.
242. Neve RL, Finch EA, Dawes LR. Expression of the Alzheimer amyloid precursor gene transcripts in the human brain. Neuron 1988; 1: 669-677.
243. Stern RA, Otvos L, Trojanowski JQ et al. Monoclonal antibodies to a synthetic peptide homologous with the first 28 amino acids of Alzheimer's disease β protein recognize amyloid and diverse glial and neuronal cell types in the central nervous system. Am J Pathol 1989; 134: 973-978.
244. Siman RL, Card JP, Welson RB et al. Expression of β-amyloid precursor protein in reactive astrocytes following neuronal damage. Neuron 1989; 3: 275-285.
245. Haass C, Hung AY, Selkoe DJ. Processing of β-amyloid precursor protein in microglia and astrocytes favors an internal localization over constitutive secretion. J Neurosci 1991; 11: 3783-3793.
246. Banati RB, Gehrmann J, Czech C et al. Early and rapid de novo synthesis of Alzheimer βA4-amyloid precursor protein (APP) in activated microglia. Glia 1993; 9: 199-210.
247. Sandbrink R, Masters CL, Beyreuther K. βA4-amyloid protein precursor mRNA isoforms without exon 15 are ubiquitously expressed in rat tissues including brain, but

not in neurons. J Biol Chem 1994; 269: 15510-15517.
248. Weidmann A, Konig G, Bunke D et al. Identification biogenesis and localization of precursors of Alzheimer's disease A4 amyloid protein. Cell 1989; 57: 115-126.
249. Oltersdorf F, Ward PJ, Beattie EC et al. In vitro mutagenesis of the β-amyloid precursor protein. Neurobiol Aging 1990; 11: 306.
250. Podlisny MB, Mammen AL, Schlossmacher MG et al. Detection of soluble forms of the β-amyloid precursor protein in human plasma. Biochem Biophys Res Commun 1990; 167: 1094-1101.
251. Palmert MR, Podlisny MB, Witker DS et al. The beta-amyloid protein precursor of Alzheimer's disease has soluble derivatives found in human brain and cerebrospinal fluid. Proc Natl Acad Sci (USA) 1989; 86: 6338-6342.
252. Koo EH, Sisodia SS, Archer DR et al. Precursor of amyloid protein in Alzheimer's disease undergoes fast anterograde axonal transport. Proc Natl Acad Sci (USA) 1990; 87: 1561-1565.
253. Schubert W, Prior R, Weidmann A et al. Localization of Alzheimer β/A4 amyloid precursor protein at central and peripheral sites. Brain Res 1991; 563: 184-194.
254. Johnson SA, McNeill T, Cordell B et al. Relation of neuronal APP-751/APP-695 mRNA ratio and neuritic plaque density in Alzheimer's disease. Science 1990; 248: 854-857.
255. Card JP, Meade RP, Davis LG. Immunocytochemical localization of the precursor protein for β amyloid in the rat central nervous system. Neuron 1988; 1: 835-846.
256. Mita S, Schon EA, Herbert J. Widespread expression of amyloid β protein precursor gene in rat brain. Am J Pathol 1989; 134: 1253-1261.
257. Monning U, Sandbrink R, Weidemann A et al. Extracellular matrix influences the biogenesis of amyloid precursor protein in microglial cells. J Biol Chem 1995; 270: 7104-7110.
258. Wasco W, Bupp K, Magendantz M et al. Identification of a mouse brain cDNA that encodes a protein related to the Alzheimer disease-associated amyloid β precursor. Proc Natl Acad Sci (USA)1992; 89: 10758-10762.
259. Wasco W, Gurubhagavatula S, Paradis M d et al. Isolation and characterization of APLP2 encoding a homologue of the Alzheimer's associated amyloid β protein precursor. Nature Genet 1993; 5: 95-99.
260. Sprecher CA, Grant FJ, Grimm G et al. Molecular cloning of the cDNA for a human amyloid precursor protein homolog: evidence for a multigene family. Biochemistry 1993; 32: 4481-4486.
261. Slunt HH, Thinakaran G, Von Koch C et al. Expression of a ubiquitous, cross-reactive homologue of the mouse β-amyloid precursor protein (APP). J Biol Chem 1994; 269: 2637-2644.
262. Oyama F, Shimada H, Oyama R. Differential expression of β amyloid precursor protein (APP) and tau m-RNA in the aged human brain: individual variability and correlation between APP-751 and four repeat tau. J Neuropathol Exp Neurol 1991; 50: 560-578.
263. Konig G, Beyreuther K, Masters CL et al. Pre-A4 RNA distribution in brain areas. Prog Clin Biol Res 1989; 317: 1027-1036.
264. Joachim CL, Mori H, Selkoe DJ. Amyloid β-protein deposition in tissues other than brain in Alzheimer's disease. Nature 1989; 341: 226-230.
265. Soininen H, Syrjanen S, Heinonen O et al. Amyloid β-protein deposition in skin of patients with dementia. Lancet 1992; 339: 245.
266. Nishimoto I, Okamoto T, Matsuura Y et al. Alzheimer amyloid protein precursor complexes with brain GTP-binding protein Go. Nature 1993; 362: 75-79.
267. Schubert D, LaCorbiere M, Saitoh T et al. Characterization of an amyloid β precursor protein that binds heparin and contains tyrosine sulfate. Proc Natl Acad Sci (USA) 1989; 86: 2066-2069.
268. Breen KC, Bruce M, Anderton BH. Beta amyloid precursor protein mediates neuronal cell-cell and cell-surface adhesion. J Neurosci Res 1991; 28: 90-100.
269. Beer J, Masters CL, Beyreuther K. Cells from peripheral tissues that exhibit high APP expression are characterized by their high

membrane fusion activity. Neurodegeneration 1995; 4: 51-59.
270. Chen M, Yankner BA. An antibody to β-amyloid and the amyloid precursor protein inhibits cell-substratum adhesion in many mammalian cell types. Neurosci Lett 1991; 125: 223-226.
271. Klier FG, Cole G, Stallcup W et al. Amyloid β-protein precursor is associated with extracellular matrix. Brain Res 1990; 515: 336-342.
272. Milward EA, Papadopoulos R, Fuller SJ et al. The amyloid protein precursor of Alzheimer's disease is a mediator of the effects of nerve growth factor on neurite outgrowth. Neuron 1992; 9: 129-137.
273. Saitoh T, Sundsmo M, Roch J-M et al. Secreted form of amyloid β protein precursor is involved in the growth regulation of fibroblasts. Cell 1989; 58: 615-622.
274. Ninomiya H, Roch JM, Sundsmo MP et al. Amino acid sequence RERMS represents the active domain of amyloid β/A4 protein precursor that promotes fibroblast growth. J Cell Biol 1993; 121: 879-886.
275. Van Nostrand WE, Wagner SL, Suzuki M et al. Protease nexin-II, a potent anti-chymotrypsin, shows identity to amyloid β precursor. Nature 1989; 341: 545-549.
276. Mattson MP, Rydel RE. β amyloid precursor protein and Alzheimer's disease: The peptide plot thickens Neurobiol Aging 1992; 13: 617-621.
277. Furukawa K, Barger SW, Blalock ER et al. Activation of K+ channels and suppression of neuronal activity by secreted β-amyloid precursor protein. Nature 1996; 379: 74-78.
278. Bush AI, Martins RN, Rumble B et al. The amyloid precursor protein of Alzheimer's disease is released by human platelets. J Biol Chem 1990; 265: 15977-15983.
279. Smith RP, Higuchi DA, Broze GJ. Platelet coagulation factor X1a-inhibitor, a form of Alzheimer's amyloid precursor protein. Science 1990; 248: 1126-1128.
280. Oltersdorf T, Fritz LC, Schenk DB et al. The secreted form of the Alzheimer's amyloid precursor protein with the Kunitz domain is protease nexin II. Nature 1989; 341: 144-147.
281. Kawarabayashi T, Shoji M, Harigaya Y et al. Expression of APP in early stages of brain damage. Brain Res 1991; 563: 334-338.
282. Kalaria RN, Bhatti SU, Palatinsky EA et al. Accumulation of β-amyloid precursor protein at sites of ischemic injury in rat brain. NeuroReport 1993; 4: 211-214.
283. Gentleman SM, Nash MJ, Sweeting CJ et al. β-amyloid precursor protein (β-APP) as a marker for axonal injury after head injury. Neurosci Lett 1993; 160: 139-144.
284. Shigematsu K, McGeer PL. Accumulation of amyloid precursor protein in damaged neuronal processes and microglia following intracerebral administration of aluminum salts. Brain Res 1992; 593: 117-123.
285. Shigematsu K, McGeer PL. Accumulation of amyloid precursor protein in neurones after intraventricular injection of colchicine. Am J Pathol 1992; 140: 787-794.
286. Goldgaber D, Harris HW, Hla T et al. Interleukin-1 regulates synthesis of amyloid β-protein precursor mRNA in human endothelial cells. Proc Natl Acad Sci (USA) 1989; 86: 7606-7610.
287. Donnelly RJ, Friedhoff AJ, Beer B et al. Interleukin-1 stimulates the beta-amyloid precursor protein promoter. Cell Mol Neurobiol 1990; 10: 485-491.
288. Buxbaum JD, Oishi M, Chen HI et al. Cholinergic agonists and interleukin-1 regulate processing and secretion of the Alzheimer β/A4 amyloid protein precursor. Proc Natl Acad Sci (USA) 1992; 89: 10075-10078.
289. Guilian D, Woodward J, Young DG et al. Interleukin-1 injected into brain stimulates astrogliosis and neovascularization. J Neurosci 1988; 8: 2485-2490.
290. Sheng JG, Boop FA, Mrak RE et al. Increased neuronal β-amyloid precursor protein expression in human temporal lobe epilepsy: Association with interleukin-1a immunoreactivity. J Neurochem 1994; 63: 1872-1879.
291. Esch ES, Keim P, Beattie EC et al. Cleavage of amyloid β peptide during constitutive processing of its precursor. Science 1990; 248: 1122-1124.
292. Zhong Z, Higaki J, Murakami K et al. Secretion of β-amyloid precursor protein

involves multiple cleavage sites. J Biol Chem 1994; 269: 627-632.
293. Sisodia SS, Koo EH, Beyreuther K et al. Evidence that β amyloid protein in Alzheimer's disease is not derived by normal processing. Science 1990; 248: 492-495.
294. Anderson JP, Esch FS, Keim PS et al. Exact cleavage site of Alzheimer amyloid precursor in neuronal PC-12 cells. Neurosci Lett 1991; 128: 126-128.
295. Estus S, Golde TE, Kunishita T et al. Potentially amyloidogenic carboxy-terminal derivatives of the amyloid protein precursor. Science 1992; 255: 726-728.
296. Busciglio J, Gabuzda DH, Matsudaira P et al. Generation of β-amyloid in the secretory pathway in neuronal and non-neuronal cells. Proc Natl Acad Sci (USA) 1993; 90: 2092-2096.
297. Buxbaum JD, Gandy SE, Cicchetti P et al. Processing of Alzheimer β/A4 amyloid precursor protein: Modulation by agents that regulate protein phosphorylation. Proc Natl Acad Sci (USA) 1990; 87: 6003-6006.
298. Suzuki T, Nairn AC, Gandy SE et al. Phosphorylation of Alzheimer amyloid precursor protein by protein kinase C. Neuroscience 1992; 48: 755-761.
299. Caporaso GL, Gandy SE, Buxbaum JD et al. Protein phosphorylation regulates secretion of Alzheimer β/A4 amyloid precursor protein. Proc Natl Acad Sci (USA) 1992; 89: 3055-3059.
300. Nitsch RM, Slack BE, Wurtman RJ et al. Release of Alzheimer amyloid precursor derivatives stimulated by activation of muscarinic acetylcholine receptors. Science 1992; 258: 304-307.
301. Kennedy H, Kametani K, Allsop D. Only Kunitz inhibitor-containing isoforms of secreted Alzheimer amyloid precursor protein show amyloid immunoreactivity in normal cerebrospinal fluid. Neurodegeneration 1992; 1: 59-64.
302. Seubert P, Vigo-Pelfrey C, Esch F et al. Isolation and quantification of soluble Alzheimer's β-peptide from biological fluids. Nature 1993; 359: 325-327.
303. Seubert P, Oltersdorf T, Lee MG et al. Secretion of β-amyloid precursor protein cleaved at the aminoterminus of the β-amyloid peptide. Nature 1992; 361: 260-263.
304. Davis D, Sinha S, Schlossmacher MG et al. Isolation and quantification of soluble Alzheimer's β-peptide from biological fluids. Nature 1992; 359: 325-327.
305. Haass C, Schlossmacher MG, Hung AY et al. Amyloid β-peptide is produced by cultured cells during normal metabolism. Nature 1992; 359: 322-325.
306. Haass C, Koo EH, Mellon A et al. Targeting of cell-surface β-amyloid precursor protein to lysosomes : alternative processing into amyloid bearing fragments. Nature 1992; 357: 500-502.
307. Shoji M, Golde TE, Ghiso J et al. Production of the Alzheimer amyloid β protein by normal proteolytic processing. Science 1992; 258: 126-129.
308. Dovey HF, Suomesaari-Chrysler S, Lieberburg I et al. Cells with a familial Alzheimer's disease mutation produce authentic β-peptide. NeuroReport 1993; 4: 1039-1042.
309. Vigo-Pelfrey C, Lee D, Keim P et al. Characterization of β-amyloid peptide from human cerebrospinal fluid. J Neurochem 1993; 61: 1965-1968.
310. Golde TE, Estus S, Younkin LH et al. Processing of the amyloid protein precursor to potentially amyloidogenic derivatives. Science 1992; 255: 728-730.
311. Knops J, Lieberburg I, Sinha S. Evidence for a nonsecretory, acidic degradation pathway for amyloid precursor protein in 293 cells. J Biol Chem 1992; 267: 16022-16024.
312. Haass C, Lemere CA, Capell A et al. The Swedish mutation causes early-onset Alzheimer's disease by β-secretase cleavage within the secretory pathway. Nature Medicine 1995; 1: 1291-1296.
313. Thomas T, Thomas G, McLendon C et al. β-amyloid-mediated vasoactivity and vascular endothelial damage. Nature 1996; 380: 168-171.
314. Jarrett JT, Berger EP, Lansbury PT. The carboxy terminus of the β-amyloid protein is critical for the seeding of amyloid formation. Implications for the pathogenesis of Alzheimer's disease. Biochemistry 1993; 32: 4693-4697.

315. Hilbich C, Kisters-Woike B, Reed J et al. Aggregation and secondary structure of synthetic amyloid βA4 peptides of Alzheimer's disease. J Molec Biol 1991; 218: 149-163.
316. Evans KC, Berger EP, Cho GG et al. Apolipoprotein E is a kinetic but not thermodynamic inhibitor of amyloid formation; implications for the pathogenesis and treatment of Alzheimer's disease. Proc Natl Acad Sci (USA) 1995; 92: 763-767.
317. Ma J, Yee A, Brewer B et al. Amyloid-associated proteins of α-1 antichymotrypsin and apolipoprotein E promote assembly of Alzheimer β protein into filaments. Nature 1994; 372: 92-94.
318. Bush AI, Pettingell WH, Multhaup G et al. Rapid induction of Alzheimer Aβ amyloid formation by zinc. Science 1994; 265: 1464-1467.
319. Mantyh PW, Ghilardi JR, Rogers S et al. Aluminum, iron, and zinc ions promote aggregation of physiological concentrations of β-amyloid peptide. J Neurochem 1993; 61: 1171-1174.
320. Snow AD, Sekiguchi R, Nochlin D et al. An important role of heparan sulphate proteoglycan (Perlecan) in a model system for the deposition and persistence of fibrillar Aβ amyloid in rat brain. Neuron 1994; 12: 219-234.
321. Tennent GA, Lovat LB, Pepys MB. Serum amyloid P component prevents proteolysis of amyloid fibrils of Alzheimer's disease and systemic amyloidosis. Proc Natl Acad Sci (USA) 1995; 92: 4299-4303.
322. Tabaton M, Nunzi MG, Xue R et al. Soluble amyloid β-protein is a marker of Alzheimer amyloid in brain but not in cerebrospinal fluid. Biochem Biophys Res Commun 1994; 200: 1598-1603.
323. Teller JK, Russo C, de Busk LM et al. Presence of soluble amyloid β peptide precedes amyloid plaque formation in Down's syndrome. Nature Medicine 1996; 2: 93-95.
324. Halverson K, Fraser PE, Kirschner DA et al. Molecular determinants of amyloid deposits in Alzheimer's disease: Conformational studies of synthetic β-protein fragments. Biochemistry 1990; 29: 2639-2644.
325. Pike CJ, Walencewicz-Wasserman AJ, Kosmoski J et al. Structure–activity analyses of β-amyloid peptides: Contributions of the β25-35 region to aggregation and neurotoxicity. J Neurochem 1995; 64: 253-265.
326. Yankner BA, Dawes LR, Fisher S et al. Neurotoxicity of a fragment of the amyloid precursor associated with Alzheimer's disease. Science 1989; 245: 417-420.
327. Shearman MS, Ragan CI, Iversen LL. Inhibition of PC12 cell redox activity is a specific, early indicator of the mechanism of β-amyloid-mediated cell death. Proc Natl Acad Sci (USA) 1994; 91: 1470-1471.
328. Shearman MS, Hawtin SR, Tailor VJ. The intracellular component of cellular 3-(4,5-dimethylthiazol-2-yl)-2,5diphenyltetrazolium bromide (MTT) reduction is specifically inhibited by β-amyloid peptides. J Neurochem 1995; 65: 218-227.
329. Loo DT, Copani A, Pike CJ et al. Apoptosis is induced by β-amyloid in cultured central nervous system neurons. Proc Natl Acad Sci (USA) 1993; 90: 7951-7955.
330. Forloni G, Chiesa R, Smiroldo S et al. Apoptosis-/mediated neurotoxicity induced by chronic application of beta amyloid fragment. NeuroReport 1993; 4: 523-526.
331. Behl C, Davis B, Klier FG et al. Amyloid beta peptide induces necrosis rather than apoptosis. Brain Res 1994; 645: 253-264.
332. Behl C, Davis JB, Lesley R et al. Hydrogen peroxide mediates amyloid β protein toxicity. Cell 1994; 77: 1-20.
333. Schubert D, Behl C, Lesley R et al. Amyloid peptides are toxic via a common oxidative mechanism. Proc Natl Acad Sci (USA) 1995; 92: 1989-1993.
334. Etcheberrigaray R, Ito E, Kim CS et al. Soluble β-amyloid induction of Alzheimer's phenotype for human fibroblast K+ channels. Science 1994; 264: 276-279.
335. Copani A, Koh J, Cotman CW. β-amyloid increases neuronal susceptibility to injury by glucose deprivation. NeuroReport 1991; 2: 763-765.
336. Meda L, Cassatella MA, Szendrei GI et al. Activation of microglial cells by β-amyloid protein and interferon γ. Nature 1995; 374: 647-650.
337. Mahley RW. Apolipoprotein E: cholesterol transport protein with expanding role in cell biology. Science 1988; 240: 622-630.

338. Diedrich JF, Minnigan H, Carp RI et al. Neuropathological changes in Scrapie and Alzheimer's disease are associated with increased expression of apolipoprotein E and cathepsin D in astrocytes. J Virol 1991; 65: 4759-4768.
339. Han S-H, Hulette C, Saunders AM et al. Apolipoprotein E is present in hippocampal neurons without neurofibrillary tangles in Alzheimer's disease and in age-matched controls. Exp Neurol 1994; 128: 13-26.
340. Poirier J, Hess M, May PC et al. Astrocytic apolipoprotein E mRNA and GFAP mRNA in hippocampus after entorhinal cortex lesioning. Mol Brain Res 1991; 11: 97-106.
341. Rebeck GW, Reiter JS, Strickland DK et al. Apolipoprotein E in sporadic Alzheimer's disease: Allelic variation and receptor interactions. Neuron 1993; 11: 575-580.
342. Namba Y, Tomonaga M, Kawasaki H et al. Apolipoprotein E immunoreactivity in cerebral amyloid deposits and neurofibrillary tangles in Alzheimer's disease and Kuru plaque amyloid in Creutzfeldt-Jakob disease. Brain Res 1991; 541: 163-166.
343. Wisniewski T, Frangione B. Apolipoprotein E: a pathological chaperone protein in patients with cerebral and systemic amyloid. Neurosci Lett 1992; 135: 235-238.
344. Schmechel D, Saunders AM, Strittmatter WJ et al. Increased amyloid β peptide deposition in cerebral cortex as a consequence of apolipoprotein E genotype in late-onset Alzheimer disease. Proc Natl Acad Sci (USA) 1993; 90: 9649-9653.
345. Yamaguchi H, Ishiguro K, Sugihara S et al. Presence of apolipoprotein E on extracellular tangles and on meningeal blood vessels precedes the Alzheimer β-amyloid deposition. Acta Neuropathol 1994; 88: 413-419.
346. Kida E, Golabek AA, Wisniewski T et al. Regional differences in apolipoprotein E immunoreactivity in diffuse plaques in Alzheimer's disease brain. Neurosci Lett 1994; 167: 73-76.
347. Saunders AM, Strittmatter WJ, Schmechel D et al. Association of Apolipoprotein E4 with late-onset familial and sporadic Alzheimer's disease. Neurology 1993; 43: 1467-1472.
348. Strittmatter WJ, Saunders AM, Schmechel D et al. Apolipoprotein E: high-avidity binding to β-amyloid and increased frequency of type 4 allelle in late-onset familial Alzheimer's disease. Proc Natl Acad Sci (USA) 1993; 90: 1977-1981.
349. Corder EH, Saunders AM, Strittmatter WJ et al. Gene dose of apolipoprotein E Type 4 allele and the risk of Alzheimer's disease in late onset families. Science 1993; 261: 921-923.
350. Houlden H, Crook R, Duff K et al. Confirmation that the apolipoprotein E4 allele is associated with late onset familial Alzheimer's disease. Neurodegeneration 1993; 2: 283-288.
351. Mayeux R, Stern Y, Ottman R et al. The Apolipoprotein E4 allele in patients with Alzheimer's disease. Ann Neurol 1993; 34: 752-754.
352. Pickering-Brown SM, Roberts D, Owen F et al. Apolipoprotein E4 alleles and non-Alzheimer forms of dementia. Neurodegeneration 1994; 3: 95-96.
353. Saunders AM, Schmader K, Breitner JCS et al. Apolipoprotein E E4 allele distribution in late onset Alzheimer's disease and in other amyloid forming diseases. Lancet 1993; 342: 710-711.
354. Wisniewski T, Golabek A, Matsubara E et al. Apolipoprotein E: Binding to soluble Alzheimer's β-amyloid. Biochem Biophys Res Commun 1993; 192: 359-365.
355. Sanan DA, Weisgraber KH, Russell SJ et al. Apolipoprotein E associates with β amyloid peptide of Alzheimer's disease to form novel monofibrils. J Clin Invest 1994; 94: 860-869.
356. Berr C, Hauw J-J, Delaere P, et al. Apolipoprotein E allele E4 is linked to increased deposition of the amyloid β-peptide (A-β) in cases with or without Alzheimer's disease. Neurosci Lett 1994; 178: 221-224.
357. Hyman B, West HL, Rebeck GW et al. Neuropathological changes in Down's syndrome, hippocampal formation: effect of age and apolipoprotein E genotype. Arch Neurol 1995; 52: 373-378.
358. Ohm TG, Kirca M, Bohl J et al. Apolipoprotein E polymorphism influences not only cerebral senile plaque load but also

Alzheimer-type neurofibrillary tangle formation. Neuroscience 1995; 66: 585-587.
359. Polvikoski T, Sulkava R, Haltia M et al. Apolipoprotein E, dementia, and cortical deposition of β-amyloid protein. N Engl J Med 1995; 333: 1242-1247.
360. LaDu MJ, Falduto MT, Manelli AM et al. Isoform-specific binding of apolipoprotein E to β-amyloid. J Biol Chem 1994; 269: 23403-23406.
361. LaDu MJ, Pederson TM, Frail DE et al. Purification of apolipoprotein E attenuates isoform-specific binding to β-amyloid. J Biol Chem 1995; 270: 9039-9042.
362. Harrington CR, Louwagie J, Rossau R et al. Influence of apolipoprotein E genotype on senile dementia of the Alzheimer and Lewy body types. Am J Pathol 1994; 145: 1472-1484.
363. Mann DMA, Iwatsubo T, Pickering-Brown SM et al. Preferential deposition of amyloid β protein (Aβ) in the form Aβ40 in Alzheimer's desease is associated with a gene dosage effect of the Apolipoprotein E E4 allele. Neurosci Lett 1996; in press.
364. Benjamin R, Leake A, Ince PG et al. Effects of apolipoprotein E genotype on cortical neuropathology in senile dementia of the Lewy body type and Alzheimer's disease. Neurodegeneration 1995; 4: 443-448.
365. Gearing M, Mori H, Mirra SS. Aβ peptide length and apolipoprotein E genotype in Alzheimer's disease. Ann Neurol 1996; 39: 395-399.
366. Heinonen O, Lehtovirta M, Soininen H et al. Alzheimer pathology of patients carrying apolipoprotein E E4 allele. Neurobiol Aging 1995; 16: 505-513.
367. Itoh Y, Yamada M. Apolipoprotein E and the neuropathology of dementia. N Engl J Med 1996; 334: 599-600.
368. Zhan S-S, Veerhuis R, Kamphorst W et al. Distribution of beta amyloid associated proteins in plaques in Alzheimer's disease and in the non-demented elderly. Neurodegeneration 1995; 4: 291-297.
369. Zhan SS, Veerhuis R, Janssen I et al. Immunohistochemical distribution of the inhibitors of the terminal complement complex in Alzheimer's disease. Neurodegeneration 1994; 3: 111-117.
370. Choi-Miura NH, Ihara Y, Fukuchi K et al. SP-40,40 is a constituent of Alzheimer's amyloid. Acta Neuropathol 1992; 83: 260-264.
371. McGeer PL, Kawamata T, Walker DG. Distribution of clusterin in Alzheimer brain tissue. Brain Res 1992; 579: 337-341.
372. Wisniewski T, Golabek A, Matsubara E et al. Apolipoprotein E: Binding to soluble Alzheimer's β-amyloid. Biochem Biophys Res Commun 1993; 192: 359-365.
373. Matsubara E, Soto C, Governale S et al. Apolipoprotein J and Alzheimer's amyloid β solubility. Biochem J 1996; 316: 671-679.
374. Golabek A, Marques M, Lalowski M et al. Amyloid β binding proteins *in vitro* and in normal human cerebrospinal fluid. Neurosci Lett 1995; 191: 79-82.
375. Koudinov A, Matsubara E, Frangione B et al. The soluble form of Alzheimer's amyloid β protein is complexed to high density lipoprotein 3 and very high density lipoprotein in normal human plasma. Biochem Biophys Res Commun 1994; 205: 1164-1171.
376. Snow AD, Mar H, Nochlin D et al. The presence of heparan sulphate proteoglycans in the neuritic plaques and congophilic angiopathy in Alzheimer's disease. Am J Pathol 1988; 133: 456-463.
377. Snow AD, Mar H, Nochlin D et al. Early accumulation of heparan sulphate in neurones and in the β amyloid protein containing lesions of Alzheimer's disease and Down's syndrome. Am J Pathol 1990; 137: 1253-1270.
378. Su JH, Cummings BJ, Cotman CW. Localization of heparan sulfate glycosaminoglycan and proteoglycan core protein in aged brain and Alzheimer's disease. Neuroscience 1992; 51: 801-813.
379. Perry G, Siedlak SL, Richey P et al. Association of heparan sulphate proteoglycan with the neurofibrillary tangles of Alzheimer's disease. J Neurosci 1991; 11: 3679-3683.
380. Buee L, Ding W, Anderson JP et al. Binding of vascular heparan sulfate proteoglycan to Alzheimer's amyloid precursor protein is mediated in part by the N-terminal region

of A4 peptide. Brain Res 1993; 627: 199-204.
381. Snow AD, Seikiguchi RT, Nochlin D et al. Heparan sulphate proteoglycan in diffuse plaques of hippocampus but not of cerebellum in Alzheimer's disease brain. Am J Pathol 1994; 144: 337-347.
382. Fraser PE, Nguyen JT, Chin DT et al. Effects of sulphate ions on Alzheimer-beta/A4 peptide assemblies–implications for amyloid fibril-proteoglycan interactions. J Neurochem 1992; 59: 1531-1540 .
383. Kisilevsky R, Lemieux LJ, Fraser PE et al. Arresting amyloidosis *in vivo* using small-molecule anionic sulphonates or sulphates: implications for Alzheimer's disease. Nature Medicine 1995; 1: 143-148.
384. Pollack SJ, Sadler IIJ, Hawtin SR et al. Sulfated glycosaminoglycans and dyes attenuate the neurotoxic effects of β-amyloid in rat PC12 cells. Neurosci Lett 1995; 184: 113-116.
385. Lorenzo A, Yankner BA. β-amyloid neurotoxicity requires fibril formation and is inhibited by Congo Red. Proc Natl Acad Sci (USA) 1994; 91: 12243-12247.
386. Threlkeld A, Adler R, Hewitt AT. Proteoglycan biosynthesis by chick embryo retina glial-like cells. Dev Biol 1989; 132: 559-568.
387. Fluharty AL, Davis LD, Trammell JL et al. Mucopolysaccharides synthesized by cultured glial cells derived from a patient with Sanfilippo A syndrome. J Neurochem 1975; 25: 429-435.
388. Snow AD, Mar H, Nochlin D et al. Peripheral distribution of dermatan sulfate proteoglycans (decorin) in amyloid-containing plaques and their presence in neurofibrillary tangles of Alzheimer's disease. J Histochem Cytochem 1992; 40: 105-113.
389. DeWitt DA, Silver J, Canning DR et al. Chondroitin sulfate proteoglycans are associated with the lesions of Alzheimer's disease. Exp Neurol 1993; 121: 149-152.
390. Snow AD, Nochlin D, Sekiguchi R et al. Identification and immunolocalization of a new class of proteoglycan (Keratan sulphate) to the neuritic plaques of Alzheimer's disease. Exp Neurol 1996; 138: 305-317.
391. Howard J, Pilkington GJ. Antibodies to fibronectin bind to plaques and other structures in Alzheimer's disease and control brain. Neurosci Lett 1990; 118: 71-76.
392. Perlmutter LS, Chui HC, Saperia D et al. Microangiopathy and the colocalization of heparan sulfate proteoglycan with amyloid in senile plaques of Alzheimer's disease. Brain Res 1990; 508: 13-19.
393. Kato T, Sasaki H, Katagiri T et al. The binding of bFGF to Alzheimer's disease neurofibrillary tangles and senile plaques. Neurosci Lett 1991; 122: 33-36.
394. Masliah E, Mallory M, Ge N et al. Amyloid precursor protein is localized in growing neurites of neonatal rat brain. Brain Res 1992; 593: 323-328.
395. Verbeek MM, Otte-Holler I, Westphal JR et al. Accumulation of intercellular adhesion molecule-1 in senile plaques in brain tissue of patients with Alzheimer's disease. Am J Pathol 1994; 144: 104-116.
396. Verbeek MM, Otte-Holler I, Wesseling P et al. Differential expression of intercellular adhesion molecule-1 (ICAM-1) in the Aβ-containing lesions in brains of patients with dementia of the Alzheimer type. Acta Neuropathol 1996; 91: 608-615.
397. Gillian AM, Brion J-P, Breen KC. Expression of the neural cell adhesion molecule (NCAM) in Alzheimer's disease. Neurodegeneration 1994; 3: 283-291.
398. Lue L-F, Rogers J. Full complement activation fails in diffuse plaques of the Alzheimer's disease cerebellum. Dementia 1992; 3: 308-313.
399. Kalaria RN, Perry G. Amyloid P component and other acute phase proteins associated with cerebellar Aβ deposits in Alzheimer's disease. Brain Res 1993; 631: 151-155.
400. Rozemuller JM, Eikelenboom P, Stam FC et al. A4 protein in Alzheimer's disease: primary and secondary cellular events in extracellular amyloid deposition. J Neuropathol Exp Neurol 1989; 48: 674-691.
401. Eikelenboom P, Hack CE, Kamphorst W et al. The sequence of neuroimmunological events in cerebral amyloid plaque formation in Alzheimer's disease. In: Corain B, Iqbal K, Nicolini M, Wisniewski H, Zatta P, eds. Alzheimer disease: Advances in Clini-

cal and Basic Research. J Wiley & Sons, Chichester. 1993; 165-170.
402. Rogers J, Cooper NR, Webster S et al. Complement activation by β-amyloid in Alzheimer disease. Proc Natl Acad Sci 1992; 89: 10016-10020.
403. Akiyama H, McGeer PL. Brain microglia constitutively express β2 integrins. J Neuroimmunol 1990; 30: 81-93.
404. McGeer PL, Akiyama H, Itagaki S et al. Activation of the classical complement pathway in brain tissue of Alzheimer's patients. Neurosci Lett 1989; 107: 341-346.
405. McGeer PL, Walker DG, Akiyama H et al. Detection of the membrane inhibitor of reactive lysis (CD59) in deceased neurons of Alzheimer brain. Brain Res 1991; 544: 315-319.
406. Akiyama H, Kawamata T, Dedhar S et al. Immunohistochemical localization of vitronectin, its receptor and beta-3 integrin in Alzheimer brain tissue. J Neuroimmunol 1991; 32: 19-28.
407. Eikelenboom P, Stam FC. Immunoglobulins and complement factors in senile plaques. An immunoperoxidase study. Acta Neuropathol 1982; 57: 239-242.
408. Ishii T, Haga S. Immuno-electron microscopic localization of complements in amyloid fibrils of senile plaques. Acta Neuropathol 1984; 63: 296-300.
409. Lampert-Etchells M, Pasinetti GM, Finch CE et al. Regional localization of cells containing complement C1q and C4 mRNAs in the frontal cortex during Alzheimer's disease. Neurodegeneration 1993; 2: 111-121.
410. Jiang H, Burdick D, Glabe CG et al. β-Amyloid activates complement by binding to a specific region of the collagen-like domain of the C1q A chain. J Immunol 1994; 152: 5050-5059.
411. Snyder SW, Wang GT, Barrett L et al. Complement C1q does not bind monomeric β-amyloid. Exp Neurol 1994; 128: 136-142.
412. Levi-Strauss M, Mallet M. Primary cultures of murine astrocytes produce C3 and factor B, two components of the alternative pathway of complement activation. J Immunol 1987; 139: 2361-2366.
413. Pasinetti GM, Johnson SA, Rozovsky I et al. Complement ClqB and C4 mMRAs responses to lesioning in rat brain. Exp Neurol 1992; 118: 117-125.
414. McGeer PL, McGeer EG. Complement proteins and complement inhibitors in Alzheimer's disease. Res Immunol 1992; 143: 620-621.
415. Eikelenboom P, Hack CE, Kamphorst W et al. Distribution pattern and functional state of complement proteins and α-1 antichymotrypsin in cerebral β/A4 deposits in Alzheimer's disease. Res Immunol 1992; 43: 617-620.
416. Abraham CR, Selkoe DJ, Potter H. Immunochemical identification of the serine protease inhibitor α-1 antichymotrypsin in the brain amyloid deposits of Alzheimer's disease. Cell 1988; 52: 487-501.
417. Shoji M, Hirai S, Yamaguchi H et al. Alpha 1-antichymotrypsin is present in diffuse senile plaques. A comparative study of β-protein and α-1-antichymotrypsin immunostaining in the Alzheimer brain. Am J Pathol 1991; 138: 247-257.
418. Rozemuller JM, Abbink JJ, Kamp AM et al. Distribution pattern and functional state of α-1-antichymotrypsin in plaques and vascular amyloid in Alzheimer's disease. Acta Neuropathol 1991; 82: 200-207.
419. Eikelenboom P, Hack CE, Kamphorst W et al. Distribution pattern and functional state of complement proteins and α-1 antichymotrypsin in cerebral β/A4 deposits in Alzheimer's disease. Res Immunol 1992; 43: 617-620.
420. Kalaria RN, Galloway PG, Perry G. Widespread serum amyloid P immunoreactivity in cortical amyloid deposits and the neurofibrillary pathology of Alzheimer's disease and other degenerative disorders. Neuropath Appl Neurobiol 1991; 17: 189-201.
421. Duong T, Pommier EC, Scheibel AB. Immunodetection of the amyloid P component in Alzheimer's disease. Acta Neuropathol 1989; 78: 429-437.
422. Coria F, Castrano E, Prelli F et al. Isolation and characterization of amyloid P component from Alzheimer's disease and other types of cerebral amyloidosis. Lab Invest 1988; 58: 454-458.

423. Bauer J, Strauss S, Schreiter-Gasser U et al. Interleukin-6 and α-2 macroglobulin indicate an acute-phase state in Alzheimer's disease cortices. FEBS Lett 1991; 285: 111-114.
424. Siedlak SL, Cras P, Kawai M et al. Basic fibroblast growth factor binding is a marker for extracellular neurofibrillary tangles in Alzheimer's disease. J Histochem Cytochem 1991; 39: 899-904.
425. Szumanska G, Vorbrodt AW, Mandybur TI et al. Lectin histochemistry of plaques and tangles in Alzheimer's disease. Acta Neuropathol 1987; 73: 1-11.
426. Mann DMA, Bonshek RE, Marcyniuk B et al. Sacchrides of senile plaques and neurofibrillary tangles in Alzheimer's disease. Neurosci Lett 1988; 85: 277-282.
427. Ji Z-S, Fazio S, Lee Y-L et al. Secretion-capture role for apolipoprotein E in remnant lipoprotein metabolism involving cell surface heparan sulfate proteoglycans. J Biol Chem 1994; 269: 2764-2772.
428. Cataldo AM, Nixon RA. Enzymatically active lysosomal proteases are associated with amyloid deposits in Alzheimer brain. Proc Natl Acad Sci (USA) 1990; 87: 3861-3865.
429. Cataldo AM, Thayer CY, Bird ED et al. Lysosomal proteinase antigens are prominently localized within senile plaques of Alzheimer's disease: evidence for a neuronal origin. Brain Res 1990; 513: 181-192.
430. Cataldo AM, Barnett JL, Mann DMA et al. Colocalization of lysosomal hydrolase and β-amyloid in diffuse plaques of the cerebellum and striatum in Alzheimer's disease and Down's syndrome. J Neuropathol Exp Neurol 1996; 55: 704-715.
431. Rifenburg RP, Perry G. Dystrophic neurites define diffuse as well as core-containing senile plaques in Alzheimer's disease. Neurodegeneration 1995; 4: 235-237.
432. Cummings BJ, Su JM, Geddes JW et al. Aggregation of the amyloid precursor protein within degenerating neurons and dystrophic neurites in Alzheimer's disease. Neuroscience 1992; 48: 763-777.
433. Cras P, Kawai M, Lowery DE et al. Senile plaque neurites in Alzheimer's disease accumulate amyloid precursor protein. Proc Natl Acad Sci (USA) 1991; 88: 7552-7556.
434. Shoji M, Hirai S, Yamaguchi H et al. Amyloid β-protein precursor accumulates in dystrophic neurites of senile plaques in Alzheimer's disease. Brain Res 1990; 512: 164-168.
435. Tate-Ostroff B, Majocha RE, Marotta CA. Identification of cellular and extracellular sites of amyloid precursor protein extracytoplasmic domain in normal and Alzheimer's disease brains. Proc Natl Acad Sci (USA) 1989; 86: 745-749.
436. Joachim CL, Games D, Morris J et al. Antibodies to non-Beta regions of the beta-amyloid precursor protein detect a subset of plaques. Am J Pathol 1991; 138: 373-384.
437. Kawai M, Cras P, Richey P et al. Subcellular localization of amyloid precursor protein in senile plaques of Alzheimer's disease. Am J Pathol 1992; 140: 947-958.
438. Masliah E, Honer WG, Mallory M et al. Topographical distribution of synaptic-associated proteins in the neuritic plaques of Alzheimer's disease hippocampus. Acta Neuropathol 1994; 87: 135-142.
439. Masliah E, Mallory M, Hansen L et al. Localization of amyloid precursor protein in GAP43-immunoreactive aberrant sprouting neurites in Alzheimer's disease. Brain Res 1992; 574: 312-316.
440. Six J, Lubke U, Lenders M-B et al. Neurite sprouting and cytoskeletal pathology in Alzheimer's disease: a comparative study with monoclonal antibodies to growth-associated protein B-50 (GAP43) and paired helical filaments. Neurodegeneration 1992; 1: 247-255.
441. Struble RG, Cork LC, Whitehouse PJ et al. Cholinergic innervation in neuritic plaques. Science 1982; 213: 413-415.
442. Armstrong DM, Bruce G, Hersh LB et al. Choline acetyltransferase immunoreactivity in neuritic plaques of Alzheimer brain. Neurosci Lett 1986; 71: 229-234.
443. Kitt CA, Struble RG, Cork LC et al. Catecholaminergic neurites in senile plaques in prefrontal cortex of aged nonhuman primates. Neuroscience 1985; 16: 691-699.

444. Struble RG, Kitt CA, Walker LC et al. Somatostatinergic neurites in senile plaques of aged non-human primates. Brain Res 1984; 324: 394-396.
445. Armstrong DM, LeRoy S, Shields D et al. Somatostatin-like immunoreactivity within neuritic plaques. Brain Res 1985; 338: 71-79.
446. Morrison JH, Rogers J, Scherr S et al. Somatostatin immunoreactivity in neuritic plaques of Alzheimer's patients. Nature 1985; 314: 90-94.
447. Nakamura S, Vincent SR. Somatostatin- and neuropeptide Y-Immunoreactive neurons in the neocortex in senile dementia of Alzheimer's type. Brain Res 1986; 370: 11-20.
448. Armstrong DM, Terry RD. Substance P immunoreactivity within neuritic plaques. Neurosci Lett 1985; 58: 139-144.
449. Dawbarn D, Emson PC. Neuropeptide Y like immunoreactivity in neuritic plaques of Alzheimer's disease. Biochem Biophys Res Commun 1985; 126: 289-294.
450. Benes FM, Farol PA, Majocha RE et al. Evidence for axonal loss in regions occupied by senile plaques in Alzheimer cortex. Neuroscience 1991; 42: 651-660.
451. Masliah E, Terry RD, Mallory M et al. Diffuse plaques do not accentuate synapse loss in Alzheimer's disease. Am J Pathol 1990; 137: 1293-1297.
452. Masliah E, Mallory M, Deerinck T et al. Re-evaluation of the structural organization of neuritic plaques in Alzheimer's disease. J Neuropathol Exp Neurol 1993; 52: 619-632.
453. Wisniewski HM, Wegiel J, Wang KC et al. Ultrastructural studies of the cells forming amyloid fibres in classical plaques. Can J Neurol Sci 1989; 16: 535-542.
454. Wisniewski HM, Vorbrodt AW, Epstein MH. Nucleoside diphosphatase (NDPase) activity associated with human β-protein amyloid fibres. Acta Neuropathol 1991; 81: 366-370.
455. Wegiel J, Wisniewski HM. The complex of microglial cells and amyloid star in three-dimensional reconstruction. Acta Neuropathol 1990; 81: 116-124.
456. Perlmutter LS, Barron E, Chui HC. Morphologic association between microglia and senile plaque amyloid in Alzheimer's disease. Neurosci Lett 1990; 119: 32-36.
457. Sheng JG, Mrak RE, Griffin WST. Microglial interleukin-1β expression in brain regions in Alzheimer's disease: correlation with neuritic plaque distribution. Neuropath Appl Neurobiol 1995; 21: 290-301.
458. Griffin WST, Sheng JG, Roberts GW et al. Interleukin-1 expression in different plaque types in Alzheimer's disease: Significance in plaque evolution. J Neuropathol Exp Neurol 1995; 54: 276-281.
459. Mann DMA, Marcyniuk B, Yates PO et al. The progression of the pathological changes of Alzheimer's disease in frontal and temporal neocortex examined both at biopsy and at autopsy. Neuropath Appl Neurobiol 1988; 14: 177-195.
460. Royston MC, Kodical NS, Mann DMA et al. Quantitative analysis of β-amyloid deposition in Down's syndrome using computerized image analysis. Neurodegeneration 1994; 3: 43-51.
461. Naidu A, Quon D, Cordell B. β-Amyloid peptide produced *in vitro* is degraded by proteinases released by cultured cells. J Biol Chem 1995; 270: 1369-1374.
462. Wisniewski HM, Barcikowska M, Kida E. Phagocytosis of β/A4 amyloid fibrils of the neuritic neocortical plaques. Acta Neuropathol 1991; 81: 588-590.
463. Dickson DW, Mattiace LA, Kure K et al. Microglia in human disease, with an emphasis on acquired immune deficiency syndrome. Lab Invest 1991; 64: 135-156.
464. Giulian D, Tapscott MJ. Immunoregulation of cells within the central nervous system. Brain Behav Immunity 1988; 2: 352-358.
465. Roberts GW, Gentleman SH, Lynch A et al. β/A4 amyloid protein deposition in the brain after head injury. Lancet 1991; 338: 1422-1423.
466. Giulian D, Baker TJ. Peptides released by amoeboid microglia regulate astroglial proliferation. J Cell Biol 1985; 101: 2411-2415.
467. Van Eldik LJ, Griffin WST. S100β in Alzheimer's disease: relation to neuropathology in brain regions. Biochim Biophys Acta 1994; 1223: 398-403.

468. Huell M, Strauss S, Volk B et al. Interleukin-6 is present in early stages of plaque formation and is restricted to the brains of Alzheimer's disease patients. Acta Neuropathol 1995; 89: 544-551.
469. Styren CD, Civin WH, Rogers J. Molecular, cellular, and pathologic characterization of HLA-DR immunoreactivity in normal elderly and Alzheimer disease brain. Exp Neurol 1990; 110: 93-104.
470. Itagaki S, McGeer PL, Akiyama H. Presence of T-cytotoxic suppressor and leukocyte common antigen positive cells in Alzheimer's disease brain tissue. Neurosci Lett 1988; 91: 259-264.
471. Boje KM, Arora PK. Microglial-produced nitric oxide and reactive nitrogen oxides mediate neuronal death. Brain Res 1992; 587: 250-256.
472. Chao CC, Molitor TW, Hu S. Neuroprotective role of IL-4 against reactive microglia. J Immunol 1993; 151: 1473-1481.
473. Mancardi GL, Liwnicz BH, Mandybur TI. Fibrous astrocytes in Alzheimer's disease and senile dementia of Alzheimer's type. Acta Neuropathol 1983; 61: 76-80.
474. Mandybur TI, Chuirazzi BA. Astrocytes and the plaques of Alzheimer's disease. Neurology 1990; 40: 635-639.
475. Schechter R, Yen S-HC, Terry RD. Fibrous astrocytes in senile dementia of the Alzheimer type. J Neuropathol Exp Neurol 1981; 40: 95-101.
476. Frederickson RCA. Astroglia in Alzheimer's disease. Neurobiol Aging 1992; 13: 239-253.
477. Card JP, Meade RP, Davis LG. Immunocytochemical localization of the precursor protein for β amyloid in the rat central nervous system. Neuron 1988; 1: 835-846.
478. Griffin WST, Stanley LC, Ling C et al. Brain interleukin 1 and S-100 immunoreactivity are elevated in Down syndrome and Alzheimer disease. Proc Natl Acad Sci (USA) 1989; 86: 7611-7615.
479. Marshak DR, Pesce SA, Stanley LC et al. Increased S100β neurotrophic activity in Alzheimer's disease temporal lobe. Neurobiol Aging 1991; 13: 1-7.
480. Sheng JG, Mrak RE, Griffin WST. S100β protein expressed in Alzheimer disease: potential role in the pathogenesis of neuritic plaques. J Neurosci Res 1994; 39: 398-404.
481. Alzheimer A. Uber eine eigenartige erkrankung der hirnrinde. Allgem Zeit Psychiat Psych-Gerich Med 1907; 64: 146-148.
482. Ashall F. Plaques and tangles in Alzheimer's disease: a historical introduction. In: Ashall F, Goate A, eds. Pathobiology of Alzheimer's Disease. Academic Press Limited, 1995: 1-7.
483. Fuller SC. A study of the neurofibrils in dementia paralytica, dementia senilis, chronic alcoholism, cerebral lues and microcephalic idiocy. Am J Insanity 1907; 63: 415-468.
484. Kidd M. Paired helical filaments in electron microscopy of Alzheimer's disease. Nature 1963; 97: 192-193.
485. Terry RD. The fine structure of neurofibrillary tangles in Alzheimer's disease. J Neuropathol Exp Neurol 1963; 22: 629-642.
486. Brion JP, Passareiro H, Nunez J et al. Mise en evidence immunologique de la proteine tau au niveau de lesions de degenerescence neurofibrillaire de la maladie d'Alzheimer. Arch Neurol 1985; 95: 229-235.
487. Jamada M, Mehraein P. Distribution of senile changes in the brain–The part of the limbic system in Alzheimer's disease and senile dementia. Neurologie 1968; 211: 308-324.
488. Morimatsu M, Hirai S, Muramatsu A et al. Senile degenerative brain lesions and dementia. J Amer Geriat Soc 1975; 23: 390-406.
489. Braak H, Braak E. Neuropathological staging of Alzheimer-related changes. Acta Neuropathol 1991; 82: 239-259.
490. Bouras C, Hof PR, Morrison JH. Neurofibrillary tangle densities in the hippocampal formation in a non-demented population define subgroups of patients with differential early pathologic changes. Neurosci Lett 1993; 153: 131-135.
491. Cras P, Smith MA, Richey PL et al. Extracellular neurofibrillary tangles reflect neuronal loss and provide further evidence of

extensive protein cross-linking in Alzheimer disease. Acta Neuropathol 1995; 89: 291-295.
492. Shibayama H, Kitoh J. Electron microscopic structure of the Alzheimer's neurofibrillary changes in case of atypical senile dementia. Acta Neuropathol 1978; 41: 229-234.
493. Yagishita S, Itoh Y, Wang N et al. Reappraisal of the fine structure of Alzheimer's neurofibrillary tangles. Acta Neuropathol 1981; 54: 239-246.
494. Metuzals J, Montpetit V, Clapin DF. Organization of the neurofilamentous network. Cell Tis Res 1981; 214: 455-482.
495. Crowther RA, Wischik CM. Image reconstruction of the Alzheimer paired helical filament. EMBO J 1985; 4: 3661-3665.
496. Crowther RA. Straight and paired helical filaments in Alzheimer's disease have a common structural unit. Proc Natl Acad Sci (USA) 1991; 88: 2292-2298.
497. Selkoe DJ, Ihara Y, Salazar FJ. Alzheimer's disease: insolubility of partially purified helical filaments in sodium dodecylsulphate and urea. Science 1982; 215: 1243-1245.
498. Delacourte A, Defossez A. Alzheimer's disease tau proteins, the promoting factors of microtubule assembly, are major components of paired helical filaments. J Neurol Sci 1986; 76: 173-186.
499. Ihara Y, Nukina N, Miura R et al. Phosphorylated tau protein is integrated into paired helical filaments in Alzheimer's disease. J Biochem Japan 1986; 99: 1807-1810.
500. Grundke-Iqbal I, Iqbal K, Tung Y-C et al. Abnormal phosphorylation of the microtubule-associated protein (tau) in Alzheimer cytoskeletal pathology. Proc Natl Acad Sci (USA) 1986; 83: 4913-4917.
501. Kosik KS, Joachim CL, Selkoe DJ. Microtubule associated protein tau is a major antigenic component of paired helical filaments in Alzheimer's disease. Proc Natl Acad Sci (USA) 1986; 83: 4044-4048.
502. Wood JG, Mirra SS, Pollock NJ et al. Neurofibrillary tangles of Alzheimer's disease share antigenic determinants with the axonal microtubule associated protein tau. Proc Natl Acad Sci (USA) 1986; 83: 4040-4043.
503. Anderton BH, Breinburg D, Downs MJ et al. Monoclonal antibodies show that neurofibrillary tangles and neurofilaments share antigenic determinants. Nature 1982; 298: 87-96.
504. Sternberger NH, Sternberger LA, Ulrich J. Aberrant neurofilament phosphorylation in Alzheimer disease. Proc Natl Acad Sci (USA) 1985; 82: 4274-4276.
505. Perry G, Rizzuto N, Autilio-Gambetti L et al. Alzheimer's paired helical filaments contain cytoskeletal components. Proc Natl Acad Sci (USA) 1985; 82: 3916-3920.
506. Ulrich J, Haugh M, Anderton BH et al. Alzheimer dementia and Pick's disease: neurofibrillary tangles and Pick bodies are associated with identical phosphorylated neurofilament epitopes. Acta Neuropathol 1987; 73: 240-246.
507. Gambetti P, Shecket G, Ghetti B et al. Neurofibrillary changes in the human brain. An immunohistochemical study with a neurofilament antiserum. J Neuropathol Exp Neurol 1983; 42: 69-79.
508. Miller CCJ, Brion J-P, Calvert R et al. Alzheimer's paired helical filaments share epitopes with neurofilament side arms. EMBO J 1986; 5: 269-276.
509. Yen S-HC, Gaskin F, Terry RD. Immunocytochemical studies of neurofibrillary tangles. Am J Pathol 1981; 104: 77-89.
510. Nukina N, Kosik KS, Selkoe DJ. Recognition of Alzheimer paired helical filaments by monoclonal neurofilament antibodies is due to cross-reaction with tau protein. Proc Natl Acad Sci (USA) 1987; 84: 3415-3419.
511. Perry G, Friedman R, Shaw G et al. Ubiquitin is detected in neurofibrillary tangles and senile plaque neurites of Alzheimer's disease brains. Proc Natl Acad Sci (USA) 1987; 84: 3033-3036.
512. Mori H, Kondo J, Ihara Y. Ubiquitin is a component of paired helical filament in Alzheimer's disease. Science 1987; 235: 1641-1644.
513. Mann DMA, Prinja D, Davies CA et al. Immunocytochemical profile of neurofibrillary tangles in Down's syndrome patients of different ages. J Neurol Sci 1989; 92: 247-260.

514. Lowe J, Blanchard A, Morrell K et al. Ubiquitin is a common factor in intermediate filament inclusion bodies of diverse type in man, including those of Parkinson's disease, Pick's disease, and Alzheimer's disease, as well as Rosenthal fibres in cerebellar astrocytomas, cytoplasmic bodies in muscle, and Mallory bodies in alcoholic liver disease. J Pathol 1988; 155: 9-15.
515. Yen S-H, Gaskin F, Fu SM. Neurofibrillary tangles in senile dementia of the Alzheimer type share an antigenic determinant with intermediate filaments of the vimentin class. Am J Pathol 1983; 113: 373-381.
516. Vogelsang GD, Zemlan FP, Dean GE. The molecular biology of Alzheimer's disease. In: Finch CE, Davies T, eds. Current Communications in Molecular Biology. New York: Cold Spring Harbor Publications, 1988: 150-158.
517. Wischik CM, Novak M, Edwards PC et al. Structural characterization of the core of the paired helical filament of Alzheimer's disease. Proc Natl Acad Sci (USA) 1988; 85: 4884-4888.
518. Wischik C, Novak M, Thagersen HC et al. Isolation of a fragment of tau derived from the core of the paired helical filament of Alzheimer's disease. Proc Natl Acad Sci (USA) 1988; 85: 4506-4510.
519. Kondo J, Honda T, Mori H et al. The carboxyl third of tau is tightly bound to paired helical filaments. Neuron 1988; 1: 827-834.
520. Kosik KS, Orecchio LD, Binder L et al. Epitopes that span the tau molecule are shared with paired helical filaments. Neuron 1988; 1: 817-825.
521. Goedert M, Wischik C, Crowther RA et al. Cloning and sequencing of the cDNA encoding a core protein of the paired helical filament of Alzheimer's disease: identification as the microtubule associated protein tau. Proc Natl Acad Sci (USA) 1988; 85: 4051-4055.
522. Wischik CM, Lai R, Harrington CR et al. Structure, biochemistry and molecular pathogenesis of paired helical filaments in Alzheimer's disease. In: Goate A, Ashall F eds. Pathobiology of Alzheimer's Disease. Academic Press Limited, London 1995; 9-39.
523. Goedert M, Spillantini MG, Jakes R et al. Multiple isoforms of human microtubule-associated protein tau: sequences and localization in neurofibrillary tangles of Alzheimer's disease. Neuron 1989; 3: 519-526.
524. Lee G, Cowan N, Kirschner M. The primary structure and heterogeneity of tau protein from mouse brain. Science 1988; 239: 285-288.
525. Neve RL, Harris P, Kosik KS et al. Identification of cDNA clones for the human microtubule associated protein tau and chromosomal localization of the genes for tau and microtubule-associated protein 2. Mol Brain Res 1986; 1: 271-280.
526. Andreadis A, Brown WM, Kosik KS. Structure and novel exons of the human tau gene. Biochemistry 1992; 31: 10626-10633.
527. Mukaetova-Ladinska EB, Harrington CR, Roth M et al. Biochemical and anatomical redistribution of tau protein in Alzheimer's disease. Am J Pathol 1993; 143: 565-578.
528. Lee VM-Y, Balin BJ, Otvos LJ et al. A major subunit of paired helical filaments and derivatized forms of normal tau. Science 1991; 251: 675-678.
529. Goedert M, Spillantini MG, Cairns NJ et al. Tau proteins of Alzheimer paired helical filaments: abnormal phosphorylation of all six brain isoforms. Neuron 1992; 8: 159-168.
530. Flament S, Delacourte A, Mann DMA. Phosphorylation of tau proteins: a major event during the process of neurofibrillary degeneration. Comparisons between Alzheimer's disease and Down's syndrome. Brain Res 1990; 516: 15-19.
531. Delacourte A, Flament S, Dibe EM et al. Pathological proteins tau 64 and 69 are specifically expressed in the somatodendritic domain of the degenerating cortical neurons during Alzheimer's disease. Acta Neuropathol 1990; 80: 111-117.
532. Buee-Scherrer V, Buee L, Hof PR et al. Neurofibrillary degeneration in amyotrophic lateral sclerosis/Parkinsonism-dementia complex of Guam. Immunochemical characterization of tau proteins. Am J Pathol 1995; 68: 924-932.

533. Vermersch P, Frigard B, David JP et al. Presence of abnormally phosphorylated tau proteins in the entorhinal cortex of aged non-demented subjects. Neurosci Lett 1992; 144: 143-146.
534. Lai RYK, Gertz HJ, Wischik DJ et al. Examination of phosphorylated tau protein as a PHF-precursor at early stage Alzheimer's disease. Neurobiol Aging 1995; 16: 433-445.
535. Mena R, Edwards P, Perez-Olvera O et al. Monitoring pathological assembly of tau and β-amyloid proteins in Alzheimer's disease. Acta Neuropathol 1995; 89: 50-56.
536. Mena R, Edwards PC, Harrington CR et al. Staging the pathological assembly of truncated tau protein into paired helical filaments in Alzheimer's disease. Acta Neuropathol 1996; 91: 633-641.
537. Lee VM-Y, Trojanowski JQ. Tau proteins and their significance in the pathobiology of Alzheimer's disease. In: Goate A, Ashall F, eds. Pathobiology of Alzheimer's Disease. Academic Press Limited, London 1995; 41-58.
538. Gustke N, Steiner B, Mandelkow E-M et al. The Alzheimer-like phosphorylation of tau proteins reduces microtubule binding and involves Ser-Pro and Thr-Pro motifs. FEBS Lett 1992; 307: 199-205.
539. Goedert M, Jakes R, Crowther RA et al. The abnormal phosphorylation of tau protein at Ser-202 in Alzheimer's disease recapitulates phosphorylation during development. Proc Natl Acad Sci (USA) 1993; 90: 5066-5070.
540. Biernat J, Mandlekow E-M, Schroter C et al. The switch of tau protein to an Alzheimer-like state includes phosphorylation of two serine-proline motifs upstream of the microtubule binding region. EMBO J 1992; 11: 1593-1597.
541. Bramblett GT, Goedert M, Jakes R et al. Abnormal tau phosphorylation at Ser^{396} in Alzheimer's disease recapitulates development and contributes to reduced microtubule binding. Neuron 1993; 10: 1089-1099.
542. Lang E, Szendrei GI, Lee VM-Y et al. Immunological and conformational characterization of a phosphorylated immunodominant epitope on the paired helical filaments found in Alzheimer's disease. Biochem Biophys Res Commun 1992; 187: 783-790.
543. Liu WK, Moore WT, Williams RT et al. Application of synthetic phospho- and unphosphopeptides to identify phosphorylation sites in a subregion of the tau molecule which is modified in Alzheimer's disease. J Neurosci Res 1993; 34: 371-376.
544. Lichtenberg-Kragg B, Mandelkow EM, Biernat J et al. Phosphorylation-dependent epitopes of neurofilament antibodies on tau protein and relationship with Alzheimer tau. Proc Natl Acad Sci (USA) 1992; 89: 5384-5388.
545. Otvos L, Feiner L, Lang E et al. Monoclonal antibody PHF-1 recognizes tau protein phosphorylated at serine residues 396 and 404. J Neurosci Res 1994; 39: 669-673.
546. Hasegawa M, Morishima-Kawashima M, Takio K et al. Protein sequence and mass spectrometric analyses of tau in the Alzheimer's disease brain. J Biol Chem 1992; 267: 17047-17054.
547. Garver TD, Lehman RAW, Lee VM-Y et al. Tau phosphorylation in human, primate and rat brain: evidence that a pool of tau is highly phosphorylated in vivo and is rapidly dephosphorylated in vitro. J Neurochem 1994; 63: 2279-2287.
548. Matsuo ES, Shin RW, Billingsley ML et al. Biopsy-derived adult human brain tau is phosphorylated at the same sites as Alzheimer's disease paired helical filaments. Neuron 1994; 13: 989-1002.
549. Sergeant N, Bussiere T, Vermersch P et al. Isoelectric point differentiates PHF-tau from biopsy-derived human brain tau proteins. NeuroReport 1995; 6: 2217-2220.
550. Morishima-Kawashima M, Hasegawa M, Takio K et al. Ubiquitin is conjugated with amino-terminally processed tau in paired helical filaments. Neuron 1993; 10: 1151-1160.
551. Biernat J, Gustke N, Drewes G et al. Phosphorylation of $serine^{262}$ strongly reduces the binding of tau protein to microtubules: distinction between PHF-like immunoreactivity and microtubule binding. Neuron 1993; 11: 153-163.

552. Watanabe A, Hasegawa M, Suzuki M et al. In vivo phosphorylation sites in fetal and adult rat tau. J Biol Chem 1993; 268: 25712-25717.
553. Litersky JM, Johnson GVW, Jakes R et al. Tau protein is phosphorylated by cyclic AMP-dependent protein kinase and calcium/calmodulin-dependent protein kinase II within its microtubule-binding domains at Ser262 and Ser356. Biochem J 1996; 316: 655-660.
554. Drewes G, Lichtenberg-Kraag B, Doring F et al. Mitogen activated protein (MAP) kinase transforms tau protein into an Alzheimer-like state. EMBO J 1992; 11: 2131-2138.
555. Greenberg SM, Koo EH, Selkoe DJ et al. Secreted β-amyloid precursor protein stimulates mitogen-activated protein kinase and enhances tau phosphorylation. Proc Natl Acad Sci (USA) 1994; 91: 7104-7108.
556. Hanger DP, Hughes K, Woodgett JR et al. Glycogen synthase kinase-3 induces Alzheimer's disease-like phosphorylation of tau: generation of paired helical filament epitopes and neuronal localization of the kinase. Neurosci Lett 1992; 147: 58-62.
557. Vulliet R, Halbran SM, Braun RK et al. Proline directed phosphorylation of human tau protein. J Biol Chem 1992; 267: 22570-22574.
558. Baumann K, Mandelkow E-M, Biernat J et al. Abnormal Alzheimer-like phosphorylation of tau protein by cyclin dependent kinases cdk2 and cdk5. FEBS Lett 1993; 336: 417-424.
559. Paudel HK, Lew J, Ali Z et al. Brain proline-directed protein kinase phosphorylates tau on sites that are abnormally phosphorylated in tau associated with Alzheimer's paired helical filaments. J Biol Chem 1993; 268: 23512-23518.
560. Lovestone S, Reynolds CH, Latimer D et al. Alzheimer's disease-like phosphorylation of the microtubule-associated protein tau by glycogen synthase kinase-3 in transfected mammalian cells. Cur Biol 1994; 4: 1077-1085.
561. Gong C-X, Singh TJ, Grundke-Iqbal I et al. Phosphoprotein phosphatase activities in Alzheimer disease brain. J Neurochem 1993; 61: 921-927.
562. Gong CX, Singh TJ, Grundke-Iqbal I et al. Alzheimer disease abnormally phosphorylated tau is dephosphorylated by protein phosphatase-2B (calcineurin). J Neurochem 1994; 62: 803-806.
563. Harris KA, Oyler GA, Doolittle GM et al. Okadaic acid induces hyperphosphorylated forms of tau protein in human brain slices. Ann Neurol 1993; 33: 77-87.
564. Billingsley ML, Ellis C, Kincaid RL et al. Calcineurin immunoreactivity in Alzheimer's disease. Exp Neurol 1994; 126: 178-184.
565. Brion JP, Couck AM, Conreur JL. Calcineurin (phosphatase 2B) is present in neurons containing neurofibrillary tangles and in a subset of senile plaques in Alzheimer's disease. Neurodegeneration 1995; 4: 13-21.
566. Chalazonitis A, Kalberg AJ, Twardzik DR et al. Transforming growth factor-β has neurotropic actions on sensory neurons in vitro and is synergistic with nerve growth factor. Dev Biol 1992; 152: 121-132.
567. Flanders KC, Lippa CF, Smith TW et al. Altered expression of TGF-β in Alzheimer's disease. Neurology 1995; 45: 1561-1569.
568. Shimohama S, Fujimoto S, Taniguchi T et al. Reduction of low-molecular-weight acid phosphatase activity in Alzheimer brains. Ann Neurol 1993; 33: 616-621.
569. Shimohama S, Fujimoto S, Chachin M et al. Alterations of low molecular weight acid phosphatase protein level in Alzheimer's disease. Brain Res 1995; 699: 125-129.
570. Wood JG, Zinsmeister P. Tyrosine phosphorylation systems in Alzheimer's disease pathology. Neurosci Lett 1991; 121: 12-16.
571. Crowther RA, Olesen OF, Jakes R et al. The microtubule binding repeats of tau protein assemble into filaments like those found in Alzheimer's disease. FEBS Lett 1992; 309: 199-202.
572. Wille H, Drewes G, Biernat J et al. Alzheimer-like paired helical filaments and antiparallel dimers formed from microtubule-associated protein tau in vitro. J Cell Biol 1992; 118: 573-584.
573. Dayan AD, Ball MJ. Histometric observations on the metabolism of tangle-bearing neurons. J Neurol Sci 1973; 19: 433-436.

574. Mann DMA, Yates PO. The relationship between formation of senile plaques and neurofibrillary tangles and changes in nerve cell metabolism in Alzheimer-type dementia. Mech Aging Dev 1981; 17: 395-401.
575. Sumpter PQ, Mann DMA, Davies CA et al. An ultrastructural analysis of the effects of accumulation of neurofibrillary tangle in pyramidal cells of the cerebral cortex in Alzheimer's disease. Neuropath Appl Neurobiol 1986; 12: 305-319.
576. Salehi A, Ravid R, Gonatas NK et al. Decreased activity of the hippocampal neurons in Alzheimer's disease is not related to the presence of the neurofibrillary tangles. J Neuropathol Exp Neurol 1995; 54: 704-709.
577. Smith MA, Taneda S, Richey PL et al. Advanced Maillard reaction end products are associated with Alzheimer disease pathology. Proc Natl Acad Sci (USA) 1994; 91: 5710-5714.
578. Smith MA, Rudnicka-Nawrot M, Richey PL et al. Carbonyl-related posttranslational modification of neurofilament protein in the neurofibrillary pathology of Alzheimer's disease. J Neurochem 1995; 64: 2660-2666.
579. Yan S-D, Chen X, Schmidt A-M et al. Glycated tau protein in Alzheimer disease: A mechanism for induction of oxidant stress. Proc Natl Acad Sci (USA) 1994; 91: 7787-7791.
580. Yan SD, Yan SF, Chen X et al. Non-enzymatically glycated tau in Alzheimer's disease induces neuronal oxidant stress resulting in cytokine gene expression and release of amyloid β-peptide. Nature Medicine 1995; 1: 693-699.
581. Lassmann H, Bancher C, Breitschopf H et al. Cell death in Alzheimer's disease evaluated by DNA fragmentation in situ. Acta Neuropathol 1995; 89: 35-41.
582. Bancher C, Brunner C, Lassmann H et al. Accumulation of abnormally phosphorylated "t" precedes the formation of neurofibrillary tangles in Alzheimer's disease. Brain Res 1989; 477: 90-99.
583. Wolozin BL, Pruchnicki A, Dickson DW et al. A neuronal antigen in the brains of Alzheimer patients. Science 1986; 232: 648-650.
584. Mena R, Wischik CM, Novak M et al. A progressive deposition of paired helical filaments (PHF) in the brain characterizes the evolution of dementia in Alzheimer's disease. J Neuropathol Exp Neurol 1991; 50: 474-490.
585. Endoh R, Ogawara M, Iwatsubo T et al. Lack of the carboxyl terminal sequence of tau in ghost tangles of Alzheimer's disease. Brain Res 1993; 601: 164-172.
586. Ikeda K, Haga C, Oyanagi S et al. Ultrastructural and immunohistochemical study of degenerate-neurite bearing ghost tangles. J Neurol 1992; 239: 191-194.
587. Bondareff W, Wischik CM, Novak M et al. Molecular analysis of neurofibrillary degeneration in Alzheimer's disease: an immunohistochemical study. Am J Pathol 1990; 137: 711-723.
588. Su JH, Cummings BJ, Cotman CW. Subpopulations of dystrophic neurites in Alzheimer brain with distinct immunocytochemical and argentophilic characteristics. Brain Res 1994; 637: 37-44.
589. Spillantini MG, Goedert M, Jakes R et al. Topographical relationship between β-amyloid and tau protein epitopes in tangle-bearing cells in Alzheimer's disease. Proc Natl Acad Sci (USA) 1990; 87: 3952-3956.
590. Tabaton M, Cammarata S, Mancardi G et al. Ultrastructural localization of β-amyloid, tau and ubiquitin epitopes in extracellular neurofibrillary tangles. Proc Natl Acad Sci (USA) 1991; 88: 2098-2102.
591. Yamaguchi H, Nakazato Y, Kawarabayashi T et al. Extracellular neurofibrillary tangles associated with degenerating neurites and neuropil threads in Alzheimer-type dementia. Acta Neuropathol 1991; 81: 603-609.
592. Cras P, Kawai M, Siedlak S et al. Neuronal and microglial involvement in β-amyloid protein deposition in Alzheimer's disease. Am J Pathol 1990; 137: 241-246.
593. Ikeda K, Akiyama H, Haga C et al. Evidence that neurofibrillary tangles undergo glial modification. Acta Neuropathol 1992; 85: 101-104.
594. Nakano I, Iwatsubo T, Otsuka N et al. Paired helical filaments in astrocytes: electron microscopy and immunohistochemis-

try in a case of atypical Alzheimer's disease. Acta Neuropathol 1992; 83: 228-232.

595. Ikeda K, Haga C, Akiyama H et al. Coexistence of paired helical filaments and glial filaments in astrocytic processes within ghost tangles. Neurosci Lett 1992; 148: 126-128.

596. Probst A, Ulrich J, Heitz PU. Senile dementia of Alzheimer type: astroglial reaction to extracellular neurofibrillary tangles in the hippocampus. Acta Neuropathol 1982; 57: 75-79.

597. Strittmatter WJ, Weisgraber K, Goedert M et al. Hypothesis: Microtubule instability and paired helical filament formation in the Alzheimer disease brain are related to Apolipoprotein E genotype. Exp Neurol 1994; 125: 163-171.

598. Pickering-Brown S, Mann DMA, Bourke JP et al. Apolipoprotein E4 and Alzheimer's disease pathology in Lewy body disease and in other β-amyloid forming diseases. Lancet 1994; 343: 1155.

599. Royston MC, Mann DMA, Pickering-Brown S et al. ApoE2 allele promotes longevity and protects patients with Down's syndrome from the development of dementia. Neuro Report 1994; 5: 2583-2585.

600. Wisniewski T, Morelli L, Wegiel J et al. The influence of Apolipoprotein E isotypes on Alzheimer's disease pathology in 40 cases of Down's syndrome. Ann Neurol 1995; 37: 136-138.

601. Buee L, Perez-Tur J, Leveugle B et al. Apolipoprotein E in Guamanian amyotrophic lateral sclerosis/ parkinsonism-dementia complex: genotype analysis and relationships to neuropathological changes. Acta Neuropathol 1996; (In press).

602. Tabaton M, Rolleri M, Masturzo P et al. Apolipoprotein E e4 allele frequency is not increased in progressive supranuclear palsy. Neurology 1995; 45: 1764-1765.

603. Renkawek K, Basman GJCGM, de Jong WW. Expression of small heat-shock protein hsp27 in reactive gliosis in Alzheimer disease and other types of dementia. Acta Neuropathol 1994; 87: 511-519.

604. Hamos JE, Oblas B, Pulaski-Salo D et al. Expression of heat shock proteins in Alzheimer's disease. Neurology 1991; 41: 345-350.

605. Smith MA, Kutty RK, Richey PL et al. Heme Oxygenase-1 associated with the neurofibrillary pathology of Alzheimer's disease. Am J Pathol 1994; 145: 42-47.

606. Smith MA, Richey PL, Kutty RK et al. Ultrastructural localization of heme oxygenase-1 to the neurofibrillary pathology of Alzheimer disease. Molec Chem Neuropathol 1995; 24: 227-230.

607. Premkumar DRD, Smith MA, Richey PL et al. Induction of heme oxygenase-1 mRNA and protein in neocortex and cerebral vessels in Alzheimer's disease. J Neurochem 1995; 65: 1399-1402.

608. Lowe J, McDermott H, Pike I et al. (B crystallin expression in non-lenticular tissues and selective presence in ubiquitinated inclusion bodies in human disease. J Pathol 1992; 166: 61-68.

609. Perry G, Cras P, Siedlak SL et al. β Protein immunoreactivity is found in the majority of neurofibrillary tangles of Alzheimer's disease. Am J Pathol 1992; 140: 283-290.

610. Hyman BT, Van Hoesen GW, Beyreuther K et al. A4 amyloid protein immunoreactivity is present in Alzheimer's disease neurofibrillary tangles. Neurosci Lett 1989; 101: 352-355.

611. Yamaguchi H, Nakazato Y, Shoji M et al. Secondary deposition of beta amyloid within extracellular neurofibrillary tangles in Alzheimer-type dementia. Am J Pathol 1991; 138: 699-705.

612. Allsop D, Haga S-I, Bruton C et al. Neurofibrillary tangles in some cases of dementia pugilistica share antigens with amyloid β-protein of Alzheimer's disease. Am J Pathol 1990; 136: 255-260.

613. Perry G, Richey PL, Siedlak SL et al. Immunocytochemical evidence that the β-protein precursor is an integral component of neurofibrillary tangles of Alzheimer's disease. Am J Pathol 1993; 143: 1586-1593.

614. Giaccone G, Pedrotti B, Migheli A et al. βPP and tau interaction: A possible link between amyloid and neurofibrillary tangles in Alzheimer's disease. Am J Pathol 1996; 148: 79-87.

615. Ledesma MD, Bonay P, Colaco C et al. Analysis of microtubule-associated protein

tau glycation in paired helical filaments. J Biol Chem 1994; 269: 21614-21619.
616. Sparkman DR, Goux WJ, Jones CM et al. Alzheimer disease paired helical filament core structures contain glycolipid. Biochem Biophys Res Commun 1991; 181: 771-779.
617. Smith MA, Rudnicka-Nawrot M, Richey PL et al. Carbonyl-related posttranslational modification of neurofilament protein in the neurofibrillary pathology of Alzheimer's disease. J Neurochem 1995; 64: 2660-2666.
618. Galloway PG, Mulvihill P, Siedlak S et al. Immunochemical demonstration of tropomyosin in the neurofibrillary pathology of Alzheimer's disease. Am J Pathol 1990; 137: 291-300.
619. Perl DP, Brody AR. Alzheimer's disease: x-ray spectrometric evidence of aluminum accumulation in neurofibrillary tangle-bearing neurons. Science 1980; 208: 297-299.
620. Good PF, Perl DP, Bierer LM et al. Selective accumulation of aluminum and iron in neurofibrillary tangles of Alzheimer's disease: a laser microprobe (LAMMA) study. Ann Neurol 1992; 31: 286-292.
621. Jacobs RW, Duong T, Jones RE et al. A reexamination of aluminum in Alzheimer's disease: analysis by energy dispersive x-ray microprobe and flameless atomic absorption spectrophotometry. Can J Neurol Sci 1989; 16: 498-503.
622. Lovell MA, Ehmann WD, Markesbery WR. Laser microprobe analysis of brain aluminum in Alzheimer's disease. Ann Neurol 1993; 33: 36-42.
623. Guy S, Jones D, Mann DMA et al. Neuroblastoma cells treated with aluminum-EDTA express an epitope associated with Alzheimer's disease neurofibrillary tangles. Neurosci Lett 1991; 121: 166-168.
624. Lewy FH. Paralysis agitans: I Pathologische anatomie. In: Handbuch der Neurologie III. Berlin: Springer, 1912: 920-933.
625. Tretiakoff MC. Contribution a l'etude d'anatomie pathologique de locus niger de soemmerling. Thesis, 1919; University of Paris.
626. Greenfield JG, Bosanquet FD. The brain stem lesions in Parkinsonism. J Neurol Neurosurg Psychiatry 1953; 16: 213-226.
627. Gibb WRG. Idiopathic Parkinson's disease and the Lewy body disorders. Neuropath Appl Neurobiol 1986; 12: 223-234.
628. Forno LS. Concentric hyaline intraneuronal inclusions of Lewy type in the brains of elderly persons (50 incidental cases); relationship to Parkinsonism. J Amer Geriat Soc 1969; 17: 557-575.
629. Forno LS, Alvord EC. The pathology of Parkinsonism. In: Recent Advances in Parkinson's disease. Blackwell Scientific Publications, Oxford 1971; 120-161.
630. Ohama E, Ikuta F. Parkinson's disease: Distribution of Lewy bodies and monoamine neuron system. Acta Neuropathol 1976; 34: 311.
631. Tomonaga M. Neuropathology of locus coeruleus: A semi-quantitative study. J Neurol 1983; 230: 231-240.
632. Forno LS, Murphy GM, Eng LF. Immunocytochemical study of Lewy bodies in sympathetic ganglia. Neurodegeneration 1992; 1: 135-144.
633. Oyanagi K, Wakabayashi K, Ohama E et al. Lewy bodies in the lower sacral parasympathetic neurons of a patient with Parkinson's disease. Acta Neuropathol 1990; 80: 558-559.
634. Wakabayashi K, Takahashi H, Takeda S et al. Parkinson's disease: the presence of Lewy bodies in Auerbach's and Meissner's plexuses. Acta Neuropathol 1988; 76: 217-221.
635. Forno LS, Langston JW. Lewy bodies and aging: Relation to Alzheimer's and Parkinson's diseases. Neurodegeneration 1993; 2: 19-24.
636. Gibb WRG, Mountjoy CQ, Mann DMA et al. A pathological study of the association between Lewy body disease and Alzheimer's disease. J Neurol Neurosurg Psychiatry 1989; 52: 701-708.
637. Woodard JS. Concentric hyaline inclusion body formation in mental disease. An analysis of 27 cases. J Neuropathol Exp Neurol 1962; 21: 442-450.
638. Perry RH, Irving D, Blessed G et al. Senile dementia of Lewy body type. A clinically and neuropathologically distinct form of Lewy body dementia in the elderly. J Neurol Sci 1990; 95: 119-139.

639. Hansen L, Salmon D, Galasko D et al. The Lewy body variant of Alzheimer's disease: A clinical and pathological entity. Neurology 1990; 40: 1-8.
640. Dickson DW, Crystal H, Mattiace LA et al. Diffuse Lewy body disease: light and electron microscopic immunocytochemistry of senile plaques. Acta Neuropathol 1989; 78: 572-584.
641. Ince P, Irving D, MacArthur F et al. Quantitative neuropathological study of Alzheimer-type pathology in the hippocampus: comparison of senile dementia of Alzheimer type, senile dementia of Lewy body type, Parkinson's disease and non-demented elderly control patients. J Neurol Sci 1991; 106: 142-152.
642. Dale GE, Probst A, Luthert P et al. Relationships between Lewy bodies and pale bodies in Parkinson's disease. Acta Neuropathol 1992; 83: 525-529.
643. Lennox G, Lowe J, Morrell K et al. Anti-ubiquitin immunocytochemistry is more sensitive than conventional techniques in the detection of diffuse Lewy body disease. J Neurol Neurosurg Psychiatry 1989; 52: 67-71.
644. Roy S, Wolman L. Ultrastructural observations in Parkinsonism. J Pathol 1969; 99: 39-44.
645. Duffy PO, Tennyson VM. Phase and electron microscopic observations of Lewy bodies and melanin granules in the substantia nigra and locus ceruleus in Parkinson's disease. J Neuropathol Exp Neurol 1965; 24: 398-414.
646. Galloway PG, Grundke-Iqbal I, Iqbal K et al. Lewy bodies contain epitopes both shared and distinct from Alzheimer neurofibrillary tangles. J Neuropathol Exp Neurol 1988; 47: 654-663.
647. Lowe J, McDermott H, Landon M et al. Ubiquitin carboxylterminal hydrolase (PGP 9.5) is selectively present in ubiquitinated inclusion bodies characteristic of human neurodegenerative diseases. J Pathol 1990; 161: 153-160.
648. Goldman J, Yen S-H, Chiu F et al. Lewy bodies of Parkinson's disease contain neurofilament antigens. Science 1983; 221: 1082-1084.
649. Bancher C, Lassmann H, Budka H et al. An antigenic profile of Lewy bodies: Immunocytochemical indication for protein phosphorylation and ubiquitination. J Neuropathol Exp Neurol 1989; 48: 81-93.
650. DeWitt DA, Richey PL, Praprotnik D et al. Chondroitin sulfate proteoglycans are a common component of neuronal inclusions and astrocytic reaction in neurodegenerative diseases. Brain Res 1994; 656: 205-209.
651. Galloway PG, Bergeron C, Perry G. The presence of tau distinguishes Lewy bodies of diffuse Lewy body disease from those of idiopathic Parkinson disease. Neurosci Lett 1989; 100: 6-10.
652. Galloway PG, Perry G. Tropomyosin distinguishes Lewy bodies of Parkinsons's disease from other neurofibrillary pathology. Brain Res 1991; 541: 347-349.
653. Galloway PG, Mulvihill P, Perry G. Filaments of Lewy bodies contain insoluble cytoskeletal elements. Am J Pathol 1992; 140: 1-14.
654. Pollanen MS, Bergeron C, Weyer L. Detergent-insoluble cortical Lewy body fibrils share epitopes with neurofilament and tau. J Neurochem 1992; 58: 1953-1956.
655. Jellinger K, Paulus W, Grundke-Iqbal I et al. Brain iron and ferritin in Parkinson's and Alzheimer's diseases. J Neural Transm Park Dis Dement Sect 1990; 2: 327-340.
656. Hirsch EC, Brandel JP, Galle P et al. Iron and aluminum increase in the substantia nigra of patients with Parkinson's disease: An X-ray microanalysis. J Neurochem 1991; 56: 446-451.
657. Hirano A. Pathology of amyotrophic lateral sclerosis. In: Gajdusek DC, Gibbs CJ, Alpers M, eds. Slow, Latent and Temperate Virus Infections. NINDB Monograph. Washington DC: US Department of Health, Education and Welfare, 1965.
658. Schochet SS, Lampert PW, Lindenberg R. Fine structure of the Pick and Hirano bodies in a case of Pick's disease. Acta Neuropathol 1968; 11: 330-337.
659. Gibson PH, Tomlinson BE. Numbers of Hirano bodies in the hippocampus of normal and demented people with Alzheimer's disease. J Neurol Sci 1977; 33: 199-206.

660. Ogata J, Budzilovitch GN, Cravioto H. A study of rod-like structures (Hirano bodies) in 240 normal and pathological brains. Acta Neuropathol 1972; 21: 61-67.
661. Hirano A, Malamud N, Elizan TS et al. Amyotrophic lateral sclerosis and parkinsonism-dementia complex on Guam. Arch Neurol 1966; 2: 225-232.
662. Laas R, Hagel C. Hirano bodies and chronic alcoholism. Neuropath Appl Neurobiol 1994; 20: 12-21.
663. Yamamoto T, Hirano A. Hirano bodies in the perikaryon of the Purkinje cell in a case of Alzheimer's disease. Acta Neuropathol 1985; 67: 167-169.
664. Nagara H, Yajima K, Suzuki K. An ultrastructural study on the cerebellum of the brindle mouse. Acta Neuropathol 1980; 52: 41-50.
665. Tomanaga M. Hirano bodies in extraocular muscle. Acta Neuropathol 1983; 60: 309-313.
666. Atsumi T, Yamamura Y, Sato T et al. Hirano bodies in the axon of peripheral nerves in a case with progressive external ophthalmoplegia with multisystem involvement. Acta Neuropathol 1980; 49: 95-100.
667. Fu Y-S, Ward J, Young HF. Unusual, rod-shaped cytoplasmic inclusions (Hirano bodies) in a cerebellar hemangioblastoma. Acta Neuropathol 1975; 31: 129-135.
668. Ho K-L, Allevato PA. Hirano body in an inflammatory cell of leptomeningeal vessel infected by fungus Paecilomyces. Acta Neuropathol 1986; 71: 159-162.
669. Schochet SSJ, McCormick EF. Ultrastructure of Hirano bodies. Acta Neuropathol 1972; 21: 50-60.
670. Gibson PH. Light and electron microscopic observations on the relationship between Hirano bodies, neuron and glial perikarya in the human hippocampus. Acta Neuropathol 1978; 42: 165-171.
671. Okamoto K, Hirai S, Hirano A. Hirano bodies in myelinated fibers of hepatic encephalopathy. Acta Neuropathol 1982; 58: 308-310.
672. Hirano A, Dembitzer HM, Kurland LT et al. The fine structure of some intraganglionic alterations. Neurofibrillary tangles, granulovacuolar bodies and 'rod-like' structures as seen in Guam amyotrophic lateral sclerosis and parkinsonism-dementia complex. J Neuropathol Exp Neurol 1968; 27: 169-182.
673. Tomonaga M. Ultrastructure of Hirano bodies. Acta Neuropathol 1974; 28: 365-366.
674. Galloway PG, Perry G, Gambetti P. Hirano body filaments contain actin and actin-associated proteins. J Neuropathol Exp Neurol 1987; 46: 185-199.
675. Goldman HE. The association of actin with Hirano bodies. J Neuropathol Exp Neurol 1983; 42: 146-152.
676. Goldman JE, Horoupian DS. An immunocytochemical study of intraneuronal inclusions of the caudate and substantia nigra. Reaction with anti-actin antiserum. Acta Neuropathol 1982; 58: 300-302.
677. Galloway PG, Perry G, Kosik KS et al. Hirano bodies contain tau protein. Brain Res 1987; 403: 337-340.
678. Peterson C, Kress Y, Vallee R et al. High molecular weight microtubule-associated proteins bind to actin lattices (Hirano bodies). Acta Neuropathol 1988; 77: 168-174.
679. Schmidt ML, Lee VM-Y, Trojanowski JQ. Analysis of epitopes shared by Hirano bodies and neurofilament proteins in normal and Alzheimer's disease hippocampus. Lab Invest 1989; 60: 513-522.
680. Manetto V, Abdul-Karim FW, Perry G et al. Selective presence of ubiquitin in intracellular inclusions. Am J Pathol 1989; 134: 505-513.
681. Simchowitz T. Histologische studien uber die senile demenz. In: Nissl F, Alzheimer A, eds. Histologische und Histopathologische Arbeiten. Jena: Gustav Fischer, 1911: 267-444.
682. Tomlinson BE, Kitchener D. Granulovacuolar degeneration of hippocampal pyramidal cells. J Pathol 1971; 106: 165-185.
683. Ball MJ, Lo P. Granulovacuolar degeneration in the aging brain and in dementia. J Neuropathol Exp Neurol 1977; 36: 474-487.
684. Xu M, Shibayama H, Kobayashi H et al. Granuolovacuolar degeneration in the hippocampal cortex of aging and demented patients–a quantitative study. Acta Neuropathol 1992; 85: 1-9.

685. Woodward JS. Clinicopathological significance of granulovacuolar degeneration in Alzheimer's disease. J Neuropathol Exp Neurol 1962; 21: 85-91.
686. Price DL, Altschuler RJ, Struble RG et al. Sequestration of tubulin in neurons in Alzheimer's disease. Brain Res 1986; 385: 305-310.
687. Kahn J, Anderton BH, Probst A et al. Immunohistological study of granulovacuolar degeneration using monoclonal antibodies to neurofilaments. J Neurol Neurosurg Psychiatry 1985; 48: 924-926.
688. Dickson DW, Ksiezak-Reding H, Davies P et al. A monoclonal antibody that recognizes a phosphorylated epitope in Alzheimer neurofibrillary tangles, neurofilaments and tau proteins immunostains granulovacuolar degeneration. Acta Neuropathol 1987; 73: 254-258.
689. Dickson DW, Liu W-K, Kress Y et al. Phosphorylated tau immunoreactivity of granulovacuolar bodies (GVB) of Alzheimer's disease: localization of two amino terminal tau epitopes in GVB. Acta Neuropathol 1993; 85: 463-470.
690. Bondareff W, Wischik CM, Novak M et al. Sequestration of tau by granulovacuolar degeneration in Alzheimer's disease. Am J Pathol 1991; 139: 641-647.
691. Okamoto K, Hirai S, Iizuka T et al. Reexamination of granulovacuolar degeneration. Acta Neuropathol 1991; 82: 340-345.
692. Mann DMA. Granulovacuolar degeneration in pyramidal cells of the hippocampus. Acta Neuropathol 1978; 42: 149-151.
693. Barden H. The histochemical relationship of neuromelanin and lipofuscin. J Neuropathol Exp Neurol 1969; 28: 419-441.
694. Hirosawa K. Electron microscopic studies on pigment granules in the substantia nigra and locus caeruleus of the Japanese monkey. Z Zellforsch 1968; 88: 187-203.
695. Sekhon SS, Maxwell DS. Ultrastructural changes in neurones of spinal anterior horns of aging mice with particular reference to accumulation of lipofuscin pigment. J Neurocytol 1974; 3: 59-72.
696. Brunk U, Ericsson JLE. Electron microscopical studies on rat brain neurons. Localization of acid phosphatase and mode of formation of lipofuscin bodies. J Ultrastruc Res 1972; 38: 1-15.
697. Mann DMA, Yates PO. Lipoprotein pigments: Their relationship to aging in the human nervous system. I–The lipofuscin content of nerve cells. Brain 1974; 97: 481-488.
698. Mann DMA, Yates PO, Stamp JE. Relationship of lipofuscin pigment to aging in the human nervous system. J Neurol Sci 1978; 35: 83-93.
699. Brody H. The deposition of aging pigment in the human cerebral cortex. J Gerontol 1960; 15: 258-261.
700. Scholtz CL, Brown A. Lipofuscin and Trans-synaptic degeneration. Virchows Arch Anat Histol 1978; 381: 35-40.
701. Samorajski T, Keefe JR, Ordy JM. Intracellular localization of lipofuscin age pigments in the nervous system. J Gerontol 1964; 19: 262-272.
702. Barden H. Relationship of Golgi thiamine pyrophosphatase and lysosomal acid phosphatase to neuromelanin and lipofuscin in cerebral neurones of aging rhesus monkey. J Neuropathol Exp Neurol 1970; 29: 225-240.
703. Bazelon M, Fenichel GM, Randall J. Studies on neuromelanin. I.A melanin system in the human adult brain stem. Neurology 1969; 17: 512-519.
704. Moses HL, Ganote CE, Beaver DL et al. Light and electron microscope studies of pigment in human and rhesus monkey substantia nigra and locus caeruleus. Anat Rec 1966; 155: 167-184.
705. Graham DG, Tiffany SM, Bell WR. Auto-oxidation versus covalent binding of quinones as the mechanism of toxicity of dopamine, 6-hydroxydopamine and related compounds towards C1300 neuroblastoma cells in vitro. Molec Pharmacol 1978; 14: 644-653.
706. Jellinger K, Kienzl E, Paulus W et al. Presence of iron in melanized dopamine neurons in Parkinson's disease. J Neurochem 1992; 59: 1168-1171.
707. Foley JM, Baxter D. On the nature of pigment granules in the cells of the locus caeruleus and substantia nigra. J Neuropathol Exp Neurol 1958; 17: 586-598.

708. Mann DMA, Yates PO. Lipoprotein pigments: Their relationship to aging in the human nervous system. II–The melanin content of pigmented nerve cells. Brain 1974; 97: 489-498.
709. Graham DG. On the origin and significance of neuromelanin. Arch Path Lab Med 1979; 103: 359-362.
710. Mann DMA, Yates PO, Barton CM. Variations in melanin content with age in the human substantia nigra. Biochem Exp Biol 1977; 13: 137-139.
711. McGeer PL, McGeer EG. Enzymes associated with the metabolism of catecholamines, acetylcholine and GABA in human controls and patients with Parkinson's disease and Huntington's Chorea. J Neurochem 1976; 26: 65-76.
712. Carlsson A, Winblad B. Influence of age and time interval between death and autopsy on dopamine and 3-methoxytyramine levels in human basal ganglia. J Neural Transm Park Dis Dement Sect 1976; 38: 271-276.
713. Winblad B, Adolfsson R, Gottfries CG et al. In: Frigerio A, ed. Recent advances in mass spectrometry in biochemistry and medicine. Academic Press, New York 1978; 253.
714. Barden H, Barret R. Localization of catecholamine fluorescence to dog hypothalamic neuromelanin bearing neurones. J Histochem Cytochem 1973; 21: 175-183.
715. Spokes EGS. An analysis of factors influencing measurements of dopamine, noradrenaline, glutamic acid decarboxylase and choline acetyl transferase in human post mortem brain tissue. Brain 1979; 102: 333-346.
716. Cross AJ, Crow TJ, Perry EK et al. Reduced dopamine-β-hydroxylase activity in Alzheimer's disease. BMJ 1981; 1: 93-94.
717. Seite R, Leonetti J, Luciani-Vuillet JL et al. Cyclic AMP and ultrastructural organization of the nerve cell nucleus: Stimulation of nuclear microtubules and microfilaments assembly in sympathetic neurons. Brain Res 1977; 124: 41-51.
718. Bouteille M, Kalifat SR, Delarue J. Ultrastructural variations of nuclear bodies in human disease. J Ultrastruc Res 1967; 19: 474-486.
719. Field EJ, Peat A. Intranuclear inclusions in neurones and glia. A study in the aging mouse. Gerontologia 1971; 17: 129-138.
720. Toper S, Bannister CM, Lincoln J et al. Nuclear inclusion bodies in Alzheimer's disease. Neuropath Appl Neurobiol 1980; 6: 245-253.
721. Zurhein GM, Chou SM. Subacute sclerosing panencephalitis. Neurology 1968; 18: 146-160.
722. Oyangi S, Rorke LB, Katz M et al. Histopathology and electronmicroscopy of 3 cases of SSPE. Acta Neuropathol 1971; 18: 58-73.
723. Martinez AJ, Oya T, Jabbour JT et al. Subacute sclerosing panencephalitis (SSPE). Reappraisal of nuclear, cytoplasmic and axonal inclusions. Ultrastructure of 8 cases. Acta Neuropathol 1974; 28: 1-13.
724. Grunnet ML. Nuclear bodies in Creutzfeldt-Jakob and Alzheimer's disease. Neurology 1975; 25: 1091-1093.
725. Shiraki H, Yamamoto T. Histochemical aspects of hepatocerebral diseases. Adv Neurol Sci 1960; 5: 73-80.
726. Ishii T, Hamada S. Histological and histochemical studies on the eosinophilic intranuclear inclusions of the pigmented cells of the substantia nigra. Adv Neurol Sci 1960; 5: 111-116.
727. Yuen P, Baxter DW. The morphology of Marinesco bodies (paranucleolar corpuscles) in the melanin pigmented nuclei in the brainstem. J Neurol Neurosurg Psychiatry 1963; 26: 178-184.
728. Leestma JE, Andrews JM. The fine structure of the Marinesco body. Arch Pathol 1969; 88: 431-437.
729. Hirai S, Okamoto K. Marinesco body. Neurol Med (Tokyo) 1986; 24: 457-462.
730. Schochet SSJ, Wyatt RB, McCormick WF. Intracytoplasmic acidophilic granules in the substantia nigra. Arch Neurol 1970; 22: 550-559.
731. Sekiya S, Tanaka M, Hayashi S et al. Light- and electron- microscopic studies of intracytoplasmic acidophilic granules in the human locus coeruleus and substantia nigra. Acta Neuropathol 1982; 56: 78-80.
732. Sasaki S, Hirano A. Study of intracytoplasmic acidophilic granules in the human

dorsal root ganglia. Neurol Med (Tokyo) 1983; 19: 263-268.
733. Culebras A, Feldman GR, Merk F. Cytoplasmic inclusion bodies within neurons of the thalamus in myotonic dystrophy. J Neurol Sci 1973; 19: 319-329.
734. Pena CE. Intracytoplasmic neuronal inclusions in the human thalamus. Acta Neuropathol 1980; 52: 157-159.
735. Cisse S, Lacoste-Royal G, Laperriere J et al. Ubiquitin is a component of polypeptides purified from corpora amylacea of aged human brain. Neurochem Res 1991; 16: 429-433.
736. Cisse S, Perry G, Lacoste-Royal G et al. Immunocytochemical identification of ubiquitin and heat-shock proteins in corpora amylacea from normal aged and Alzheimer's disease brains. Acta Neuropathol 1993; 85: 233-240.
737. Austin JH, Sakai M. Corpora amylacea. In: Minckler J, ed. Pathology of the nervous system. McGraw-Hill, New York 1972; 29-61.
738. Sakai M, Austin J, Witmer F et al. Studies of corpora amylacea. I. Isolation and preliminary characterization by chemical and histochemical techniques. Arch Neurol 1969; 21: 526-544.
739. Liu HM, Anderson K, Caterson B. Demonstration of a keratan sulfate proteoglycan and a mannose-rich glycoconjugate in corpora amylacea of the brain by immunocytochemical and lectin-binding methods. J Neuroimmunol 1987; 14: 49-60.
740. Steyaert A, Cisse S, Merhi Y et al. Purification and polypeptide composition of corpora amylacea from aged human brain. J Neurosci Methods 1990; 31: 59-64.
741. Stam FC, Roukema PA. Histochemical and biochemical aspects of corpora amylacea. Acta Neuropathol 1973; 25: 95-102.
742. Cisse S, Schipper HM. Experimental induction of corpora amylacea-like inclusions in rat astroglia. Neuropath Appl Neurobiol 1995; 21: 423-431.
743. Singhrao SK, Neal JW, Newman GR. Corpora amylacea could be an indicator of neurodegeneration. Neuropath Appl Neurobiol 1993; 19: 269-276.
744. Ramsey HJ. Ultrastructure of corpora amylacea. J Neuropathol Exp Neurol 1965; 24: 29-39.
745. Takahashi K, Agari M, Nakamura H. Intra-axonal corpora amylacea in ventral and lateral horns of the spinal cord. Acta Neuropathol 1975; 31: 151-158.
746. Anzil AP, Herrlinger H, Blinzinger J et al. Intraneuritic corpora amylacea. Virchows Archives 1974; 364: 297-301.
747. Yagashita S, Itoh Y. Corpora amylacea in the peripheral nerve axon. Acta Neuropathol 1977; 37: 73-76.
748. Singhrao SK, Neal JW, Piddlesden SJ et al. New immunocytochemical evidence for a neuronal/oligodendroglial origin for corpora amylacea. Neuropath Appl Neurobiol 1994; 20: 66-73.
749. Martin JE, Mather K, Swash M et al. Heat shock protein expression in corpora amylacea in the central nervous system: clues to their origin. Neuropath Appl Neurobiol 1991; 17: 113-119.
750. Prabhakar S, Kurien E, Gupta RS et al. Heat shock protein immunoreactivity in CSF: Correlation with oligoclonal banding and demyelinating disease. Neurology 1994; 44: 1644-1648.
751. Singhrao SK, Morgan BP, Neal JW et al. A functional role for corpora amylacea based on evidence from complement studies. Neurodegeneration 1995; 4: 335-345.
752. Tokutake S, Nagase H, Morisaki S et al. X-ray microprobe analysis of corpora amylacea. Neuropath Appl Neurobiol 1995; 21: 269-273.
753. Lowenthal A, Bruyn GW. Calcification of the striopallidodentate system. In: Vinken PJ, Bruyn GW eds. Handbook of Clinical Neurology. North Holland Publishing Company, Amsterdam 1968; 703-729.
754. Hurst EW. On the so-called calcification in the basal ganglia of the brain. J Pathol 1926; 24: 65-84.
755. Neumann MA. Iron and calcium dysmetabolism in the brain. J Neuropathol Exp Neurol 1963; 22: 148-163.
756. Strassman G. Iron and calcium deposits in the brain. J Neuropathol Exp Neurol 1949; 8: 428-435.

757. Slager UT, Wagner JA. The incidence, composition, and pathological significance of intracerebral vascular deposits in the basal ganglia. J Neuropathol Exp Neurol 1956; 15: 417-431.
758. Takashima S, Becker LE. Basal ganglia calcification in Down's syndrome. J Neurol Neurosurg Psychiatry 1985; 48: 61-64.
759. Wisniewski KE, French JH, Rosen JF et al. Basal Ganglia Calcification (BGC) in Down's Syndrome (DS)–Another Manifestation of Premature Aging. Ann NY Acad Sci 1982; 396: 179-189.
760. Mann DMA. Calcification of the basal ganglia in Down's syndrome and Alzheimer's disease. Acta Neuropathol 1988; 76: 595-598.
761. Escourolle R, Poirer J. Manual of Basic Neuropathology. Philadelphia: Saunders, WB. 1973.
762. Beck E, Matthews WB, Stevens DL et al. Creutzfeldt-Jakob disease. The neuropathology of a transmission experiment. Brain 1969; 92: 699-716.
763. Mann DMA, Stamp JE, Yates PO et al. The fine structure of the axonal torpedo in Purkinje cells of the human cerebellum. Neurol Res 1980; 1: 369-378.
764. Carpenter S. Proximal axonal enlargement in motor neuron disease. Neurology 1968; 18: 841-851.
765. Gertz HJ, Cervos-Navarro J, Frydl V et al. Glycogen accumulation of the aging human brain. Mech Aging Dev 1985; 31: 25-35.
766. Mann DMA, Sumpter PQ, Davies CA et al. Glycogen accumulations in the cerebral cortex in Alzheimer's disease. Acta Neuropathol 1987; 73: 181-184.
767. Nichols NR, Day JR, Laping NJ et al. GFAP mRNA increases with age in rat and human brain. Neurobiol Aging 1993; 14: 421-429.
768. Hansen LA, Armstrong DM, Terry RD. An immunohistochemical quantification of fibrous astrocytes in the aging human cerebral cortex. Neurobiol Aging 1987; 8: 1-6.
769. Hansen LA, De Teresa R, Davies P et al. Neocortical morphometry, lesion counts, and choline acetyltransferase levels in the age spectrum of Alzheimer's disease. Neurology 1988; 38: 48-54.
770. Sakai Y, Rawson C, Lindburg K et al. Serum and transforming growth factor beta regulate glial fibrillary acidic protein in serum-free-derived mouse embryo cells. Proc Natl Acad Sci (USA) 1990; 87: 8378-8382.
771. Arai N. Grumose or foamy spheroid bodies involving astrocytes in the human brain. Neuropath Appl Neurobiol 1995; 21: 238-245.
772. McGeer PL, Itagaki S, Tago H et al. Reactive microglia in patients with senile dementia of the Alzheimer-type are positive for the histocompatibility glycoprotein HLA-DR. Neurosci Lett 1987; 79: 195-200.
773. Rozemuller JM, Eikelenboom P, Pals ST et al. Microglial cells around amyloid plaques in Alzheimer's disease express leucocyte adhesion molecules of the LFA-1 family. Neurosci Lett 1989; 101: 288-292.
774. Kaneko Y, Kitamoto T, Tateishi J et al. Ferritin immunohistochemistry as a marker for microglia. Acta Neuropathol 1989; 79: 129-136.
775. Dickson DW, Crystal H, Mattiace LA et al. Diffuse Lewy body disease: light and electron microscopic immunocytochemistry of senile plaques. Acta Neuropathol 1989; 78: 572-584.
776. Suzuki H, Franz H, Yamamoto T et al. Identification of the normal microglial population in human and rodent nervous tissue using lectin histochemistry. Neuropath Appl Neurobiol 1988; 14: 221-227.
777. Mattiace LA, Davies P, Dickson DW. Detection of HLA-DR on microglia in the human brain is a function of both clinical and technical factors. Am J Pathol 1990; 36: 1101-1114.
778. Bell MA, Ball MJ. Morphometric comparison of hippocampal microvasculature in aging and demented people: diameter and densities. Acta Neuropathol 1981; 53: 299-318.
779. Mann DMA, Eaves NR, Marcyniuk B et al. Quantitative changes in cerebral cortical microvasculature in aging and dementia. NeurobiolAging 1986; 7: 321-330.
780. Buee L, Hof PR, Bouras C et al. Pathological alterations of the cerebral microvasculature in Alzheimer's disease and related dementing disorders. Acta Neuropathol 1994; 87: 469-480.

781. Hassler D. Arterial deformities in senile brains. Acta Neuropathol 1967; 8: 219-229.
782. Ravens JR. Vascular changes in the human senile brain. In: Cervos-Navarra J, ed. Pathology of Cerebrospinal Microcirculation. New York: Raven Press, 1978: 487-501.
783. Mann DMA, Purkiss MS, Bonshek RE et al. Lectin histochemistry of cerebral microvessels in aging, Alzheimer's disease and Down's syndrome. NeurobiolAging 1992; 13: 137-143.
784. Divry P. Etude histo-chimique des plaques seniles. J Belge Neurol Psychiat 1927; 27: 643-657.
785. Scholz W. Studien zur Pathologie der Hirngefässe: II Die drüsige Entartung der Hirnarterien und-Capillaren. Z Gesamte Neurol Psychiatr 1938; 162: 694-715.
786. Glenner GG. On causative theories in Alzheimer's disease. Hum Pathol 1985; 16: 433-435.
787. Vinters HV. Cerebral amyloid angiopathy. A critical review. Stroke 1987; 18: 311-324.
788. Mountjoy CQ, Tomlinson BE, Gibson PH. Amyloid and senile plaques and cerebral blood vessels. J Neurol Sci 1982; 57: 89-103.
789. Castano EM, Frangione B. Biology of disease: Human amyloidosis, Alzheimer's disease and related disorders. Lab Invest 1988; 58: 122-132.
790. Glenner GG, Wong CW. Alzheimer's disease and Down's syndrome: sharing of a unique cerebrovascular amyloid fibril. Biochem Biophys Res Commun 1984; 122: 1131-1135.
791. Tagliavini F, Ghiso J, Timmers WF et al. Co-existence of Alzheimer's amyloid precursor protein and amyloid protein in cerebral vessel walls. Lab Invest 1990; 62: 791-797.
792. Ko L, Sheu K-FR, Blass JP. Immunohistochemical co-localization of amyloid precursor protein with cerebrovascular amyloid of Alzheimer's disease. Am J Pathol 1991; 139: 523-533.
793. Shoji M, Hirai S, Harigaya Y et al. The amyloid β precursor protein is localized in smooth muscle cells of leptomeningeal vessels. Brain Res 1990; 530: 113-116.
794. Kawai M, Kalaria RN, Cras P et al. Degeneration of amyloid precursor protein containing smooth muscle cells in cerebral amyloid angiopathy. Brain Res 1993; 623: 142-146.
795. Yamaguchi H, Yamazaki T, Lemere CA et al. Beta amyloid is focally deposited within the outer basement membrane in the amyloid angiopathy of Alzheimer's disease. Am J Pathol 1992; 141: 249-259.
796. Wisniewski HM, Wegiel J, Wang KC et al. Ultrastructural studies of the cells forming amyloid in the cortical vessel wall in Alzheimer's disease. Acta Neuropathol 1992; 84: 117-127.
797. Wisniewski HM, Frackowiak J, Mazur-Kolecka B. In vitro production of β-amyloid in smooth muscle cells isolated from amyloid angiopathy-affected vessels. Neurosci Lett 1995; 183: 120-123.
798. Wisniewski HM, Wegiel J. β-amyloid formation by myocytes of leptomeningeal vessels. Acta Neuropathol 1994; 87: 233-241.
799. Premkumar DRD, Cohen DL, Hedera P et al. Apolipoprotein E-e4 alleles in cerebral amyloid angiopathy and cerebrovascular pathology associated with Alzheimers's disease. Am J Pathol 1996; 148: 2083-2095.
800. Anglade P, Mouatt-Prigent M, Agid Y et al. Synaptic plasticity in the caudate nucleus of patients with Parkinson's disease. Neurode generation 1996; 5: 121-128.
801. Geddes JW, Monaghan D, Cotman CW et al. Plasticity of hippocampal circuitry in Alzheimer's disease. Science 1985; 230: 1179-1181.
802. Hyman BT, Kromer LJ, Van Hoesen GW. Reinnervation of the hippocampal perforant pathway zone in Alzheimer's disease. Ann Neurol 1987; 21: 259-267

CHAPTER 3

Pathological Changes in Neurodegenerative Disease

3.1. INTRODUCTION

From what has been said so far it should be apparent that it is not really possible to draw clear cut qualitative distinctions, either in clinical or pathological terms, between the effects of so-called normal aging and those of the common neurodegenerative diseases of later life (i.e. AD, PD). Pathological diagnoses of AD and PD are made on the basis of the amount, rather than the type, and distribution of damage imposed upon the brain. In clinical terms this degree of damage may be represented by a subthreshold scoring on rating systems designed to test powers of memory or reasoning, or motor function. Such observations imply that aging and neurodegeneration might exist on a sliding scale of change the one inevitably leading, with time, into the other, with a certain and perhaps rather arbitrary "cut off point" separating the two.

In this chapter it is not necessary to further review the morphological and structural aspects of the destructive lesions that accompany AD and PD since it has already been made clear that in these terms such changes are essentially the same as those that can be present in the non-diseased elderly. Distinctions between aged but healthy persons and diseased individuals reside only in the quantity of damage present. Accordingly, the pathology of these latter disorders will be only briefly commented upon, mainly to establish these commonalities. Instead, emphasis will be placed in this chapter upon the mechanisms that cause such changes to develop since it is likely that amyloid plaques, neurofibrillary tangles, Lewy bodies etc 'simply' represent the common endpoints of pathological cascade processes, each of which may have (perhaps many) different causes. Factors which drive or trigger the cascade in aging alone may differ from those in disease, not only with respect of their actual nature, but also in the effectiveness with which they operate. It might perhaps be variations at this etiological level that determine

(i) how much damage will eventually be caused and (ii) the time scale over which this might take place, thereby dictating whether clinical dysfunction and disease will result or not. In both health and disease the same pathogenetic routes leading to the eventual tissue appearance of these pathological tombstones will be followed.

3.2. THE PATHOLOGICAL CHANGES OF ALZHEIMER'S DISEASE

3.2.1. Gross Changes in the Brain

3.2.1.1. Imaging studies

CT scanning of the brains of patients with clinically suspected AD generally provides little positive diagnostic information and serves mainly to detect other disorders involving traumatic, cerebrovascular or neoplastic damage that may mimic or confound the disease.[1-5] While in some instances a normal scan is recorded, most patients show a variable cerebral cortical atrophy with a widening of the sulci, especially in the region of the Sylvian fissure and an enlargement of the lateral ventricles, particularly at their posterior extent and extension into the temporal lobes.[1-5] Since both cerebral atrophy and ventricular enlargement are often observed in non-demented elderly persons,[2,3,6-11] quantitative measures of the cerebrospinal fluid containing spaces in and around the brain have been used in order to gain objective criteria for the presence of AD. Although, in general, a greater atrophy (and increase in the cerebrospinal fluid containing space) occurs in AD there is much overlap and no clear distinctions between diagnostic groups have emerged. Hence, the usefulness of CT scanning for the detection of tissue changes specific to AD is limited. Nevertheless, recent work[12-14] has suggested that measurement of the thickness of the medial temporal lobe may effectively discriminate those patients with AD from those without.

On MRI scan the brains of patients with AD display many of the same structural characteristics as are visible on CT scanning though the precision of this particular methodology produces additional and more useful diagnostic information. The ventricular enlargement and cerebral atrophy seen on CT are likewise apparent on MRI in most individuals, predominantly within the temporal lobe structures[15-21] but sometimes involving also the basal ganglia,[22] though again such changes do not effectively discriminate between persons with AD and the normal elderly. However, the extent of medial temporal lobe (hippocampal) atrophy, measurable either as total tissue volume, tissue thickness, mean interuncal distance, or ratio of cross-sectional area to that of the globus pallidus has, like in CT scanning, been acclaimed[17-21] as a potentially valuable discriminant with at least 85% of patients with subsequently proven AD having hippocampal measures lower than the highest recorded non-demented control value.

The improved resolution of MRI means that brain grey and white matter areas are well separated and individual structures easily recognized and quantifiable. In this way Rusinek et al[23] demonstrated a small, though significant, preferential loss of grey than white matter in AD, this being reduced to 45% of the total tissue from an overall norm of about 50%. Proportionately greater reductions, however, were seen in the grey matter of the frontal, temporal and occipital lobes and in the basal ganglia, while no (differential) changes were seen in the parietal lobes and posterior fossa (cerebellum). Such data imply that AD preferentially damages the grey matter regions of the neocortex.

White matter hyperintensities (akin to the periventricular translucencies (leukoaraiosis) seen on CT scanning) associated with arteriosclerosis and microinfarction, or vascular ectasia and dilated perivascular space (etat lacunaire), are prominent in many patients with AD[24-27] but not so much more than is seen among elderly non-demented subjects.[24,28-30] Indeed, such changes are absent in the brains of those AD patients free from cerebrovascular risk factors. [26,31] Conversely, a demented patient with prominent white matter lesions is more likely to have suffered, in pathological terms, from multi-infarct dementia (subcortical arteriosclerotic encephalopathy) than AD, though a combination of the two disorders is possible.

In PET studies,[32-47] AD is characterized by a bilateral, sometimes asymmetric, reduction in glucose metabolism in the parieto-temporo-occipital association cortex–a change that is not seen in the non-demented elderly. Indeed, in subsequently proven cases of AD, this metabolic change was observed to precede the onset of cognitive impairment by 2-3 years[42,43] with the extent of cognitive impairment subsequently matching further reductions in glucose metabolism. In late stages of disease the frontal lobes may become involved, especially in disinhibited patients. [44] Patients with language disturbances show a reduced glucose metabolism in left fronto-temporal regions, those with dyspraxia in right parietal areas.[42] The metabolic rate in the cerebellum, basal ganglia, thalamus and primary sensory cortex remains normal, or nearly so.

The pathophysiological basis for the glucose PET changes is unclear. On one hand it could relate to a reduction in the number of functional neurones due to neurofibrillary degeneration. On the other hand, there might be a reduced function of individual cells, while retaining a normal cell complement, perhaps because of a decreased afferent input into the affected region. Some combination of both is possible. The posterior hemisphere changes in AD are greater than what the amount of pathology usually present (as plaque and tangle density) would perhaps predict and this exaggerated decline may be due to the concomitant loss of input from a degenerated cholinergic projection system.[45]

Hence, both PET and MRI provide data that is not only of potential diagnostic importance but which may also help to pinpoint critically damaged, or functionally impaired, regions in the brain at an early stage in the illness.

3.2.1.2. Autopsy changes

At autopsy, the brain is often severely and globally atrophic, weighing about 1000g or less, particularly in patients who become affected before the age of 65.[48-51] In older individuals, however, there is often a relative sparing of many cortical regions and atrophy is focused upon the temporal lobe.[48] In such individuals brain weight may be near normal or fall within the normal range of 1200-1450g.[52,53] Although overshadowed by the more obvious cerebral cortical atrophy, a mild reduction in the size of the brain stem, basal ganglia and cerebellum also occurs in many patients, especially the younger ones.

Atherosclerosis within the large extracerebral arteries is frequently, and quite impressively, absent, even in old age, though in about 20% of (mostly elderly) cases a moderate to severe degree of atherosclerosis can be present and overt infarction of the cerebral cortex may also be seen. The overlying meninges are commonly thickened and opaque, but not so much different from those of unaffected persons of similar age.

The decrease in brain weight is reflected by volumetric changes with the meninges often coming to lie loosely over the surface of the brain and much cerebrospinal fluid being accommodated between them and the atrophied cortex. This becomes particularly obvious if the meninges are peeled away from the underlying cortex. Atrophy of the left and right sides of the cerebral cortex tends to be roughly equivalent, averaging about 10-24%, and affects the frontal and temporal regions particularly though in most cases no region is entirely spared.[48,50,51,54,55] The overall extent of atrophy is roughly age-dependent, this tending to be widespread and severe in younger patients but becoming less extensive and often restricted to the temporal lobe in older subjects.[48] However, even in some young individuals atrophy may be slight, yet the histopathological changes of the disease (i.e. the plaques and tangles) may still be rampant throughout the brain. At post mortem, the hippocampus and amygdala are obviously atrophied in most instances;[51] indeed a reduction in the cross-sectional area of the hippocampus has been observed on brain imaging even at early clinical stages, with further atrophy paralleling an increasing impairment of brain function as the disease progresses.[12-14,17-21]

The lateral ventricles of the brain are normally and often considerably enlarged, especially at their posterior extent and within the temporal horn, though as with cerebral atrophy the size of these, particularly in older subjects, often appears within normal age limits.[1,56] Usually the III and IV ventricles and the aqueduct are enlarged, roughly in proportion to the dilatation of the lateral ventricles. Ventricular dilatation reflects the cerebral atrophy, though the extent of the enlargement does not usually match that which might be expected on the basis of tissue loss alone, suggesting that some collapse or contraction of the tissue may accompany the atrophy in later stages of the disease.[50]

3.2.2. Patterns of Nerve Cell Loss

3.2.2.1. Cerebral cortex

While it has been widely assumed that cerebral atrophy should reflect a thinning of the cortical mantle, quantitative studies have not always borne this out. As with changes in brain size, studies which have examined mostly younger patients have indeed usually found a substantial reduction in cortical thickness[49,51,53,57] whereas in older patients a lesser, and sometimes nonsignificant, decrease has been found.[52,55,58] Other studies[51,58] have shown that the overall length of the cortical ribbon is also reduced. Hence, as might be expected, the cortical atrophy of AD is 3-dimensional with reductions in both the length of the cortex, due (presumably) to a fallout of neurones and processes in a columnar fashion, and its thickness, due to a regression of axons and dendrites, contributing to the overall tissue change.[51,55]

In hematoxylin-eosin or Nissl stained sections the cerebral cortex often fails to show any obvious abnormalities (the pathognomonic plaques and tangles are usually only poorly visualized, if at all, by these methods) yet experienced microscopists have long concluded that the number (density) of nerve cells is reduced in the disease, particularly in younger cases where the loss of nerve cells may sometimes be so severe as to confer a microvacuolation to the tissue, especially in the outer cortical laminae. Collapse and disintegration of the cortical ribbon following cell loss may lead to a breakdown or blurring of the normal laminar arrangement of cells within the cortex.

These predictions of nerve cell loss have, in general, been borne out by quantitative studies. In these a decrease in neuronal density (beyond that of 'normal aging') occurs, particularly with respect to the large pyramidal cells, over many cortical regions but especially in the frontal, temporal and cingulate cortices with relative preservation of parietal and occipital regions.[49,52,53,57,59-64] The extent of nerve cell loss so reported is highly variable, ranging from 0-80% according to which particular brain region was investigated or the age distribution of the patients studied, with greatest losses being recorded in the frontal and temporal lobes of younger patients. Furthermore, since only a few of these studies attempted to compensate for the 'packing down' of surviving cells in the (often grossly) atrophic tissue it is likely that in many instances the losses quoted are underestimates of the true neuronal fall out; a consideration particularly relevant in those instances where no, or a minimal, cell loss (reduced cell density) was reported. Nevertheless, not all investigators have reported a significant loss of nerve cells from the cerebral cortex even when allowing for cortical atrophy. Regeur et al[55] found a non-significant cell loss of 6% in a group of very elderly (i.e. over 80 years of age) patients. It is however possible that in such individuals, where the extent of tissue atrophy was slight (about 14%), a shrinkage (atrophy) of nerve cells, but without overt loss of cell bodies, may have been sufficient to produce the clinical deficit.

The extent of nerve cell loss (beyond that of age alone) from the cerebral cortex in AD correlates inversely with patient age though the total loss of cells (when old patients are compared with young ones) does not change according to age.[53] This suggests that dementia will occur when a certain threshold to maintenance of function has been crossed and that this might, at least in part, be represented by a critical density of surviving nerve cells. In younger patients the required loss of cells would be brought about solely by the presence of AD whereas in older subjects, in whom some cell loss would likely have already taken place because of their age (Chapter 2), a lesser degree of Alzheimer-related damage would be needed to cross this threshold to change. In this way, the wide variations in extent of cerebral cortical nerve cell loss reported in AD can be reconciled. Older persons with AD may need the additional loss of only relatively few nerve cells to tip the balance in favor of clinical dysfunction; [55] comparative studies of nerve cell counts between these elderly demented subjects and other apparently healthy individuals, who might be just about maintaining normal function, would thus be expected to show only little (significant) difference.

3.2.2.2. Hippocampus and amygdala

There seems to be little dispute that the hippocampal atrophy of AD results from a severe loss of nerve cells both within, and projecting to, the hippocampal formation. In many studies a preference for areas such as CA1 and subiculum has been noted[53,65-70] and an involvement of the entorhinal cortex, especially the large stellate cells of layer II has been emphasized.[67,71] Indeed, the extent of damage to all these areas is similar irrespective of patient age underscoring the critical importance of the hippocampal formation in the pathogenesis of the disease. While a severe plaque and tangle formation within many parts of the amygdala has long been recognized[72-82] and gross atrophy of this region is commonplace,[51] only a single study[73] has actually quantified the extent of nerve cell loss from this region. In this, a reduction in nerve cell density was noted in all subdivisions, though greatest changes occurred in the cortical and medial nuclei where, having taken into account the overall tissue shrinkage, losses of neurones of between 60 and 70% were estimated. Such changes, together with the usual heavy plaque and tangle formation probably make the amygdala one of the most, if not the most, damaged areas of brain in AD.

3.2.2.3. Cerebellum

Although a mild to moderate atrophy of the cerebellar hemispheres is commonly seen in AD, especially in younger or more severely affected cases, no changes in Purkinje or dentate cell number or density have been found. Indeed, the very stability of this cell population has often been remarked upon.[83,84]

3.2.2.4. Subcortical regions

An extensive loss of nerve cells from the cholinergic basal forebrain region has been reported by many workers on numerous occasions.[74,85-104] However, as is seen in the cerebral cortex, less change often occurs in this region in some elderly demented persons where little or no actual loss of cells may take place though surviving cells are much shrunken.[105] The noradrenergic locus caeruleus is likewise consistently and severely damaged, especially in younger subjects where a virtually complete loss of cells is commonplace.[92,93,102,106-121] The serotonergic raphe nuclei are also usually affected, but less severely than the nucleus basalis and locus caeruleus.[92,93,102,114,118,120-125] The dopaminergic cells of the ventral tegmentum are damaged and lost[125-127] while those of the substantia nigra are usually well preserved.[126-129] Some patients can however show a severe cell loss from this latter region[118,124,129] but these may be suffering from the related disorder CLBD rather than AD itself. Loss of nerve cells from the globus pallidus, [130] ventral striatum[131] and anterodorsal nucleus of the thalamus[132] has been reported, together with that of the (large) cholinergic neurones of the corpus striatum.[133-135] In addition a nerve cell loss from other brain regions has been inferred on the basis of the presence of an extensive NFT formation within such cells or a heavy tissue amyloid deposition, though whether any actual cell loss has occurred has not always been confirmed by quantitation. Among such regions are the peripeduncular and parabrachial nuclei,[118,125,129] as well as other diverse nerve cell types in brainstem areas,[118,129] certain medially located thalamic nuclei[129,136-138] and particular hypothalamic nuclei.[74,139-141]

The extent of nerve cell loss within sub-regions of certain vulnerable cell groups also seems to vary greatly. For example, in the nuclei basalis complex the nucleus of Meynert seems more affected than the diagonal band of Broca and the septal nuclei[96] and in the locus caeruleus, rostral and dorsal parts of the nucleus are affected more than caudal and ventral regions.[113,117] These observations suggest that the pattern of cell loss in such regions might relate to a retrograde damage occasioned at the nerve terminals in those cortical projection areas (frontal and temporal cortex) that are most severely affected in AD.[74,113,117-119]

Nonetheless, some subcortical brain regions apparently remain spared in AD, these being the olivary and pontine nuclei in the brainstem (Mann unpublished observations), the mammillary bodies[142] and the lateral thalamic nuclei,[129] though this conclusion is inferred from an absence of NFT and remains to be confirmed by cell counting.

With damage and loss of function, changes in markers of neurotransmission, or energy metabolism will result. Some of these changes will reflect a loss of structural elements, such as synapses,[143-148] though here compensatory changes may occur[143,144] and these could help to offset some of the losses. It is not the purpose of this chapter to review the wealth of neurotransmitter-related changes that take place in AD; these in any case have been well documented elsewhere.[149,150] Likewise, alterations in energy metabolism, involving mitochondrial enzymes[151-153] and their expression[154] or oxidative phosphorylation and glucose utilization,[155] may arise from actual damage to, and loss of, nerve cells or functional deafferentation, or both; again these have been reviewed elsewhere.[156] At present, there is no firm evidence to suggest that these neurochemical changes reflect anything but secondary or adaptive responses to the primary disease process going on within neurones (i.e. the neurofibrillary degeneration) that leads directly to their demise. Neurotransmitter alterations are thus unlikely to take part in triggering the pathological cascade though they may influence the rate or progress of the disorder. They are, however, of clear interest in terms of our understanding and assessment of what any residual capacity the brain might have for correct or corrective function, since at present it is only potentially possible to treat (some of) the symptoms of AD by modulating or stimulating the remains of these failing systems. On the other hand, changes in energy metabolism might favor the production of reactive oxygen species which, if unscavenged, could mediate or exacerbate the underlying pathological process. (See later).

3.2.3. The Degenerative Changes of AD

Degenerative changes in surviving nerve cells in AD are well recognized. Of particular importance is the formation and accumulation of NFT since this is likely to be the proximal cause of the neuronal dysfunction, atrophy and death of nerve cells, or if not it is, at the very least, closely related to this. In affected cells accumulation of NFT is associated with decreases in the amount of rough endoplasmic reticulum and ribosomes;[157] changes in these organelles may reflect the widespread loss of cytoplasmic RNA from neurones in regions where NFT are profuse.[53,68,91,158-164] Although early studies[165-167] demonstrated a massive reduction in the extent of dendrites of individual neurones in the cerebral cortex, the scale of this was challenged by later work by Coleman and Flood.[168-170] These latter studies emphasized a failure of compensatory sprouting rather than an overall loss of dendrites, which in any case was found to be mild. These same workers[171] reported a lack of dendritic change in pyramidal cells of CA2/3 of the hippocampus, though because this region usually appears histologically normal in AD[69] the absence of any regressive changes would perhaps be expected. Nonetheless, widespread dendritic abnormalities do exist within the neuropil of the cerebral cortex in the form of 'neuropil threads'; these are thought to represent an abnormal sprouting of dendrites from those nerve cells involved in neurofibrillary degeneration.[172,173]

3.2.3.1. Synaptic changes

A major loss of synapses from the cerebral cortex in AD has been demonstrated quantitatively under the electron microscope in biopsy specimens from both frontal cortex[144,147] and temporal cortex.[143,145,148] This synapse loss outweighs that of neuronal fallout and leads to a decline in the ratio between neuronal and synaptic densities.[145,148] This might infer that all surviving cells, regardless of type, share a net reduction in synaptic contacts or it may simply reflect a preferential loss of the large pyramidal cells which would, presumably because of their size, have a disproportionately greater number of synapses than the smaller pyramidal cells and non-pyramidal interneurones. This reduction in the number of synapses in AD, as determined by electron microscopy, has also been detected on numerous occasions by immunohistochemistry using antibodies directed against synapse associated proteins.[174-180] Although the extent of the synapse loss may be fairly even throughout different parts of the cortex, local accentuation seems to occur in regions of neuritic plaques though, interestingly, no change in synaptic density is seen in areas of diffuse amyloid.[181] Hence, neuritic plaques, associated as they are with aggregated β-pleated amyloid, show damage to the neuropil whereas the non-aggregated diffuse plaques are seemingly not associated with any such damage. The former type might therefore be considered as malignant plaques, the latter as benign deposits.[182] As a result of synapse loss, neighboring or surviving synapses enlarge their contact surface area and may thereby off-set, at least partially, some of the functional (neurotransmission) consequences of actual synaptic fallout.[144,145] The magnitude of the synapse loss does not seem to be affected by variations in ApoE genotype,[180] though since NFT density does not so vary[180,183-186] this finding might have been anticipated.

3.2.3.2. Neurofibrillary tangles

The basic structural characteristics of NFT, their mode of formation and the cellular affects of their accumulation have already been described in Chapter 2 and need no further elaboration here. At autopsy, an established case of AD will show widespread NFT throughout the brain. In the neocortex NFT are most common within areas of association cortex, principally within the large pyramidal cells of layers III and V, but are much less common in primary somatic and sensory (auditory, visual) cortex.[187-192] However, in older subjects with AD, NFT may be much less extensively present in the neocortex, occurring mainly in the temporal cortex, especially the inferior and middle temporal gyri. In the hippocampal formation the large stellate cells of layer II are most severely and most consistently affected[193-197] though pyramidal neurones in areas CA1 and subiculum are also considerably damaged.[53,65-71,193-197] Neurones in the other hippocampal fields are much less involved; indeed those in area CA2 are usually resistant to NFT formation and only occasional cells in the dentate gyrus may show such changes.[69] The amygdala is also consistently and severely damaged with neurones in cortical and medial nuclei being those most affected[72-82] as are those of the olfactory bulb and tracts and the associated olfactory (piriform) cortex.[188,198-201] Many subcortical structures show NFT formation, principal among which are neurones of the nucleus basalis complex,[85-104] the locus caeruleus[106-121] and the raphe nuclei,[120-125] though many other diverse cell types throughout the periventricular and periaqueductal grey matter can also contain NFT, as do cells in the mid-line thalamic nuclei.[116,117,120,121,129,133,136-138] Neurones of the cerebellum (Purkinje cells and those of the dentate nucleus) and brainstem (inferior olivary and pontine nuclei), the motor neurones of the cranial nerve nuclei and the anterior horn cells of the spinal cord, and the neurones of the supraoptic and paraventricular nuclei of the hypothalamus are all conspicuous by a consistent absence of NFT.

Braak and colleagues[193] have developed a pathological staging system for AD based on the distribution and intensity of NFT formation throughout the brain. In this, stages 1 and 2 are associated with either a complete lack of NFT anywhere in the brain (Stage 1) or the presence of only a very occasional NFT in cells of layer II of the entorhinal cortex (Stage 2). In stages 3 and 4 this latter region is consistently (Stage 3) and severely (maximally) (Stage 4) affected and NFT may also be seen in other hippocampal regions (CA1 and subiculum) and in the amygdala (Stage 3), then spreading into the temporal neocortex (Stage 4). Stages 5 and 6 represent the familiar advanced pathological end-stages of AD seen at autopsy where NFT are widespread throughout the temporal neocortex and other areas of association cortex and many subcortical nuclei are also involved. By this scheme, NFT formation would originate within the entorhinal cortex, spreading through other regions of the hippocampus and amygdala into the temporal neocortex, then out into remaining neocortical association areas and subcortical structures such as the nucleus basalis, locus caeruleus and the raphe complex. Clinical change is absent at stages 1 and 2 (preclinical AD), mild at stages 3 and 4 but profound at stages 5 and 6. Studies of the temporal progression of AD pathology in the non-demented elderly[194,197] or in DS,[196] showing a similar chronological sequence of events with increasing age reinforce this scheme. Pathological changes in terms of NFT formation in the non-demented elderly[82,194,195,197] would correspond to Braak stages 1 and 2.

Hence, the pattern of neurofibrillary degeneration in AD proceeds in a hierarchical fashion, probably commencing in the entorhinal cortex but rapidly involving other hippocampal and amygdaloid regions, spreading into the temporal cortex and thence into other neocortical regions and involving the olfactory bulbs and tract and various subcortical substructures. This pattern of spread would be consistent with a cortico-cortical progress of pathology along defined pathways through the brain[188,202,203] though whether this proceeds in a retrograde (feed forward) or anterograde (feedback) manner, or even some combination of the two, remains uncertain.[202,203]

3.2.3.3. Amyloid plaques

As with NFT the general structural characteristics of plaques have been given in Chapter 2 and hence only those aspects of plaque formation particular to AD are noted here. In AD, Aβ deposits, in the form of diffuse plaques, are widespread throughout the hippocampus (involving chiefly CA1 and subiculum, presubiculum and dentate gyrus), entorhinal cortex, amygdala (mainly cortical and medial nuclei), the neocortex (association areas), basal ganglia (caudate, putamen), thalamus (medial nuclei), hypothalamus, geniculate bodies, periaqueductal and periventricular grey matter, and the cerebellar cortex.[118,204-226] Such diffuse deposits are conspicuously absent from the deep nuclei of the cerebellum, the brainstem and the spinal cord. Neuritic plaques have a much more restricted distribution being essentially confined to the neocortex, hippocampus and amygdala.[82,195,197,213,227] Cored amyloid plaques follow a similar distribution to that of neuritic plaques within cerebral cortical regions, and in most instances comprise the same structural entity. However, in the cerebellum cored plaques are occasionally seen, usually in the Purkinje cell layer but also sometimes in the molecular cell layer[211,218,219,221,225,228,229] along with the more common diffuse plaques. In contrast, however, to cerebral cortical plaques, tau-immunoreactive neurites are never seen within any of the cerebellar deposits, even the cored type.[221] Within the cerebral cortex layers II and III are generally most severely affected by plaques and it is here where the majority of both the diffuse and the cored plaques are seen; the deeper cortical layers tend to be less populated by plaques though, when present, they are often large and usually of a cored type.[186-192] In the primary visual cortex there is often a particular predilection for plaques to be located in layer IV[189,190] though on other occasions plaques are more evenly spread through all layers.[222] In the cerebral cortex plaques often appear clustered and sometimes, when Aβ deposition is not too severe, an obviously vertical orientation can be seen.[188,230] This distribution pattern arises because the plaques lie between, but seemingly do not involve, the parallel ascending apical dendrites of adjacent pyramidal cells.[231,232] The sulcal depths are usually more affected than the gyral crests both in terms of plaque density[233,234] and the proportion of cored plaques present.[235]

3.2.3.4. Cerebrovascular changes

Although hypertensive or atherosclerotic disease of large and medium arteries in AD is no more frequent nor more severe than is seen in aged non-demented individuals (indeed the cerebrovasculature is often conspicuously normal in these respects), microvascular changes, involving a looping and coiling or an atrophy of vessels[236-239] and producing a net reduction in the overall extent of the capillary bed beyond that of age alone, have been recorded.[238-241] As with normal aging,[240,241] this additional decrease in microvascular density in AD most likely represents the reduced metabolic demands of a depleted and dysfunctional neuronal population, though changes in endothelial cells or alterations involving a thickening of the basement membrane[242-245] or an increase in astrocytic processes beyond that seen in aging alone,[246-247] could compromise neuronal viability perhaps through alterations in blood brain barrier function. Nonetheless, studies suggesting that cerebrospinal fluid serum ratios for albumin are not increased over control values in AD patients lacking cerebrovascular changes[248,249]–but see Elovaara et al[250]– imply an intact blood brain barrier function. The presence of serum proteins, like albumin, within senile plaques in some cases of AD[251-253] may reflect a binding to Aβ following transudation across locally damaged vessels, this being associated with concomitant cerebrovascular disease (hypertension). The presence of albumin within large inflated astrocytes surrounding blood vessels in those cases of AD showing a similar plaque staining[251,252] would accord with this. Studies on patients with multi-infarct dementia,[254,255] in whom a defective blood brain barrier is known to be present, have shown that an increase in cerebrospinal fluid/serum albumin ratio, indicative of vascular leakage and tissue permeation by serum proteins, can indeed follow damage to the small blood vessels of the brain.

Hence in cases of AD uncomplicated by cerebrovascular disease blood brain barrier function is likely to be generally well maintained. Amyloid angiopathy is present in most (perhaps even all) cases of AD[256-258] and although variable in extent and distribution it seems to most consistently and most severely affect the leptomeningeal arteries supplying the occipital cortex and cerebellum. In these regions parenchymal arteries are often also affected and a capillary involvement sometimes leads to dyshoric change. The histological and immunohistochemical profile of affected vessels in AD is identical to that (when present) in non-demented elderly subjects (see chapter 2).

Brain scanning showing some degree of leukoaraiosis may be present in many cases of AD,[1-5,24-27] particularly in the white matter of the posterior hemisphere, and this is usually more severe and occurs more frequently in the very elderly subjects with AD than in those patients dying before 80 years of age. However, the extent of such changes does not seem to be more severe than is commonly encountered in non-demented subjects of a similar age and may thus represent an incidental age-related pathology of the cerebral blood vessels due perhaps to hypertension.

3.2.3.5. Other changes

Granulovacuolar degeneration of nerve cells,[259-264] and the presence of Hirano bodies,[53,265,266] in the CA1 region of the hippocampus is a constant aspect of the histopathology of AD. The morphological and histochemical properties of both these pathologies have been described in Chapter 2 and although their frequency in AD grossly outweighs that seen in the non-demented elderly their structural features seem identical.

3.3. THE INITIAL SITE OF DAMAGE

A consideration of those nerve cell types affected in AD leads to the conclusion that although the loss of such cells is likely to be mediated by NFT accumulation, the actual capability to form NFT is not dictated by nerve cell morphology (i.e. pyramidal vs non-pyramidal), topography (cortical vs subcortical) or neurotransmitter characteristics (cholinergic vs aminergic vs amino acid). The pattern of cell loss seems to be dictated by connectivity. Virtually all nerve cells lost in AD are known (or suspected) to connect directly with the plaque-rich regions of the association cortex (including the hippocampus and amygdala) or to be located within these regions. In this context observations[113,115,117] that ventrally located neurones in the locus caeruleus which project to the basal ganglia, cerebellum and spinal cord are spared at the expense of dorsally and centrally located neurones projecting to cerebral (association) cortical regions are of particular relevance. Indeed, within such cortically projecting cells, those with connections to the frontal and temporal regions are more severely damaged than those with occipital connections. Dopaminergic neurones of the substantia nigra, projecting to the corpus striatum, are more or less undamaged when compared with nearby cortically projecting ventral tegmental neurones.[126,127,129] Neurones of the midline nuclei of the thalamus projecting to frontal cortex are damaged,[129,136] whereas those of the dorsal and lateral nuclei which project to (relatively) plaque and tangle free somatosensory cortex are preserved. Finally, a greater loss of neurones from the nucleus basalis complex is seen in those subregions projecting to the temporal cortex where plaque densities are usually highest.[96]

However, apparent exceptions to this pattern are seen in the hypothalamus. Despite showing extensive cell loss and NFT formation,[139] neurones of the suprachiasmatic nucleus do not project to the cerebral cortex but connect principally with the supraoptic and paraventricular nuclei,[267] regions which are usually devoid of both NFT and plaques[110,224] and from which no loss of nerve cells occurs either.[268] However, cells of the supraoptic and paraventricular nuclei do have collateral projections to the periventricular (medial) thalamus and to other hypothalamic regions[267] where Aβ containing plaques[207,224,269] and NFT[117,120,121,269-272] are widespread. Similarly, the projections of the lateral tuberal nucleus (that is severely damaged in AD) [140] are not to the cortex but to other local hypothalamic nuclei where again there are Aβ plaques and dystrophic neurites.[140,224] Hence, in this context the hypothalamic grey matter could be classed as 'cortex', in the same way as regions like the hippocampus and amygdala.

Therefore, the 'cerebral cortex' may be the initial site of the lesions of AD. This argument is strengthened by 'longitudinal' studies of patients with DS[84,196,273-284] in whom plaque formation, and specifically the deposition of Aβ, within the cerebral cortex seems to be the earliest pathological change associated with the development of AD.

3.4. THE PRESENCE OF ALZHEIMER-TYPE PATHOLOGY IN SITUATIONS OTHER THAN AD

3.4.1. Down's Syndrome

It has long been known[273,274,285,286] that all persons with DS, or at least probably all of those in whom a full trisomy 21 is present, will develop the histopathological changes of AD should they live into middle age and beyond. Established cases of AD in elderly DS subjects show an identical pathology to that in AD (especially the younger) cases in the general population whether this be in terms of the morphological form, number or distribution of plaques and tangles,[273,274,285,286] or the nature and extent of nerve cell loss from cerebral cortical and subcortical regions.[287] Hence, there appears to be nothing, in histopathological terms at least, to differentiate AD in an elderly DS subject from that in anyone else coming from the general population. This very similitude has allowed DS to become perhaps the best known and most widely used 'model' of the pathological process of AD, since the predictability with which such persons acquire the disease by middle age sets them apart from members of the general population in whom, except perhaps for those (relatively rare) cases where the disease relates to mutations on chromosomes 1, 14 or 21, (see Chapter 4) it is still not possible to predict who will and who will not get the disease. Neither can it be said without reservation that those non-demented persons dying with minimal AD pathology would indeed have gone on to develop the full disease had they lived longer even though this is highly probable.

Therefore, by examining the brains of persons with DS dying at different ages it becomes possible to 'reconstruct', from cross-sectional observations in such individuals, the time course of events relating to the development and progress of the pathological changes of AD. By doing this, the process of AD is seen to start in a constant way and to progress in an orderly and sequential fashion. The initial histopathology takes place, sometimes as early as the late teens or early twenties but always by the age of 30 years, in neocortical association areas, especially temporal neocortex, and involves a deposition of Aβ.[84,196,275-279,281,283,288] At this stage the hippocampus and amygdala may or may not show amyloid plaques. Deposition of Aβ does not commence in the cerebellum and corpus striatum until the late thirties or early forties.[84,280] The Aβ in these early diffuse plaques exists solely as $A\beta_{42}$ and this is so both in the cerebral cortex[278,281,288] and in the corpus striatum or cerebellum.[280] Many, but not all, of these early cerebral cortical plaques quickly come to contain ApoE,[278,289] HSPG[290] and complement factors[291] though the cerebellar deposits throughout their lifetime contain little ApoE[281,289] or complement (Mann–unpublished) and no HSPG.[292] Plaques with neurites and cores, containing $A\beta_{40}$ and microglial cells do not usually appear in the cerebral cortex until well into the late thirties or early forties.[196,275,278,281,287,288] At this same time NFT can be seen, initially in the layer II stellate cells of the entorhinal cortex and in some CA1 pyramidal cells,[196,275] but later occur throughout the hippocampus, entorhinal cortex and amygdala, progressing by the age of 45-55 into the temporal neocortex and other neocortical regions. NFT are never seen in the cerebellum,[84,280] and only rarely in the basal ganglia, at any age. After the age of 50 a pathology entirely typical of AD is present in most patients though some can lag a little in this respect whereas in others this stage may be reached a little earlier in life.

Microvascular changes involving either a decrease in the overall capillary bed[239] or a thickening and change in glycan composition of the basement membrane[245] occur in elderly DS subjects, as in those with AD. However, because such changes in the basement membrane are not seen in younger DS individuals[245] they are unlikely to directly contribute to the formation of plaques or tangles by, for example, conferring some defect to the cerebrovasculature that may make the early development of Alzheimer-type changes more likely.

Hence, in DS the process of AD commences with a deposition of Aβ as $A\beta_{42}$ (perhaps even as $A\beta_{1-42}$ or $A\beta_{3-42}$)[278,288] within the neocortex. Indeed, analysis of soluble Aβ extracted from the brain tissue of young persons with DS has shown this to exist as $A\beta_{1-42}$ and that high (but presumably not high enough for fibrillization to occur) levels of this are present in the tissue of such individuals even before Aβ deposits (plaques) can be detected within the tissue by immunohistochemistry.[293] In DS, therefore, the process of AD takes at least 20 years to complete with an initial prodromal phase of Aβ deposition occupying the first half of the process. This, however, does not necessarily mean that the pathological process of AD in the general population runs over this same time period; the differing etiologies responsible for causing AD in DS and in the general population (see Chapter 4) may mean that a much more aggressive course is run in the latter with the whole process in most instances being compressed into about a decade or less. Nonetheless, studies of DS have made it absolutely clear that the deposition of Aβ is fundamental to the development of the pathology of AD and may even be critical in triggering the whole disease process. Despite etiological differences, and concerns regarding the uncertain progress of changes, studies on nondemented persons in the general population show close pathological parallels to younger DS subjects, whether this be in terms of the form and distribution of plaques and tangles, or in respect of the peptide composition of the deposited Aβ.[294] Hence, a pathogenetic cascade process for AD based on the initial deposition of Aβ has been put forward and it is argued that all cases of AD, irrespective of underlying etiology, should comply with this.

3.4.2. Cortical Lewy Body Disease

Cortical Lewy body disease (CLBD) has been considered to represent, after AD, the second most common cause of dementia among the very old. The cardinal clinical signs are those of a fluctuating confusional state with hallucinations and behavioral disorder; parkinsonism may or may not occur and when present usually has onset late in the course of the illness.[295,296] In addition to a loss of nerve cells from the substantia nigra, and the presence of typical Lewy bodies within some of the surviving neurones in this region, the cerebral cortex also shows a widespread presence of Lewy bodies (see Chapter 2) together with, in the great majority of cases, many, mostly diffuse, amyloid plaques in the association cortical regions, particularly the frontal, temporal, insular and cingulate cortices.[297-304] NFT are generally absent or sparse in the neocortex, but are usually abundant in hippocampal and entorhinal regions, though in some instances these are widespread throughout both the hippocampus and the neocortex and an additional diagnosis of AD in merited.[297-304] The overall pattern of cortical atrophy in CLBD,[304] in keeping with the presence of an Alzheimer-type pathology, is similar to that in AD itself.[50,51] Although the numerical density of amyloid deposits in CLBD may be roughly similar to that in AD[299,300,303] the Aβ within such plaques exists mostly as $A\beta_{42}$ and much less $A\beta_{40}$ (as compared to AD) is present (Mann and Iwatsubo–unpublished data).

It is still not clear whether CLBD and AD represent separate disorders of the very elderly that share similar pathological features, perhaps because of overlapping etiological factors. Cases of CLBD show an increase, similar to that in AD, in the frequency of the E4 allelic variant of the apolipoprotein E gene.[305-307] The two disorders may thus provide a spectrum of disease, with cases of "pure AD" and "pure CLBD" (i.e. cortical Lewy bodies without Alzheimer-type pathology) being "poles" and those of mixed pathology occupying the broad mid-ground. Certainly, the roughly equivalent pattern of atrophy in AD and CLBD[50,51,304] and the similar pattern of deposition of Aβ[299,300,303] would be consistent with such a spectrum of change; even the absence, or mildness, of neurofibrillary degeneration in many instances could be equated with a less advanced pathological process as evidenced by the lower $A\beta_{40}$ levels. However, there are some important distinctions between the two disorders and the differing distribution of hippocampal NFT and neurites with CA2/CA3 regions being more affected than CA1 and subiculum in CLBD, but conversely so in AD,[298,303] might perhaps argue for separate etiologies.

3.4.3. Hereditary Cerebral Hemorrhage with Amyloidosis (Dutch Variant)

The Dutch variant of hereditary cerebral hemorrhage with amyloidosis (HCHWA-D) is an autosomally dominant inherited disorder affecting certain families of this ethnic origin. The disorder is clinically characterized by the occurrence of (usually fatal) brain hemorrhages in middle age accompanied by a variable degree of dementia. The pathological hallmark is a severe deposition of amyloid within leptomeningeal and intraparenchymal blood vessels which leads to infarction or hemorrhage.[308] Additionally, amyloid plaques are present within the cerebral cortex[309-314] though a neurofibrillary pathology (NFT, neuritic plaques or neuropil threads) is absent.[310,312] It is of interest that while the Aβ deposited within plaques and blood vessel walls is chemically the same as that in AD,[315] the plaques in HCHWA-D are nearly always of a diffuse form and are composed (almost) entirely of $A\beta_{42}$, with very little or no $A\beta_{40}$, even though the blood vessel walls contain much $A\beta_{40}$ (and $A\beta_{42}$).[314] Furthermore, such diffuse plaques, apart from not containing tau positive neurites, are also free from a microglial reaction and astrocytic changes.[311,313]

The cause of HCHWA-D relates presumably to the point mutation at codon 693 on the APP gene which replaces glutamic acid by glutamine.[309,316,317] The effect of this amino acid substitution may be to alter the microchemistry of the soluble Aβ produced through APP catabolism in favor of a more hydrophobic species with increased potential for fibrillization.[318-320] Why the diffuse plaques in the cerebral cortex in HCHWA-D fail to mature into neuritic plaques, like those in AD, is unknown though here the absence of microglial cells[311,313] and $A\beta_{40}$[314] may be relevant . A further mutation at codon 692 of the APP gene, replacing alanine by glycine, occurs in a family of Flemish origin[321] and leads to a clinical and pathological phenotype similar to that in the APP_{693} mutation. However, here the amino acid substitution does not apparently increase the propensity of the Aβ peptide to fibrilize[318] and the mechanism whereby this genetic change produces the disease remains uncertain.

3.4.4. Head Injury

Epidemiological studies[322,323] have suggested that a prior head injury with loss of consciousness may be a risk factor for the subsequent development of AD. Furthermore, occasional case histories have been documented that show a pathology typical of AD, which has seemingly arisen some time after a single episode of head trauma.[324-326] Moreover, Alzheimer-type pathological changes have been described in the brains of about one-third of patients who have suffered serious head injury during road traffic, or other, accident and have survived for hours, days or weeks following the incident.[327-329] In these latter patients an increased expression of APP by neurones is seen[330] and diffuse Aβ deposits are present within the cerebral cortex at an age where none would normally be anticipated.[327-329] The deposited Aβ is widely scattered and bears no obvious correlation to the site of contusion.[329] Such diffuse plaques contain only $A\beta_{1-42}$[328] and no NFT are seen. Alterations like these are strongly reminiscent of the early pathological changes of DS. It is of much interest (see Chapter 4) that the frequency of the ApoE E4 allele is increased in those head injured patients showing Aβ deposition to a level similar to that in AD,[328] thereby further supporting a role for ApoE E4 protein in the pathogenesis of Aβ deposition.

Nonetheless, accidental injuries of this kind are not the only cause of head trauma. It has long been known that (retired) amateur and professional boxers develop a progressive encephalopathy termed "Punch-drunk syndrome" or dementia pugilistica[331-334] and a similar kind of syndrome has been described in a repeatedly battered wife.[335] This is characterized pathologically by the widespread presence of diffuse Aβ deposits in the cerebral cortex,[336,337] these again comprising mostly of $A\beta_{42}$. Although NFT, structurally and chemically typical of AD, are common in mid-brain and brain stem structures and in the hippocampus[337-341] they are generally sparse within the cerebral neocortex and neuritic plaques are not seen.[341] Hence, NFT formation in dementia pugilistica does not always topographically coincide with the Aβ deposits and the distribution of NFT through the brain is untypical of AD.[342] Furthermore, Aβ deposition and NFT formation do not always concur; Geddes et al[343] have described the case of a young (23 year old) boxer dying accidentally during a fight following subdural hemorrhage, in whom NFT were widely present in the inferior and lateral neocortical areas (temporal and frontal lobes), but not in hippocampal or subcortical regions, and no Aβ deposition whatsoever had taken place. Hof et al[342] have also noted the absence of any Aβ deposition in two of three cases of dementia pugilistica even though NFT were numerous in all. Hence, although both acute and chronic repeated head trauma can recapitulate (some of) the pathological changes of AD, neither the topographic distribution nor the morphological characteristics of these are in most instances (entirely) typical. The relevance of this 'model' to the pathogenesis of AD still remains uncertain.

3.4.5. Chronic Renal Dialysis

About one-third of patients dying following prolonged renal dialysis show many diffuse Aβ deposits within their brains,[344] and an increased immunostaining for APP within pyramidal cells,[344,345] again at an age where none would normally be expected. Nonetheless, while conventional histological (silver staining) and immunohistochemical methods (tau immunostaining) usually fail to detect NFT in such patients,[344,346] (but see ref. 345) biochemical studies[345] have shown in most patients a lack of normal tau and the presence of abnormal (hyperphosphorylated) tau with protease resistant PHF protein (PHF-tau) also occurring in some of these. Patients who had undergone chronic dialysis in this way were at the time subjected to high circulating levels of aluminum because of the aluminum containing phosphate binding compounds present in the dialysis medium used to control their hyperphosphatemia. Consequently high levels of aluminum were accumulated in their brains.[346] Because aluminum has been associated in AD with NFT[347] or plaque cores[348,349] (but see ref. 350,351) it is possible that this element might play a role in the development of AD-type pathological changes not only in patients who may have been subject to prolonged medicinal exposure, as in those undergoing renal dialysis, but perhaps even in the general population too (see Chapter 4).

3.4.6. Other Conditions

NFT, apparently histologically and immunohistochemically, typical of those seen in AD and DS have been reported to be present, sometimes in high numbers, in diverse disorders such as subacute sclerosing panencephalitis (late measles infection),[339,352-355] postencephalitic Parkinsonism and the Parkinsonism-dementia amyotrophic lateral sclerosis complex on Guam,[356-358] myotonic dystrophy,[359]

Hallervorden-Spatz disease,[339] Gerstmann-Straüssler-Scheinker syndrome among the Indiana kindred,[360] as well as in leprosy[361] and in chronic alcoholism.[362] In all these conditions Aβ deposits are absent (or if present they occur no more frequently than would be expected for the age of the patient), though in the Indiana kindred amyloid deposits in the form of prion plaques are widespread.[360]

3.5. THE PATHOGENESIS OF ALZHEIMER'S DISEASE

From studies in DS (particularly) and from those on non-demented, or only mildly demented, individuals in the general population, it seems likely that the pathological cascade of AD is initiated by a deposition of Aβ, probably in the form of $A\beta_{1-42}$, this perhaps originating from changes in the way, or sites at which, APP is broken down in the brain or how the subsequently formed soluble Aβ is handled by the tissue. In inherited forms of AD, mutational changes may directly favor the formation of $A\beta_{1-42}$ (see Chapter 4) whereas in other situations such as DS or in renal dialysis or head injury increased (local) expression of APP may simply 'overload' conventional degradation routes with the excess APP becoming diverted towards amyloid forming pathways. Elevation of the extracellular amount of $A\beta_{1-42}$ will mean that threshold levels to fibrillogenesis might be met sooner with $A\beta_{42}$ containing deposits forming at an uncharacteristically early age. In 'normal' old age a failure to clear newly formed, though soluble, $A\beta_{1-42}$ may be responsible for the build up of this peptide, though changes in expression level of certain APP species, shifts in use of catabolic pathways, or diminished requirements for secretory (non-amyloid forming) products of APP could all play a part.

Nonetheless, once Aβ is deposited in the tissue a similar route towards the formation of $A\beta_{42}$ containing diffuse plaques is followed in all instances. What subsequently happens appears to be regionally or disease specific, differing according to whether the Aβ deposits occur in the cerebral cortex or elsewhere, in AD/DS or otherwise. The critical distinguishing feature of AD (and DS) is that, here the cerebral cortical $A\beta_{42}$ containing diffuse plaques transform over time into cored plaques containing both $A\beta_{42}$ and $A\beta_{40}$. This generally does not happen in the cerebellum or the corpus striatum in AD (and DS)[280] nor does it happen in head injury,[326,327,336] HCHWA-D [314] and renal dialysis,[344] nor to any great extent in the non-demented elderly. Why this process of transformation should so favor the cerebral cortex in AD is not clear though it may involve a selective action of non-neuronal cells such as microglial cells or astrocytes which might modify the already deposited Aβ or add to it from their own sources (see Chapter 2). Nonetheless, transformation of plaques in this way is associated with the development of the more 'malignant' pathological aspects of the disease, namely the neuritic changes that occur within cored, microglial cell containing plaques and which are not seen when such cored plaques are absent, and the formation of NFT. In the cerebellum, for example, plaques always remain as $A\beta_{42}$ containing diffuse plaques[280] and NFT and neuritic changes involving tau pathology never occur. Similarly, in boxers[336-343] or in renal dialysis[344,345] NFT either do not occur, or if they are present they do not usually coincide with regions of Aβ deposition. Furthermore, NFT, apparently identical in structure to those in AD (and DS), can be found in many other disease situations where (excessive) Aβ deposition is not seen.[352-362] Perhaps the best way of regarding Aβ deposition and NFT formation is that they both represent the end-points (markers) of pathological cascade processes and reflect (part of) the limited repertoire of response that the brain may show in the face of multiple and diverse etiologies; neither are, per se, specific to AD. In many patients (i.e. those whom by convention we say suffer from AD) these two processes go hand in hand though even here, the process of Aβ deposition need not necessarily lead directly to that of NFT formation despite observations that the latter chronologically follows the former. In many other instances each pathology can exist independently. Hence, it is possible that in the cerebral cortex in AD these two separate cascade processes may be occurring coincidentally, but concurrently, with factors that promote Aβ deposition not necessarily having any direct influence over the formation of NFT. Nevertheless Aβ promoting factors, or even Aβ itself, could alter the microenvironment of the brain in such a way as to facilitate neurofibrillary damage (see chapter 2).

3.6. SYSTEMIC CHANGES IN AD

It is well recognized that many patients with AD undergo considerable physical deterioration, particularly during the later stages of the illness. Much body weight is lost, despite adequate nourishment, with terminal body weights of around 30Kg being commonplace, especially among elderly females. The reasons for this body wasting are unclear but may relate to the hypothalamic pathology.[224,268-272] AD, however, may be a systemic illness that albeit principally involves the brain but also damages other tissues or organs.

In this context, an Alzheimer-like pathology has been widely sought in other body tissues in AD. Infrequent deposits of Aβ have been noted in the skin of some patients[363-365] and tau-like changes in the olfactory mucosa reported in others.[366,367] Such changes are not seen in the non-demented elderly. Local sources, such as fibroblasts, might be the source of Aβ within the skin, since such cells are known to express APP and secrete Aβ in culture.[368,369] However, it is also possible that a circulating source, such as platelets,[370,371] might be responsible for these body deposits of Aβ. Yet it is not known, even in carriers of the AD causing mutations, whether such changes exert any pathological effects in the (skin) tissue. Despite the (normally) high expression levels of APP within the skin and other peripheral tissues such as kidney,[372] Aβ deposits remain low or (usually) absent. This implies that APP is predominantly, or even exclusively, metabolized in extracerebral tissues along nonamyloidogenic pathways or that whatever Aβ is produced is quickly removed, perhaps because peripheral tissues lack the amyloid associated factors (chaperones) that promote its deposition in brain.

Apart from their capability for secretion of Aβ, fibroblasts in AD have been reported to suffer from an altered DNA repair capacity and increased X-ray sensitivity,[373] a reduced cholinergic activity,[374,375] a reduced intracellular, Ca^{2+},[376] a decreased adhesiveness,[377] an impaired oxidative metabolism[155] and an increased susceptibility to free radical damage.[378] Whether these changes reflect a differential aging of fibroblasts in culture from AD patients compared to those from normals or are fundamental to the presence of disease is uncertain, though other studies[379] suggest that the replicative features of fibroblasts in culture do not differ in AD from those of elderly non-demented individuals.

It has long been known that platelets, as well as having high APP expression levels and the capacity to secrete cleaved APP products, also contain many active enzymes, mitochondria and certain receptors in common with neurones. In 1984, Zubenko and colleagues[380] demonstrated an increase in platelet membrane fluidity in AD, findings confirmed in subsequent studies.[381-383] Such changes are of interest since they contradict those of aging alone where membranes become less fluid with time. Furthermore, alterations in fluidity similar to those in AD are not seen in multi-infarct dementia.[381] A study of platelets may therefore help to clarify some of the cellular mechanisms or membrane vulnerabilities that may increase the risk, or extent, of cellular damage in AD occasioned by plaque and tangle formation.

Hence, AD may be a more complex systemic disorder than has so far been recognized with cell and tissue changes extending far beyond the confines of the CNS. The study of non-neural tissues such as skin (fibroblasts) or blood (platelets) in model systems may help to unravel some of the tissue or cellular problems ongoing in the disease not only in the brain but also throughout the rest of the body.

3.7. THE PATHOLOGICAL CHANGES OF PARKINSON'S DISEASE

3.7.1. IMAGING STUDIES

In PD, both CT and MRI[384-388] scanning are relatively uninformative in diagnostic terms with patients showing either a normal scan or a degree of ventricular enlargement and cerebral atrophy not dissimilar to that commonly seen in aged, but healthy subjects. Although the corpus striatum represents the pathophysiological site of origin of many of the clinical symptoms of the disease no structural abnormalities are evident on scan, except occasional signal hyperintensities are seen both here and in the cerebral cortical white matter and internal capsule. These probably reflect minor cerebrovascular lesions but again are no more than would be anticipated for age.

MRI however does provide additional information on the tissue region containing the substantia nigra which may be of diagnostic relevance. A narrowing of the width of the tissue between the red nucleus and the pars reticulata of the substantia nigra is seen together with a lowering of its normally high signal intensity,[384-388]–changes which are not seen in elderly non-Parkinsonian subjects. A reduction in the band width between the pars reticulata and the red nucleus might reflect an atrophy of the pars compacta, where the degenerating pigmented nigral cells are located. The iron released from degenerated nerve cells and accumulated in the nigral region[389-395] would produce this lowering of signal intensity.

PET imaging, using fluorodopa uptake, produces vivid images of impaired dopamine metabolism throughout all areas of the corpus striatum, but particularly so in the putamen.[396-399] Such images are consistent with the loss of dopaminergic nerve terminals (or dopamine uptake from the same) due to the death of nigral neurones, especially those in the ventrolateral tier that project to the putamen. Global glucose metabolism is usually depressed in PD, the severity depending upon the extent of motor impairment and the presence or otherwise of dementia. If the latter is present, bilateral posterior hemisphere defects typical of those in AD are seen.[400,401] This is probably not unexpected since such demented patients are likely to have suffered from CLBD–a condition in which Alzheimer-type pathology with a distribution similar to that of AD itself is commonplace.[304]

3.7.2. HISTOPATHOLOGY

Parkinson's disease is characterized pathologically by a loss of the neuromelanin containing nerve cells of certain critical regions of the mid-brain and brainstem with the formation of the pathognomonic change–the Lewy body (see chapter 2)–within (some of) the surviving cells.[402-419] The classic areas of involvement are thus the substantia nigra, the locus caeruleus and the dorsal motor vagus nucleus. However, it would be a mistake to consider PD as a disorder solely of these pigmented, catecholaminergic cell types since many other non-catecholaminergic, non-neuromelanin pigmented cells like the cholinergic neurones of the nucleus basalis of Meynert[414,415] are affected in the disease. Furthermore, in many other regions in the brainstem, mid-brain, spinal cord, basal ganglia, thalamus, amygdala and cerebral cortex Lewy bodies can be present, even if nerve cell loss is not apparent.[408,413,416] Conversely, many other dopaminergic neurones in the mesencephalic grey matter and in the hypothalamus escape destruction.[406,420,421] The sympathetic and parasympathetic nervous systems may also be involved[422,423] and Lewy bodies can even be found in the neural networks of the gastrointestinal tract.[424] The clinical signs of PD emerge only when at least half the neurones of the substantia nigra[403,405] and 80% of the striatal dopamine[425] have been lost. Within the degenerating substantia nigra, many microglial cells can be seen[426] and while these may be responsible for the phagocytosis of neuromelanin pigment, iron or other debris from dead neurones their presence in end-stage tissues at autopsy indicates that the disease process is still ongoing, even during the terminal phases of the illness.[426]

3.7.3. Patterns of Cell Loss

The pattern of nerve cell loss from the substantia nigra in PD is non-random[403,405,427-429] with neuronal loss (averaging 75% overall) starting in the lateroventral tier then spreading to other regions, particularly the medioventral tier. In normal aging neuronal losses average 33% overall, but these are greatest in the dorsal tier followed by medioventral and lateroventral tiers.[403,405] In the locus caeruleus, cell loss occurs uniformly throughout the entire rostrocaudal extent of the nucleus.[115,117] However, other workers[119] contend that this cell loss is only apparent; a shrinkage and depigmentation of these large pigmented neurones in PD make them become small and non-pigmented and as such could have escaped counting in previous studies[115,117] where counts were restricted to the larger pigmented neurones.

3.7.4. Aging and Parkinson's Disease

A characteristic feature of cells of the substantia nigra and locus caeruleus (and certain other brainstem neurones–but to a lesser extent) is their pigmentation by neuromelanin.[430,431] The amount of pigment in these cell types increases with age, reaching a peak at around 60 years of age, then falling thereafter.[432,433] This suggests that cell loss in later life might select out those cells with highest quantities of pigment.[434] Indeed the dorsal tier of the substantia nigra contains the most heavily pigmented neurones and it is this very group of cells that suffers the greatest loss with 'normal' aging.[403,405]

These observations thus argue for a relationship between the extent of neuromelanin pigmentation and that of cell loss in later life. Neuromelanin is generated through the incorporation of the autooxidated products of catecholamines into lipofuscin granules (see Chapter 2).[435] Free radicals or other toxins may be produced during this process and these may in turn adversely affect the function and viability of the nigral cells, especially those with the highest pigment levels. Neurotoxins such as MPTP may affect nigral cells by binding to their pigment once having been taken up into such neurones via the dopamine uptake system (see Chapter 4).

However, while this might be a possible explanation for the progressive cell loss from the substantia nigra seen in normal[405,436-438] it cannot adequately explain the cause of PD since in this disorder non-neuromelanin containing, non-catecholaminergic cells, such as those of the cholinergic nucleus basalis, are also seriously damaged yet the dopamine containing tubero-infundibular neurones of the hypothalamus, among other such cell types, are little involved.[439] Furthermore, in the substantia nigra it is the least pigmented ventral tier that sustains the greatest cell loss.[403,405,427-429] In the locus caeruleus the pattern of cell loss in PD[115,117] is also different to that of normal aging[117,440] (but see ref. 441) which resembles, though to a much lesser extent, that seen in AD[113,115,117] and DS.[442] Hence, PD may represent more than just a simple 'acceleration' of those damaging changes that are in any case taking place as part of the normal aging of these regions of brain. Nonetheless, such a 'weakening' of cells of the substantia nigra and locus caeruleus with aging may act as a pathogenetic 'springboard' upon which other and PD specific changes may become superimposed. In this way aging would facilitate the development of clinical change by making the initial inroads into these vulnerable cell populations thereby paving the way for subsequent disease-specific changes to "finish off" the remaining cells.

It is likely that asymptomatic elderly persons with occasional (incidental) Lewy bodies[404,443-445] do in fact represent preclinical cases of PD, since in these the pattern of nerve cell loss in the substantia nigra (although slight and averaging 20%) closely resembles that of fully developed PD and is distinct from that seen in normally aged cases without Lewy bodies.[405] PD is therefore probably a specific disease of old age rather than being one that simply represents an acceleration, or a particularly aggressive form, of those pathological changes that are taking place in these vulnerable brain regions as we all grow older.

References

1. Hubbard BM, Anderson JM. Age, senile dementia and ventricular enlargement. J Neurol Neurosurg Psychiatry 1981; 44: 631-635.
2. George AE, de Leon MJ, Ferris SH et al. Parenchymal CT correlates of senile dementia (Alzheimer's disease): Loss of gray-white matter discriminability. Am J Neuroradiol 1981; 2: 205-213.
3. Huckman MS, Fox J, Topel J. The validity of criteria for the evaluation of cerebral atrophy by computed tomography. Radiology 1975; 116: 85-92.
4. Koller WC, Glatt SL, Fox JH et al. Cerebellar atrophy: relationship to aging and cerebral atrophy. Neurology 1981; 31: 1486-1488.
5. Drayer BP, Heyman A, Wilkinson W et al. Early onset Alzheimer's disease: An analysis of CT findings. Ann Neurol 1985; 17: 407-410.
6. Gado M, Hughs CP, Danziger W et al. Volumetric measurements of the cerebrospinal fluid spaces in demented subjects and controls. Radiology 1982; 144: 535-538.
7. LeMay M. Radiologic changes of the aging brain and skull. Am J Radiol 1984; 143: 383-389.
8. Schwartz M, Creasey H, Grady CL et al. Computed tomographic analysis of brain morphometrics in 30 healthy men, aged 21 to 81 years. Ann Neurol 1985; 17: 146-157.
9. Pfefferbaum A, Zatz LM, Jernigan TL. Computer-interactive method for quantifying cerebrospinal fluid and tissue in brain CT scans: Effects of aging. J Comput Assist Tomogr 1986; 10: 571-578.
10. Drayer BP. Imaging of the aging brain. Radiology 1988; 166: 785-796.
11. Creasey H, Rapoport SI. The aging human brain. Ann Neurol 1985; 17: 2-10.
12. Jobst KA, Smith AD, Szatmari M et al. Detection in life of confirmed Alzheimer's disease using a simple measurement of medial temporal lobe atrophy by computed tomography. Lancet 1992; 340: 1179-1183.
13. Jobst KA, Smith AD, Szatmari M et al. Rapidly progressing atrophy of medial temporal lobe in Alzheimer's disease. Lancet 1994; 343: 829-830.
14. Pasquier F, Bail L, Lebert F et al. Determination of medial temporal lobe atrophy in early Alzheimer's disease with computed tomography. Lancet 1994; 343: 861-862.
15. Tanna NK, Kohn MI, Horwich DN et al. Analysis of brain and cerebrospinal fluid volumes with MR imaging: Impact on PET data correction for atrophy. Radiology 1991; 178: 123-130.
16. Seab JP, Jagust WJ, Wong STS et al. Quantitative NMR measurements of hippocampal atrophy in Alzheimer's disease. Magn Reson Med 1988; 8: 200-208.
17. Dahlbeck SW, McCluney KW, Yeakley JW et al. The interuncal distance: A new MR measurement for the hippocampal atrophy of Alzheimer disease. Am J Neuroradiol 1991; 12: 931-932.
18. Kesslak JP, Nalcioglu O, Cotman CW. Quantification of magnetic resonance scans for hippocampal and parahippocampal atrophy in Alzheimer's disease. Neurology 1991; 41: 51-54.
19. Jack CR, Petersen RC, O'Brien PC et al. MR-based hippocampal volumetry in the diagnosis of Alzheimer's disease. Neurology 1992; 42: 183-188.
20. Erkinjuntti T, Lee DH, Gao F et al. Temporal lobe atrophy on magnetic resonance imaging in the diagnosis of early Alzheimer's disease. Arch Neurol 1993; 50: 305-310.
21. Laakso MP, Soininen H, Helkala E-L et al. Volumes of hippocampus and amygdala in the magnetic resonance imaging based diagnosis of early Alzheimer's disease. In: Iqbal

K, Mortimer JA, Winblad B et al., eds. Research Advances in Alzheimer's Disease and Related Disorders. ed. : John Wiley & Sons Ltd, 1995: 181-188.
22. Jernigan TL, Salmon DP, Butters N et al. Cerebral structures on MRI: Part II. Specific changes in Alzheimer's and Huntington's diseases. Biol Psychiat 1991; 29: 68-81.
23. Rusinek H, de Leon MJ, George AE et al. Alzheimer disease: Measuring loss of cerebral gray matter with MR imaging. Radiology 1991; 178: 109-114.
24. Fazekas F, Chawluk JB, Alavi A et al. MR signal abnormalities at 1.5 T in Alzheimer's dementia and normal aging. Am J Neuroradiol 1987; 8: 421-426.
25. Harrell LE, Duvall E, Folks DG et al. The relationship of high-intensity signals on magnetic resonance images to cognitive and psychiatric state in Alzheimer's disease. Arch Neurol 1991; 48: 1136-1140.
26. Kumar A, Yousem D, Souder E et al. High-intensity signals in Alzheimer's disease without cerebrovascular risk factors: A magnetic resonance imaging evaluation. Am J Psychiatry 1992; 149: 248-250.
27. Bradley WG, Waluch V, Brant-Zawadzki M et al. Patchy, periventricular white matter lesions in the elderly: A common observation during NMR imaging. Noninvas Med Imaging 1984; 1: 35-41.
28. Brant-Zawadski M, Fein G, Van Dyke C et al. MR imaging of the aging brain: patchy white-matter lesions and dementia. Am J Neuroradiol 1985; 6: 675-682.
29. George AE, de Leon MJ, Kalnin A et al. Leukoencephalopathy in normal and pathologic aging: II MRI of brain lucencies. Am J Neuroradiol 1986; 7: 561-570.
30. Rezek DL, Morris JC, Fulling KH et al. Periventricular white matter lucencies in senile dementia of the Alzheimer type and in normal aging. Neurology 1987; 37: 1365-1368.
31. Leys D, Soetaert G, Petit H et al. Periventricular and white matter magnetic resonance imaging hyperintensities do not differ between Alzheimer's disease and normal aging. Arch Neurol 1990; 47: 524-527.
32. Ferris SH, De Leon MJ, Wolf AP et al. Positron emission tomography in dementia. Adv Neurol 1983; 38: 123-129.
33. Foster NL, Chase TN, Fedio P et al. Alzheimer's disease: Focal cortical changes shown by positron emission tomography. Neurology 1983; 33: 961-965.
34. Friedland RP, Budinger TF, Koss E et al. Alzheimer's disease: anterior-posterior and lateral hemispheric alterations in cortical glucose utilization. Neurosci Lett 1983; 53: 235-240.
35. De Leon MJ, Ferris SH, George AE et al. Positron emission tomographic studies of aging and Alzheimer disease. Am J Neuroradiol 1983; 4: 568-571.
36. Grady CL, Haxby JV, Schlageter NL et al. Stability of metabolic and neuropsychological asymmetries in dementia of the Alzheimer type. Neurology 1986; 36: 1390-1392.
37. Grady CL, Haxby JV, Horwitz B et al. Neuropsychological and cerebral metabolic function in early vs late onset dementia of the Alzheimer type. Neuropsychologia 1987; 25: 807-816.
38. Haxby JV, Grady CL, Koss E et al. Longitudinal study of cerebral metabolic asymmetries and associated neuropsychological patterns in early dementia of the Alzheimer type. Arch Neurol 1990; 47: 753-760.
39. Azari NP, Rapoport SI, Grady CL et al. Patterns of interregional correlations of cerebral glucose metabolic rates in patients with dementia of the Alzheimer type. Neurodegeneration 1992; 1: 101-111.
40. Chawluk JB, Alavi A, Dann R et al. Positron emission tomography in aging and dementia: Effect of cerebral atrophy. J Nucl Med 1987; 28: 431-437.
41. DeCarli C, Atack JR, Ball MJ et al. Post-mortem regional neurofibrillary tangle densities but not senile plaque densities are related to regional cerebral metabolic rates for glucose during life in Alzheimer's disease patients. Neurodegeneration 1992; 1: 113-121.
42. Haxby JV, Duara R, Grady CL et al. Relations between neuropsychological and cerebral metabolic asymmetries in early Alzheimer's disease. J Cereb Blood Flow Metab 1985; 5: 193-200.

43. Grady CL, Haxby JV, Horwitz B et al. A longitudinal study of the early neuropsychological and cerebral changes in dementia of the Alzheimer type. J Clin Exp Neuropsychol 1988; 10: 576-596.
44. Kumar A, Schapiro MB, Haxby JV et al. Cerebral metabolic and cognitive studies in dementia with frontal lobe behavioural features. J Psychiatr Res 1990; 24: 97-109.
45. Friedland RP, Brun A, Budinger TF. Pathological and positron emission tomographic correlations in Alzheimer's disease. Lancet 1985; i: 228-230.
46. Cutler NR, Haxby JV, Duara R et al. Clinical history, brain metabolism, and neuropsychological function in Alzheimer's disease. Ann Neurol 1985; 18: 298-309.
47. Duara R, Grady C, Haxby J et al. Positron emission tomography in Alzheimer's disease. Neurology 1986; 36: 879-887.
48. Hubbard BM, Anderson JM. A quantitative study of cerebral atrophy in old age and senile dementia. J Neurol Sci 1981; 50: 135-145.
49. Hansen LA, Teresa R, Davies P et al. Neocortical morphometry, lesion count and choline acetyl transferase levels in the age spectrum of Alzheimer's disease. Neurology 1988; 38: 48-54.
50. De la Monte S. Quantitation of cerebral atrophy in preclinical and end-stage Alzheimer's disease. Ann Neurol 1989; 25: 450-459.
51. Mann DMA. The topographic distribution of brain atrophy in Alzheimer's disease. Acta Neuropathol 1991; 83: 81-86.
52. Terry RD, Peck A, De Teresa R et al. Some morphometric aspects of the brain in senile dementia of the Alzheimer type. Ann Neurol 1981; 10: 184-192.
53. Mann DMA, Yates PO, Marcyniuk B. Some morphometric observations on the cerebral cortex and hippocampus in presenile Alzheimer's disease, senile dementia of Alzheimer type and Down's syndrome in middle age. J Neurol Sci 1985; 69: 139-159.
54. Miller AKH, Alston RL, Corsellis JAN. Variation with age in the volumes of grey and white matter in the cerebral hemispheres of man: measurements with an image analyser. Neuropath Appl Neurobiol 1980; 6: 119-132.
55. Regeur L, Jensen GB, Pakkenberg H et al. No global neocortical nerve cell loss in brains from patients with senile dementia of Alzheimer's type. Neurobiol Aging 1994; 15: 347-352.
56. Tomlinson BE, Blessed G, Roth M. Observations on the brains of demented old people. J Neurol Sci 1970; 11: 205-242.
57. Neary D, Snowden JS, Mann DMA et al. Alzheimer's disease: a correlative study. J Neurol Neurosurg Psychiatry 1986; 49: 229-237.
58. Duyckaerts C, Hauw JJ, Piette F et al. Cortical atrophy in senile dementia of Alzheimer type is mainly due to a decrease in cortical length. Acta Neuropathol 1985; 66: 72-74.
59. Shefer VF. Absolute numbers of neurones and thickness of the cerebral cortex during aging, senile and vascular dementia, and Pick's and Alzheimer's diseases. Neurosci Behav Physiol 1973; 6: 319-324.
60. Colon EJ. The cerebral cortex in presenile dementia. A quantitative analysis. Acta Neuropathol 1973; 23: 281-290.
61. Mountjoy CQ, Roth M, Evans NJR et al. Cortical neuronal counts in normal elderly controls and demented patients. Neurobiol Aging 1983; 4: 1-11.
62. Brun A, Englund E. Regional pattern of degeneration in Alzheimer's disease: neuronal loss and histopathological grading. Histopathology 1981; 5: 549-564.
63. Hubbard BM, Anderson JM. Age-related variations in the neurone content of the cerebral cortex in senile dementia of Alzheimer type. Neuropath Appl Neurobiol 1985; 11: 369-382.
64. Tomlinson BE, Henderson G. Some quantitative cerebral findings in normal and demented old people. In: Terry RD, Gerhon S, eds. Neurobiology of Aging. ed. New York: Raven Press, 1976: 183-204.
65. Ball MJ. Neuronal loss, neurofibrillary tangles and granulovacuolar degeneration in the hippocampus with aging and dementia. A quantitative study. Acta Neuropathol 1977; 37: 111-118.
66. Shefer VF. Hippocampal pathology as a possible factor in the pathogenesis of senile dementias. Neurosci Behav Physiol 1977; 8: 236-239.

67. Hyman BT, Damasio AR, van Hoesen GW et al. Alzheimer's disease: cell-specific pathology isolates the hippocampal formation. Science 1984; 225: 1168-1170.
68. Doebler JA, Markesbery WR, Anthony A et al. Neuronal RNA in relation to neuronal loss and neurofibrillary pathology in the hippocampus in Alzheimer's disease. J Neuropathol Exp Neurol 1987; 46: 28-39.
69. Davies DC, Horwood N, Isaacs SL et al. The effect of age and Alzheimer's disease in pyramidal neurone density in the individual fields of the hippocampal formation. Acta Neuropathol 1992; 83: 510-517.
70. Cras P, Smith MA, Richey PL et al. Extracellular neurofibrillary tangles reflect neuronal loss and provide further evidence of extensive protein cross-linking in Alzheimer disease. Acta Neuropathol 1995; 89: 291-295.
71. Braak H, Braak E. On areas of transition between entorhinal allocortex and temporal isocortex within human brain. Normal morphology and laminar specific pathology. Acta Neuropathol 1985; 68: 325-332.
72. Hooper MW, Vogel FS. The limbic system in Alzheimer's disease. Am J Pathol 1976; 85: 1-13.
73. Herzog AG, Kemper TL. Amygdaloid changes in aging and dementia. Arch Neurol 1980; 37: 625-629.
74. Saper CB, German DC, White CL. Neuronal pathology in the nucleus basalis and associated cell groups in senile dementia of Alzheimer type: possible role in cell loss. Neurology 1985; 35: 1089-1095.
75. Unger JW, McNeill TH, Lapham LL et al. Neuropeptides and neuropathology in the amygdala in Alzheimer's disease: relationship between somatostatin, neuropeptide Y and subregional distribution of neuritic plaques. Brain Res 1988; 452: 293-302.
76. Brashear HR, Godec MS, Carlsen J. The distribution of neuritic plaques and acetylcholinesterase staining in the amygdala in Alzheimer's disease. Neurology 1988; 38: 1694-1699.
77. Tsuchiya K, Kosaka K. Neuropathological study of the amygdala in presenile Alzheimer's disease. J Neurol Sci 1990; 100: 165-173.
78. Murphy GM, Eng LF, Ellis WG et al. Antigenic profile of plaques and neurofibrillary tangles in the amygdala in Down's syndrome: a comparison with Alzheimer's disease. Brain Res 1990; 537: 102-108.
79. Hyman BT, Van Hoesen GW, Damasio AR. Memory related neural systems in Alzheimer's disease; An anatomic study. Neurology 1990; 40: 1721-1730.
80. Scott SA, DeKosky ST, Sparks DL et al. Amygdala cell loss and atrophy in Alzheimer's disease. Ann Neurol 1992; 32: 555-563.
81. Kromer-Vogt LJ, Hyman BT, van Hoesen GW et al. Pathologic alterations in the amygdala in Alzheimer's disease. Neuroscience 1990; 37: 377-385.
82. Mann DMA, Tucker CM, Yates PO. The topographic distribution of senile plaques and neurofibrillary tangles in the brains of non-demented persons of different ages. Neuropath Appl Neurobiol 1987; 13: 123-139.
83. Li Y-T, Woodruff-Pak DS, Trojanowski JQ. Amyloid plaques in the cerebellar cortex and the integrity of Purkinje cell dendrites. Neurobiol Aging 1993; 15: 1-9.
84. Mann DMA, Jones D, Prinja D et al. The prevalence of amyloid (A4) protein deposits within the cerebral and cerebellar cortex in Down's syndrome and Alzheimer's disease. Acta Neuropathol 1990; 80: 318-327.
85. Nakano I, Hirano A. Loss of large neurones of the medial septal nucleus in an autopsy case of Alzheimer's disease. J Neuropathol Exp Neurol 1981; 41: 341.
86. Whitehouse PJ, Price DL, Struble RG et al. Alzheimer's disease and senile dementia: loss of neurons in the basal forebrain. Science 1982; 215: 1237-1239.
87. Perry RH, Candy JM, Perry EK et al. Extensive loss of choline acetyl transferase activity is not related to neuronal loss in the nucleus basalis of Meynert in Alzheimer's disease. Neurosi Lett 1982; 33: 311-315.
88. Candy JM, Perry RH, Perry EK et al. Pathological changes in the nucleus of Meynert in Alzheimer's and Parkinson's diseases. J Neurol Sci 1983; 59: 277-289.
89. Wilcock GK, Esiri MM, Bowen DM et al. The nucleus basalis in Alzheimer's disease;

cell counts and cortical biochemistry. Neuropath Appl Neurobiol 1983; 9: 175-179.
90. Tagliavini F, Pilleri G. Basal nucleus of Meynert. J Neurol Sci 1983; 62: 243-260.
91. Mann DMA, Yates PO, Marcyniuk B. Changes in nerve cells of the nucleus basalis of Meynert in Alzheimer's disease and their relationship to aging and the accumulation of lipofuscin pigment. Mech Aging Dev 1984; 25: 189-204.
92. Mann DMA, Yates PO, Marcyniuk B. A comparison of changes in the nucleus basalis and locus caeruleus in Alzheimer's disease. J Neurol Neurosurg Psychiatry 1984; 47: 201-203.
93. Mann DMA, Yates PO, Marcyniuk B. Alzheimer's presenile dementia, senile dementia of Alzheimer type and Down's syndrome in middle age form an age-related continuum of pathological changes. Neuropath Appl Neurobiol 1984; 10: 185-207.
94. Arendt T, Bigl V, Arendt A et al. Loss of neurones in the nucleus basalis of Meynert in Alzheimer's disease, paralysis agitans and Korsakoff's disease. Acta Neuropathol 1983; 61: 101-108.
95. Arendt T, Bigl V, Tennstedt A et al. Correlation between cortical plaque count and neuronal loss in nucleus basalis in Alzheimer's disease. Neurosci Lett 1984; 48: 81-85.
96. Arendt T, Bigl V, Tennstedt A et al. Neuronal loss in different parts of the nucleus basalis is related to neuritic plaque formation in cortical target areas in Alzheimer's disease. Neuroscience 1985; 14: 1-14.
97. Nagai T, McGeer PL, Peng JH et al. Choline acetyl transferase immunohistochemistry in brains of Alzheimer's patients and controls. Neurosci Lett 1983; 36: 195-199.
98. McGeer PL, McGeer EG, Suzuki JS et al. Aging, Alzheimer's disease and the cholinergic system of the basal forebrain. Neurology 1984; 34: 741-745.
99. Rogers JD, Brogan D, Mirra SS. The nucleus basalis of Meynert in neurological disease: A quantitative morphological study. Ann Neurol 1985; 17: 163-170.
100. Doucette R, Fisman M, Hachinsky VC et al. Cell loss from the nucleus basalis of Meynert in Alzheimer's disease. Can J Neurol Sci 1986; 13: 435-440.
101. Doucette R, Ball MJ. Left-right symmetry of neuronal cell counts in the nucleus basalis of Meynert of control and of Alzheimer diseased brains. Brain Res 1987; 422: 357-360.
102. Ichimiya Y, Arai H, Kosaka K et al. Morphological and biochemical changes in the cholinergic and monoaminergic systems in Alzheimer type dementia. Acta Neuropathol 1986; 70: 112-116.
103. Allen SJ, Dawbarn D, Wilcock GK. Morphometric immunochemical analysis of neurones in the nucleus basalis of Meynert in Alzheimer's disease. Brain Res 1988; 454: 275-281.
104. De Lacalle S, Iraizoz I, Ma Gonzalo L. Differential changes in cell size and number in topographic subdivisions of human basal nucleus in normal aging. Neuroscience 1991; 43: 445-456.
105. Pearson RCA, Sofroniew MV, Cuello AC et al. Persistence of cholinergic neurons in the basal nucleus in a brain with senile dementia of the Alzheimer's type demonstrated by immunohistochemical staining for choline acetyltransferase. Brain Res 1983; 289: 375-379.
106. Tomlinson BE, Irving D, Blessed G. Cell loss in the locus caeruleus in senile dementia of Alzheimer type. J Neurol Sci 1981; 49: 419-428.
107. Bondareff W, Mountjoy CQ, Roth M. Loss of neurones of origin of the adrenergic projection to cerebral cortex (nucleus locus caeruleus) in senile dementia. Neurology 1992; 32: 164-168.
108. Mann DMA, Yates PO, Hawkes J. The noradrenergic system in Alzheimer and multi-infarct dementias. J Neurol Neurosurg Psychiatry 1982; 45: 113-119.
109. Mann DMA, Yates PO, Hawkes J. The pathology of the human locus caeruleus. Clin Neuropathol 1983; 2: 1-7.
110. Mann DMA, Yates PO, Marcyniuk B. Changes in Alzheimer's disease in the magnocellular neurones of the supraoptic and paraventricular nuclei of the hypothala-

mus and their relationship to the noradrenergic deficit. Clin Neuropathol 1985; 4: 127-134.
111. Iversen LL, Rossor MN, Reynolds GP et al. Loss of pigmented dopamine β hydroxylase positive cells from locus caeruleus in senile dementia of Alzheimer's type. Neurosci Lett 1983; 39: 95-100.
112. Chui HC, Mortimer JA, Slager UT et al. Pathological correlates of dementia in Parkinson's disease. ArchNeurol 1986; 43: 991-995.
113. Marcyniuk B, Mann DMA, Yates PO. The topography of cell loss from locus caeruleus in Alzheimer's disease. J Neurol Sci 1986; 76: 335-345.
114. Zweig RM, Ross CA, Hedreen JC et al. The neuropathology of aminergic nuclei in Alzheimer's disease. Ann Neurol 1988; 24: 233-242.
115. Chan-Palay V, Asan E. Alterations in catecholamine neurons of the locus coeruleus in senile dementia of the Alzheimer type and in Parkinson's disease with and without dementia and depression. J Comp Neurol 1989; 287: 373-392.
116. Strong R, Huang JS, Huang SS et al. Degeneration of the cholinergic innervation of the locus caeruleus in Alzheimer's disease. Brain Res 1991; 542: 23-28.
117. German DC, Manaye KF, White CL et al. Disease specific patterns of locus caeruleus cell loss: Parkinson disease, Alzheimer's disease and Down's syndrome. Ann Neurol 1992; 32: 667-676.
118. Giess R, Schlote W. Localisation and association of pathomorphological changes at the brainstem in Alzheimer's disease. Mech Aging Dev 1995; 84: 209-226.
119. Hoogendijk WJG, Pool CW, Troost D et al. Image analyser-assisted morphometry of the locus coeruleus in Alzheimer's disease, Parkinson's disease and amyotrophic lateral sclerosis. Brain 1995; 118: 131-143.
120. Ishii T. Distribution of Alzheimer's neurofibrillary changes in brain stem and hypothalamus of senile dementia. Acta Neuropathol 1966; 38: 181-187.
121. Hirano A, Zimmerman HM. Alzheimer's neurofibrillary changes. A topographic study. Arch Neurol 1962; 7: 73-88.
122. Curcio CA, Kemper T. Nucleus raphe dorsalis in dementia of the Alzheimer type: neurofibrillary changes and neuronal packing density. J Neuropathol Exp Neurol 1985; 43: 359-368.
123. Yamamoto T, Hirano A. Nucleus raphe dorsalis in Alzheimer's disease: neurofibrillary tangles and loss of large neurons. Ann Neurol 1985; 17: 573-577.
124. Tabaton M, Schenone A, Romagnoli P et al. A quantitative and ultrastructural study of substantia nigra and nucleus centralis superior in Alzheimer's disease. Acta Neuropathol 1985; 68: 213-223.
125. Jellinger K. Quantitative changes in some subcortical nuclei in aging, Alzheimer's disease and Parkinson's disease. Neurobiol Aging 1987; 8: 556-561.
126. Mann DMA, Yates PO, Marcyniuk B. Dopaminergic neurotransmitter systems with Alzheimer's disease and Down's syndrome at middle age. J Neurol Neurosurg Psychiatry 1987; 50: 341-344.
127. Gibb WRG, Mountjoy CQ, Mann DMA et al. The substantia nigra and ventral tegmental area in Alzheimer's disease and Down's syndrome. J Neurol Neurosurg Psychiatry 1989; 52: 193-200.
128. Mann DMA, Yates PO, Marcyniuk B. Monoaminergic neurotransmitter systems in presenile Alzheimer's disease and in senile dementia of Alzheimer type. Clin Neuropathol 1984; 3: 199-205.
129. German DC, White CL, Sparkman DR. Alzheimer's disease: neurofibrillary tangles in nuclei that project to the cerebral cortex. Neuroscience 1987; 21: 305-312.
130. Lehericy S, Hirsch EC, Hersh LB et al. Cholinergic neuronal loss in the globus pallidus of Alzheimer's disease patients. Neurosci Lett 1991; 123: 152-155.
131. Lehericy S, Hirsch EC, Cervera P et al. Selective loss of cholinergic neurones in the ventral striatum of patients with AD. Proc Natl Acad Sci (USA) 1989; 86: 8580-8584.
132. Xuereb JH, Perry EK, Candy JM et al. Parameters of cholinergic neurotransmission in the thalamus in Parkinson's disease and Alzheimer's disease. J Neurol 1990; 99: 185-197.

133. Oyanagi K, Takahashi H, Wakabayashi K et al. Selective involvement of large neurones in the neostriatum of Alzheimer's disease and senile dementia: A morphometric investigation. Brain Res 1987; 411: 205-211.
134. Oyanagi K, Takahashi H, Wakayashi K et al. Correlative decrease of large neurones in the neostriatum and basal nucleus of Meynert in Alzheimer's disease. Brain Res 1989; 504: 354-357.
135. Oyanagi K, Takahashi H, Wakabayashi K et al. Large neurons in the neostriatum in Alzheimer's disease and progressive supranuclear palsy: a topographic, histologic and ultrastructural investigation. Brain Res 1991; 544: 221-226.
136. Masliah E, Terry RD, Buzsaki G. Thalamic nuclei in Alzheimer's disease–evidence against the cholinergic hypothesis of plaque formation. Brain Res 1989; 493: 240-246.
137. Tourtellotte WG, Van Hoesen GW, Hyman BT et al. Alz-50 immunoreactivity in the thalamic reticular nucleus in Alzheimer's disease. Brain Res 1989; 515: 227-234.
138. Braak H, Braak E. Alzheimer's disease affects limbic nuclei of the thalamus. Acta Neuropathol 1991; 81: 261-268.
139. Swaab DF, Fliers E, Partiman TS. The suprachiasmatic nucleus of the human brain in relation to sex, age and senile dementia. Brain Res 1985; 342: 37-44.
140. Kremer B, Swaab D, Bots G et al. The hypothalamic lateral tuberal nucleus in Alzheimer's disease. Ann Neurol 1991; 29: 279-284.
141. Zhou J-N, Hofman MA, Swaab DF. VIP neurons in the human SCN in relation to sex, age, and Alzheimer's disease. Neurobiol Aging 1995; 16: 571-576.
142. Wilkinson A, Davies I. The influence of age and dementia on the neurone population of the mammillary bodies. Age Aging 1978; 7: 151-160.
143. Scheff SW, Price DA. Synapse loss in the temporal lobe in Alzheimer's disease. Ann Neurol 1993; 33: 190-199.
144. Scheff SW, DeKosky ST, Price DA. A quantitative assessment of cortical synaptic density in Alzheimer's disease. Neurobiol Aging 1990; 11: 29-37.
145. Davies CA, Mann DMA, Sumpter PQ et al. A quantitative analysis of the neuronal and synaptic content of the frontal and temporal cortex in patients with Alzheimer's disease. J Neurol Sci 1987; 78: 151-164.
146. Gibson PH. EM Study of the numbers of cortical synapses in the brains of aging people and people with Alzheimer-type dementia. Acta Neuropathol 1983; 62: 127-133.
147. DeKosky ST, Scheff SW. Synpase loss in frontal cortex biopsies in Alzheimer's disease: correlation with cognitive severity. Ann Neurol 1990; 27: 457-464.
148. Masliah E, Hansen T, Mallory M et al. Immunoelectron microscopic study of synaptic pathology in Alzheimer's disease. Acta Neuropathol 1991; 81: 428-433.
149. Mann DMA, Yates PO. Neurotransmitter deficits in Alzheimer's disease and in other dementing disorders. Hum Neurobiol 1986; 5: 147-158.
150. Hedera P, Whitehouse PJ. Neurotransmitters in neurodegeneration. In: Calne DB, ed. Neurodegenerative Diseases. Philadelphia, London, Toronto, Montreal, Sydney, Tokyo: WB Saunders Co, 1994: 97-117.
151. Parker WD, Parks J, Filley CM et al. Electron chain transfer defects in Alzheimer brain. Neurology 1994; 44: 1090-1096.
152. Simonian NA, Hyman BT. Functional alterations in Alzheimer's disease: Diminution of cytochrome oxidase in the hippocampal formation. J Neuropathol Exp Neurol 1993; 52: 580-585.
153. Mutisya EM, Bowling AC, Walker LC et al. Impaired energy metabolism in aging and Alzheimer's disease. Soc Neurosci Abs 1993; 19: 1474.
154. Simonian NA, Hyman BT. Functional alterations in Alzheimer's disease: selective loss of mitochondrial-encoded cytochrome oxidase mRNA in the hippocampal formation. J Neuropathol Exp Neurol 1994; 53: 508-512.
155. Sims NR, Finegan JM, Blass JP. Altered metabolic properties of cultured skin fibroblasts in Alzheimer's disease. Ann Neurol 1987; 21: 451-457.
156. Blass JP, Gibson GE, Sheu K-FR et al. Mitochondria, aging and neurological dis-

eases. In: Zatta P, Nicolini M, eds. Non-Neuronal Cells in Alzheimer's Disease. Singapore, New Jersey, London, Hong Kong: World Scientific, 1995: 95-107.
157. Sumpter PQ, Mann DMA, Davies CA et al. An ultrastructural analysis of the effects of accumulation of neurofibrillary tangle in pyramidal cells of the cerebral cortex in Alzheimer's disease. Neuropath Appl Neurobiol 1986; 12: 305-319.
158. Mann DMA, Yates PO, Barton CM. Cytophotometric mapping of neuronal changes in senile dementia. J Neurol Neurosurg Psychiatry 1977; 40: 299-302.
159. Mann DMA, Sinclair KGA. The quantitative assessment of lipofuscin pigment, cytoplasmic RNA and nucleolar volume in senile dementia. Neuropath Appl Neurobiol 1978; 4: 129-135.
160. Mann DMA, Yates PO. The relationship between formation of senile plaques and neurofibrillary tangles and changes in nerve cell metabolism in Alzheimer-type dementia. Mech Aging Dev 1981; 17: 395-401.
161. Doebler JA, Markesbery WR, Anthony A et al. Neuronal RNA in relation to Alz-50 immunoreactivity in Alzheimer's disease. Ann Neurol 1988; 23: 20-24.
162. Doebler JA, Rhoads RE, Anthony A et al. Neuronal RNA in Pick's and Alzheimer's diseases. Arch Neurol 1989; 46: 134-137.
163. Mann DMA, Neary D, Yates PO et al. Neurofibrillary pathology and protein synthetic capability in nerve cells in Alzheimer's disease. Neuropath Appl Neurobiol 1981; 7: 37-47.
164. Mann DMA, Neary D, Yates PO et al. Alterations in protein synthetic capability in nerve cells in Alzheimer's disease. J Neurol Neurosurg Psychiatry 1981; 44: 97-102.
165. Scheibel ME, Lindsay RD, Tomiyasu U et al. Progressive dendritic changes in the aging human cortex. Exp Neurol 1975; 47: 392-403.
166. Scheibel ME, Lindsay RD, Tomiyasu U et al. Progressive dendritic changes in the aging human limbic system. Exp Neurol 1976; 53: 420-430.
167. Scheibel ME, Tomiyasu U, Scheibel AB. The aging human Betz cell. Exp Neurol 1977; 56: 598-609.
168. Buell SJ, Coleman PD. Dendritic growth in the aged human brain and failure of growth in senile dementia. Science 1979; 206: 854-856.
169. Buell SJ, Coleman PD. Quantitative evidence for selective dendritic growth in normal human aging but not in senile dementia. Brain Res 1981; 214: 23-41.
170. Flood DG, Buell SJ, Horwitz GJ et al. Dendritic extent in human dentate gyrus granule cells in normal aging and senile dementia. Brain Res 1987; 402: 205-216.
171. Flood DG, Guarnaccia M, Coleman PD. Dendritic extent in human CA2/3 hippocampal pyramidal neurones in normal aging and senile dementia. Brain Res 1987; 409: 88-96.
172. Braak H, Braak E, Grundke-Iqbal I et al. Occurrence of neuropil threads in the senile human brain and in Alzheimer's disease: a third location of paired helical filaments outside of neurofibrillary tangles and neuritic plaques. Neurosci Lett 1986; 65: 351-355.
173. Ihara Y. Massive somatodendritic sprouting of cortical neurons in Alzheimer's disease. Brain Res 1988; 459: 138-144.
174. Terry RD, Masliah E, Salmon P et al. Physical basis of cognitive alterations in Alzheimer's disease: synapse loss is the major correlate of cognitive impairment. Ann Neurol 1991; 30: 572-580.
175. Masliah E, Terry RD, DeTeresa RM et al. Immunohistochemical quantification of the synapse-related protein synaptophysin in Alzheimer disease. Neurosci Lett 1989; 103: 234-239.
176. Honer WG, Dickson DW, Gleeson J et al. Regional synaptic pathology in Alzheimer's pathology. Neurobiol Aging 1992; 13: 375-382.
177. Hamos JE, DeGennaro LJ, Drachman DA. Synaptic loss in Alzheimer's disease and other dementias. Neurology 1989; 39: 355-361.
178. Zhan S-S, Beyreuther K, Schmitt HP. Quantitative assessment of the synaptophysin immuno-reactivity of the cortical neuropil in various neurodegenerative disorders with

dementia. Dementia 1993; 4: 66-74.
179. Brun A, Liu X, Erikson C. Synapse loss and gliosis in the molecular layer of the cerebral cortex in Alzheimer's disease and in frontal lobe degeneration. Neurodegeneration 1995; 4: 171-177.
180. Heinonen O, Lehtovirta M, Soininen H et al. Alzheimer pathology of patients carrying apolipoprotein E E4 allele. Neurobiol Aging 1995; 16: 505-513.
181. Masliah E, Terry RD, Mallory M et al. Diffuse plaques do not accentuate synapse loss in Alzheimer's disease. Am J Pathol 1990; 137: 1293-1297.
182. Wisniewski HM, Wegiel J, Kotula L. Some neuropathological aspects of Alzheimer's disease and its relevance to other disciplines. Neuropath Appl Neurobiol 1996; 22: 3-11.
183. Rebeck GW, Reiter JS, Strickland DK et al. Apolipoprotein E in sporadic Alzheimer's disease: Allelic variation and receptor interactions. Neuron 1993; 11: 575-580.
184. Schmechel D, Saunders AM, Strittmatter WJ et al. Increased amyloid β peptide deposition in cerebral cortex as a consequence of apolipoprotein E genotype in late-onset Alzheimer disease. Proc Natl Acad Sci (USA) 1993; 90: 9649-9653.
185. Benjamin R, Leake A, Ince PG et al. Effects of apolipoprotein E genotype on cortical neuropathology in senile dementia of the Lewy body type and Alzheimer's disease. Neurodegeneration 1995; 4: 443-448.
186. Harrington CR, Louwagie J, Rossau R et al. Influence of apolipoprotein E genotype on senile dementia of the Alzheimer and Lewy body types. Am J Pathol 1994; 145: 1472-1484.
187. Esiri MM, Pearson RCA, Powell TPS. Thecortex of the primary auditory area in Alzheimer's disease. Brain Res 1986; 366: 385-387.
188. Pearson RCA, Hiorns RW, Wilcock GK et al. Anatomical correlates of the distribution of the pathological changes in the neocortex in Alzheimer's disease. Proc Natl Acad Sci (USA) 1985; 82: 4531-4534.
189. Beach TG, McGeer EG. Lamina-specific arrangement of astrocytic gliosis and senile plaques in Alzheimer's disease visual cortex. Brain Res 1988; 463: 357-361.
190. Braak H, Braak E, Kalus P. Alzheimer's disease: areal and laminar pathology in the occipital isocortex. Acta Neuropathol 1989; 77: 494-506.
191. Rogers J, Morrison JH. Quantitative morphology and regional and laminar distributions of senile plaques in Alzheimer's disease. J Neurosci 1985; 5: 2801-2808.
192. Lewis DA, Campbell MJ, Terry RD et al. Laminar and regional distributions of neurofibrillary tangles and neuritic plaques in Alzheimer's disease: A quantitative study of visual and auditory cortices. J Neurosci 1987; 7: 1799-1808.
193. Braak H, Braak E. Neuropathological staging of Alzheimer-related changes. Acta Neuropathol 1991; 82: 239-259.
194. Bouras C, Hof PR, Morrison JH. Neurofibrillary tangle densities in the hippocampal formation in a non-demented population define subgroups of patients with differential early pathologic changes. Neurosci Lett 1993; 153: 131-135.
195. Giannakopoulos P, Hof PR, Mottier S et al. Neuropathological changes in the cerebral cortex of 1258 cases from a geriatric hospital: retrospective clinicopathological evaluation of a 10-year autopsy population. Acta Neuropathol 1994; 87: 456-468.
196. Mann DMA, Younis N, Jones D et al. The time course of pathological events concerned with plaque formation in Down's syndrome with particular reference to the involvement of microglial cells. Neurodegeneration 1992; 1: 201-215.
197. Arriagada PV, Marzloff K, Hyman BT. Distribution of Alzheimer-type pathologic changes in non-demented elderly individuals matches the pattern in Alzheimer's disease. Neurology 1992; 42: 1681-1688.
198. Mann DMA, Tucker CM, Yates PO. Alzheimer's disease: an olfactory connection? Mech Aging Dev 1988; 42: 1-15.
199. Reyes PF, Golden GT, Fagel PL et al. The prepiriform cortex in dementia of the Alzheimer type. Arch Neurol 1987; 44: 644-645.
200. Esiri MM, Wilcock GK. The olfactory bulbs in Alzheimer's disease. J Neurol Neurosurg Psychiatry 1984; 47: 56-60.
201. Ohm TG, Braak H. Olfactory bulb changes

in Alzheimer's disease. Acta Neuropathol 1987; 73: 365-369.
202. Armstrong RA, Slaven A. Does the neurodegeneration of Alzheimer's disease spread between visual cortical regions B17 and B18 via the feedforward or feedback short cortico-cortical projections? Neurodegeneration 1994; 3: 191-196.
203. De LaCoste M-C, White CL. The role of cortical connectivity in Alzheimer's disease pathogenesis: A review and model system. Neurobiol Aging 1993; 14: 1-16.
204. Ikeda S-I, Allsop D, Glenner GG. The morphology and distribution of plaque and related deposits in the brains of Alzheimer's disease and control cases: an immunohistochemical study using amyloid β protein antibody. Lab Invest 1989; 60: 113-122.
205. Bugiani O, Giaccone G, Frangione B et al. Alzheimer patients: preamyloid deposits are more widely distributed than senile plaques throughout the central nervous system. Neurosci Lett 1989; 103: 262-268.
206. Tagliavini F, Giaccone G, Frangione B et al. Preamyloid deposits in the cerebral cortex of patients with Alzheimer's disease and non demented individuals. Neurosci Lett 1988; 93: 191-196.
207. Ogomori K, Kitamoto T, Tateishi J et al. β amyloid protein is widely distributed in the central nervous system of patients with Alzheimer's disease. Am J Pathol 1989; 134: 243-251.
208. Yamaguchi H, Hirai S, Morimatsu M et al. Diffuse type of senile plaques in the brains of Alzheimer-type dementia. Acta Neuropathol 1988; 77: 113-119.
209. Yamaguchi H, Hirai S, Morimatsu M et al. A variety of cerebral amyloid deposits with brains of Alzheimer-type dementia demonstrated by β-protein immunostaining. Acta Neuropathol 1988; 76: 541-549.
210. Yamaguchi H, Hirai S, Morimatsu M et al. Diffuse type of senile plaques in the cerebellum of Alzheimer-type dementia detected by β-protein immunostaining. Acta Neuropathol 1989; 77: 314-319.
211. Wisniewski HM, Bancher C, Barcikowska M et al. Spectrum of morphological appearance of amyloid deposits in Alzheimer's disease. Acta Neuropathol 1989; 78: 337-347.
212. Barcikowska M, Wisniewski HM, Bancher C et al. About the presence of paired helical filaments in dystrophic neurites participating in the plaque formation. Acta Neuropathol 1989; 78: 225-231.
213. Price JL, Davis PB, Morris JC et al. The distribution of plaques, tangles and related immunohistochemical markers in healthy aging and Alzheimer's disease. Neurobiol Aging 1991; 12: 295-312.
214. Giaccone G, Tagliavini F, Linoli G et al. Down patients: extracellular preamyloid deposits precede neuritic degeneration and senile plaques. Neurosci Lett 1989; 97: 232-238.
215. Brilliant M, Elble RJ, Ghobrial M et al. Distribution of amyloid in the brainstem of patients with Alzheimer disease. Neurosci Lett 1992; 148: 23-26.
216. Suenaga T, Hirano A, Llena JF et al. Modified Bielschowsky and immunocytochemical studies on cerebellar plaques in Alzheimer's disease. J Neuropathol Exp Neurol 1990; 49: 31-40.
217. Suenaga T, Hirano A, Llena JF et al. Modified Bielschowsky staining and immunohistochemical studies on striatal plaques in Alzheimer's disease. Acta Neuropathol 1990; 80: 280-286.
218. Cole G, Williams P, Alldrick D et al. Amyloid plaques in the cerebellum in Alzheimer's disease. Clin Neuropathol 1989; 4: 188-191.
219. Cole G, Neal JW, Singrao SK et al. The distribution of amyloid plaques in the cerebellum and brain stem in Down's syndrome and Alzheimer's disease: a light microscopical analysis. Acta Neuropathol 1993; 85: 542-552.
220. Joachim CL, Morris JH, Selkoe DJ. Diffuse senile plaques occur commonly in the cerebellum in Alzheimer's disease. Am J Pathol 1989; 135: 309-319.
221. Mann DMA, Iwatsubo T, Snowden JS. Atypical amyloid (Aβ) deposition in the cerebellum in Alzheimer's disease: an immunohistochemical study using end-specific Aβ monoclonal antibodies. Acta Neuropathol 1996; 91: 647-653.
222. Leuba G, Saini K. Pathology of subcortical visual centres in relation to cortical degen-

eration in Alzheimer's disease. Neuropath Appl Neurobiol 1995; 21: 410-422.
223. Iseki E, Matsushita M, Kosaka K et al. Distribution and morphology of brain stem plaques in Alzheimer's disease. Acta Neuropathol 1989; 78: 131-136.
224. Standaert DG, Lee VM-Y, Greenberg BD et al. Molecular features of hypothalamic plaques in Alzheimer's disease. Am J Pathol 1991; 139: 681-691.
225. MacKenzie IRA, McKelvie PA, Beyreuther K et al. βA4 amyloid protein deposition in the cerebellum in Alzheimer's disease and Down's syndrome. Dementia 1991; 2: 237-242.
226. Braak H, Braak E, Bohl J et al. Alzheimer's disease: amyloid plaques in the cerebellum. J Neurol Sci 1989; 93: 277-287.
227. McKee AC, Kosik KS, Kowall NW. Neuritic pathology and dementia in Alzheimer's disease. Ann Neurol 1991; 30: 156-165.
228. Pro JD, Smith CH, Sumi SM. Presenile Alzheimer disease: amyloid plaques in the cerebellum. Neurology 1980; 30: 820-825.
229. Morioka E. Senile amyloid changes in the cerebellum with special reference to senile plaques and amyloid angiopathy. Neuropathol 1985; 6: 313-323.
230. Akiyama H, Yamada T, McGeer PL et al. Columnar arrangement of β-amyloid protein deposits in the cerebral cortex of patients with Alzheimer's disease. Acta Neuropathol 1993; 85: 400-403.
231. Kosik KS, Rogers J, Kowall NW. Senile plaques are located between apical dendritic clusters. J Neuropathol Exp Neurol 1987; 46: 1-11.
232. Duyckaerts C, Hauw J-J, Bastenaire F et al. Laminar distribution of neocortical senile plaques in senile dementia of the Alzheimer type. Acta Neuropathol 1986; 70: 249-256.
233. Gentleman SM, Allsop D, Bruton CJ et al. Quantitative differences in the deposition of βA4 protein in the sulci and gyri of frontal temporal isocortex in Alzheimer's disease. Neurosci Lett 1992; 136: 27-30.
234. Clinton J, Roberts GW, Gentleman SM et al. Differential pattern of β-amyloid protein deposition within cortical sulci and gyri in Alzheimer's disease. Neuropath Appl Neurobiol 1993; 19: 277-281.
235. McKenzie JE, Gentleman SM, Royston MC et al. Quantification of plaque types in sulci and gyri of medial frontal lobe in patients with Alzheimer's disease. Neurosci Lett 1992; 143: 23-26.
236. Hassler D. Arterial deformities in senile brains. Acta Neuropathol 1967; 8: 219-229.
237. Ravens JR. Vascular changes in the human senile brain. In: Cervos-Navarra J, ed. Pathology of Cerebrospinal Microcirculation. ed. New York: Raven Press, 1978: 487-501.
238. Fischer VW, Siddiqi A, Yusufaly Y. Altered angioarchitecture in selected areas of brains with Alzheimer's disease. Acta Neuropathol 1990; 79: 672-679.
239. Buee L, Hof PR, Bouras C et al. Pathological alterations of the cerebral microvasculature in Alzheimer's disease and related dementing disorders. Acta Neuropathol 1994; 87: 469-480.
240. Mann DMA, Eaves NR, Marcyniuk B et al. Quantitative changes in cerebral cortical microvasculature in aging and dementia. Neurobiol Aging 1986; 7: 321-330.
241. Bell MA, Ball MJ. Morphometric comparison of hippocampal microvasculature in aging and demented people: diameter and densities. Acta Neuropathol 1981; 53: 299-318.
242. Kidd M. Alzheimer's Disease–An electron microscopical study. Brain 1964; 87: 307-318.
243. Mancardi GL, Perdelli F, Rivano C et al. Thickening of the basement membrane of cortical capillaries in Alzheimer's disease. Acta Neuropathol 1980; 49: 79-83.
244. Scheibel AB, Duong T, Jacobs R. Alzheimer's disease as a capillary dementia. Ann Med 1989; 21: 103-107.
245. Mann DMA, Purkiss MS, Bonshek RE et al. Lectin histochemistry of cerebral microvessels in aging, Alzheimer's disease and Down's syndrome. Neurobiol Aging 1992; 13: 137-143.
246. Mancardi GL, Liwnicz BH, Mandybur TI. Fibrous astrocytes in Alzheimer's disease and senile dementia of Alzheimer's type. Acta Neuropathol 1983; 61: 76-80.
247. Schechter R, Yen S-HC, Terry RD. Fibrous astrocytes in senile dementia of the Alzheimer type. J Neuropathol Exp Neurol 1981; 40: 95-101.

248. Alafuzoff I, Adolfsson R, Bucht G et al. Albumin and immunoglobulin in plasma and cerebrospinal fluid, and blood-cerebrospinal fluid barrier function in patients with dementia of Alzheimer type and multi-infarct dementia. J Neurol Sci 1983; 60: 465-472.
249. Blennow K, Wallin A, Fredman P et al. Blood-brain barrier disturbance in patients with Alzheimer's disease is related to vascular factors. Acta Neurol Scand 1990; 81: 323-326.
250. Elovaara I, Icen A, Palo J et al. CSF in Alzheimer's disease: Studies on blood-brain barrier function and intrathecal protein synthesis. J Neurol Sci 1985; 70: 73-80.
251. Mann DMA, Davies JS, Yates PO et al. Immunohistochemical staining of senile plaques. Neuropath Appl Neurobiol 1982; 8: 55-61.
252. Wisniewski HM, Kozlowski PB. Evidence for blood-brain barrier changes in senile dementia of the Alzheimer type (SDAT). Ann NY Acad Sci 1982; 396: 119-129.
253. Ishii T, Haga S. Immuno-electron microscopic localization of immunoglobulins in amyloid fibrils of senile plaques. Acta Neuropathol 1976; 36: 243-249.
254. Alafuzoff I, Adolfsson R, Grundke-Iqbal I et al. Perivascular deposits of serum proteins in cerebral cortex in vascular dementia. Acta Neuropathol 1985; 66: 292-298.
255. Wallin A, Blennow K, Fredman P et al. Blood-brain barrier function in vascular dementia. Acta Neurol Scand 1990; 81: 318-322.
256. Glenner GG. On causative theories in Alzheimer's disease. Hum Pathol 1985; 16: 433-435.
257. Vinters HV. Cerebral amyloid angiopathy. A critical review. Stroke 1987; 18: 311-324.
258. Esiri M, Wilcock GK. Cerebral amyloid angiopathy in dementia and old age. J Neurol Neurosurg Psychiatry 1986; 49: 1221-1226.
259. Simchowitz T. Histologische studien uber die senile demenz. In: Nissl F, Alzheimer A, eds. Histologischc und Histopathologische Arbeiten. Jena: Gustav Fischer, 1911: 267-444.
260. Tomlinson BE, Kitchener D. Granulovacuolar degeneration of hippocampal pyramidal cells. J Pathol 1971; 106: 165-185.
261. Ball MJ. Topographic distribution of neurofibrillary tangles and granulovacuolar degeneration in the hippocampal cortex of aging and demented patients. A quantitative study. Acta Neuropathol 1978; 42: 73-80.
262. Ball MJ, Lo P. Granulovacuolar degeneration in the aging brain and in dementia. J Neuropathol Exp Neurol 1977; 36: 474-487.
263. Xu M, Shibayama H, Kobayashi H et al. Granuolovacuolar degeneration in the hippocampal cortex of aging and demented patients–a quantitative study. Acta Neuropathol 1992; 85: 1-9.
264. Woodward JS. Clinicopathological significance of granulovacuolar degeneration in Alzheimer's disease. J Neuropathol Exp Neurol 1962; 21: 85-91.
265. Gibson PH, Tomlinson BE. Numbers of Hirano bodies in the hippocampus of normal and demented people with Alzheimer's disease. J Neurol Sci 1977; 33: 199-206.
266. Ogata J, Budzilovitch GN, Cravioto H. A study of rod-like structures (Hirano bodies) in 240 normal and pathological brains. Acta Neuropathol 1972; 21: 61-67.
267. Berk ML, Finkelstein JA. An autoradiographic determination of the afferent projections of the suprachiasmatic nucleus of the hypothalamus. Brain Res 1981; 226: 1-13.
268. Goudsmit E, Hopman MA, Fliers E et al. The supraoptic and paraventricular nuclei of the human hypothalamus in relation to sex, age and Alzheimer's disease. Neurobiol Aging 1990; 11: 529-536.
269. Rudelli RD, Ambler MW, Wisniewski HM. Morphology and distribution of Alzheimer neuritic (senile) and amyloid plaques in striatum and diencephalon. Acta Neuropathol 1984; 64: 273-281.
270. Simpson J, Yates CM, Watts AG et al. Congo Red birefringent structures in the hypothalamus in senile dementia of the Alzheimer type. Neuropath Appl Neurobiol 1988; 14: 381-393.

271. McDuff T, Sumi SM. Subcortical degeneration in Alzheimer's disease. Neurology 1985; 35: 123-126.
272. Saper CB, German DC. Hypothalamic pathology in Alzheimer's disease. Neurosci Lett 1987; 74: 364-370.
273. Mann DMA. The pathological association between Down syndrome and Alzheimer disease. Mech Aging Dev 1988; 43: 99-136.
274. Mann DMA. Alzheimer's disease and Down's syndrome. Histopath 1988; 13: 125-138.
275. Mann DMA, Esiri MM. Regional acquisition of plaques and tangles in Down's syndrome patients under 50 years of age. J Neurol Sci 1989; 89: 169-179.
276. Allsop D, Haga S-I, Haga C et al. Early senile plaques in Down's syndrome brains show a clear relationship with cell bodies of neurones. Neuropath Appl Neurobiol 1989; 15: 531-542.
277. Rumble B, Retallack R, Hilbich C et al. Amyloid (A4) protein and its precursor in Down's syndrome and Alzheimer's disease. N Engl J Med 1989; 320: 1446-1452.
278. Lemere CA, Blusztajn JK, Yamaguchi H et al. Sequence of deposition of heterogeneous amyloid β-peptides and APO E in Down syndrome: Implications for initial events in amyloid plaque formation. Neurobiol Dis 1996; 3: 16-32.
279. Royston MC, Kodical NS, Mann DMA et al. Quantitative analysis of β-amyloid deposition in Down's syndrome using computerized image analysis. Neurodegeneration 1994; 3: 43-51.
280. Mann DMA, Iwatsubo T. Diffuse plaques in the cerebellum and corpus striatum in Down's syndrome contain amyloid β protein (Aβ) only in the form of Aβ42(43). Neurodegeneration 1996; 5: 115-120.
281. Kida E, Wisniewski KE, Wisniewski HM. Early amyloid-β deposits show different immunoreactivity to the amino- and carboxy-terminal regions of β-peptide in both Alzheimer's disease and Down's syndrome brain. Neurosci Lett 1995; 193: 1-4.
282. Motte J, Williams RS. Age-related changes in the density and morphology of plaques and neurofibrillary tangles in Down's syndrome brains. Acta Neuropathol 1989; 77: 535-546.
283. Mann DMA, Brown AMT, Prinja D et al. An analysis of the morphology of senile plaques in Down's syndrome patients of different ages using immunocytochemical and lectin histochemical methods. Neuropath Appl Neurobiol 1989; 15: 317-329.
284. Mann DMA, Iwatsubo T, Fukumoto H et al. Microglial cells and amyloid β protein (Aβ) deposition; association with Aβ40 containing plaques. Acta Neuropath 1995; 90: 472-477.
285. Wisniewski HM, Rabe A. Discrepancy between Alzheimer type neuropathology and dementia in persons with Down's syndrome. Ann NY Acad Sci 1986; 477: 247-259.
286. Oliver C, Holland AJ. Down's syndrome and Alzheimer's disease: a review. Psychol Med 1986; 16: 307-322.
287. Mann DMA, Yates PO, Marcyniuk B et al. Loss of nerve cells from cortical and subcortical areas in Down's syndrome patients at middle age: quantitative comparisons with younger Down's patients and patients with Alzheimer's disease. J Neurol Sci 1987; 80: 79-89.
288. Iwatsubo T, Mann DMA, Odaka A et al. Amyloid β protein (Aβ) deposition: Aβ42(43) precedes Aβ40 in Down syndrome. Ann Neurol 1995; 37: 294-299.
289. Davies CA, Mann DMA. Co-localization of apolipoprotein E and amyloid β protein in Down's syndrome. Ann Neurol 1996; (in press):.
290. Snow AD, Mar H, Nochlin D et al. Early accumulation of heparan sulphate in neurones and in the β amyloid protein containing lesions of Alzheimer's disease and Down's syndrome. Am J Pathol 1990; 137: 1253-1270.
291. Kalaria RN, Perry G. Amyloid P component and other acute phase proteins associated with cerebellar Aβ deposits in Alzheimer's disease. Brain Res 1993; 631: 151-155.
292. Snow AD, Seikiguchi RT, Nochlin D et al. Heparan sulphate proteoglycan in diffuse plaques of hippocampus but not of cerebellum in Alzheimer's disease brain. Am J Pathol 1994; 144: 337-347.

293. Teller JK, Russo C, de Busk LM et al. Presence of soluble amyloid β peptide precedes amyloid plaque formation in Down's syndrome. Nature Medicine 1996; 2: 93-95.
294. Fukumoto H, Asami-Odaka A, Suzuki N et al. Amyloid β protein (Aβ) deposition in normal aging has the same characteristics as that in Alzheimer's disease: predominance of Aβ42(43) and association of Aβ40 with cored plaques. Am J Pathol 1996; 148: 259-265.
295. Byrne EJ, Lennox G, Godwin-Austen RB et al. Dementia associated with cortical Lewy bodies: proposed clinical diagnostic criteria. Dementia 1991; 2: 283-284.
296. McKeith IG, Perry RH, Fairbairn AF et al. Clinical diagnostic criteria for Lewy body dementia. Dementia 1992; 3: 251-252.
297. Dickson DW, Crystal H, Mattiace LA et al. Diffuse Lewy body disease: light and electron microscopic immunocytochemistry of senile plaques. Acta Neuropathol 1989; 78: 572-584.
298. Dickson DW, Ruan D, Crystal H et al. Hippocampal degeneration differentiates diffuse Lewy body disease (DLDB) from Alzheimer's disease: light and electron microscopic immunocytochemistry of CA2-3 neurites specific to DLDB. Neurology 1991; 41: 1402-1409.
299. Gentleman SM, Williams B, Royston MC et al. Quantification of β/A4 protein deposition in the medial temporal lobe: a comparison of Alzheimer's disease and senile dementia of the Lewy body type. Neurosci Lett 1992; 142: 9-12.
300. Ince P, Irving D, MacArthur F et al. Quantitative neuropathological study of Alzheimer-type pathology in the hippocampus: comparison of senile dementia of Alzheimer type, senile dementia of Lewy body type, Parkinson's disease and non-demented elderly control patients. J Neurol Sci 1991; 106: 142-152.
301. Perry RH, Irving D, Blessed G et al. Senile dementia of Lewy body type. A clinically and neuropathologically distinct form of Lewy body dementia in the elderly. J Neurol Sci 1990; 95: 119-139.
302. Hansen L, Salmon D, Galasko D et al. The Lewy body variant of Alzheimer's disease: A clinical and pathological entity. Neurology 1990; 40: 1-8.
303. Lippa CF, Smith TW, Swearer JM. Alzheimer's disease and Lewy body disease: A comparative clinicopathological study. Ann Neurol 1994; 35: 81-88.
304. Mann DMA, Snowden JS. The topographic distribution of brain atrophy in cortical Lewy body disease: comparison with Alzheimer's disease. Acta Neuropathol 1995; 89: 178-183.
305. Benjamin R, Leake A, Edwardson JA et al. Apolipoprotein E genes in Lewy body and Parkinson's disease. Lancet 1994; 343: 1565.
306. Pickering-Brown S, Mann DMA, Bourke JP et al. Apolipoprotein E4 and Alzheimer's disease pathology in Lewy body disease and in other β-amyloid forming diseases. Lancet 1994; 343: 1155.
307. Galasko D, Saitoh T, Xia Y et al. The apolipoprotein E allele E4 is over-represented in patients with the Lewy body variant of Alzheimer's disease. Neurology 1994; 44: 1950-1951.
308. Luyendijk W, Bots GTAM, Vegtner-van der Vlis M et al. Hereditary cerebral haemorrhage caused by cortical amyloid angiopathy. J Neurol Sci 1988; 85: 267-280.
309. Van Duinen SG, Castano EM, Prelli F et al. Hereditary cerebral haemorrhage with amyloidosis in patients of the Dutch origin is related to Alzheimer disease. Proc Natl Acad Sci (USA) 1987; 84: 5991-5994.
310. Maat-Schieman MLC, Van Duinen SG, Haan J et al. Morphology of cerebral plaque-like lesions in hereditary cerebral haemorrhage with amyloidosis (Dutch). Acta Neuropathol 1992; 84: 674-679.
311. Maat-Schieman MLC, Rozemuller AJ, Van Duinen SG et al. Microglia in diffuse plaques in hereditary cerebral haemorrhage with amyloidosis (Dutch). An immunohistochemical study. J Neuropathol Exp Neurol 1994; 53: 483-491.
312. Maat-Schieman MLC, Radder CM, Van Duinen SG et al. Hereditary cerebral haemorrhage with amyloidosis (Dutch): a model for argyrophilic plaque formation without neurofibrillary pathology. Acta Neuropathol 1994; 88: 371-378.

313. Rozemuller JM, Bots GTAM, Roos RAC et al. Acute phase proteins but not activated microglial cells are present in parenchymal β/A4 deposits in the brains of patients with hereditary cerebral haemorrhage with amyloidosis-Dutch type. Neurosci Lett 1992; 140: 137-140.
314. Mann DMA, Iwatsubo T, Ihara Y et al. Predominant deposition of Aβ42(43) in plaques in cases of Alzheimer's disease and hereditary cerebral haemorrhage associated with mutations in the amyloid precursor protein gene. Am J Pathol 1996; 148: 1257-1266.
315. Wisniewski T, Frangione B. Molecular biology of Alzheimer's amyloid–Dutch variant, Mol. Neurobiol 1992; 6: 75-86.
316. Levy E, Carman MD, Fernandez-Madrid IJ et al. Mutation of the Alzheimer's disease amyloid gene in hereditary cerebral haemorrhage, Dutch type. Science 1990; 248: 1124-1126.
317. Van Broeckhoven C, Haan J, Bakker E et al. Amyloid β protein precursor gene and hereditary cerebral hemorrhage with amyloidosis (Dutch). Science 1990; 248: 1120-1128.
318. Clements A, Walsh DM, Williams CH et al. Effects of the mutations Glu^{22} to Gln and Ala^{21} to Gly on the aggregation of a synthetic fragment of the Alzheimer's amyloid β/A4 peptide. Neurosci Lett 1993; 161: 17-20.
319. Fabian H, Szendrei GI, Mantsch HH et al. Comparative analysis of human and Dutch-type Alzheimer β-amyloid peptides by infrared spectroscopy and circular dichroism. Biochem Biophys Res Commun 1993; 191: 232-239.
320. Wisniewski T, Ghiso J, Frangione B. Peptides homologous to the amyloid protein of Alzheimer's disease containing a glutamine for glutamic acid substitution have accelerated amyloid fibril formation. Biochem Biophys Res Commun 1991; 179: 1247-1254.
321. Hendriks L, Van Duijn CM, Cras P et al. Presenile dementia and cerebral haemorrhage linked to a mutation at codon 692 of the β-amyloid precursor protein gene. Nature Genet 1992; 1: 218-221.
322. Mortimer JA, French LR, Hutton JT et al. Head injury as a risk factor for Alzheimer's disease. Neurology 1985; 35: 264-267.
323. Mayeux R, Ottman R, Tang M-X et al. Genetic susceptibility and head injury as risk factors for Alzheimer's disease among community-dwelling elderly persons and their first-degree relatives. Ann Neurol 1993; 33: 494-501.
324. Corsellis JAN, Brierley JB. Observations on the pathology of insidious dementia following head injury. J Ment Sci 1959; 105: 714-720.
325. Rudelli R, Strom JO, Welch PT et al. Posttraumatic premature Alzheimer's disease: Neuropathologic findings and pathogenetic considerations. Arch Neurol 1982; 39: 570-575.
326. Clinton J, Ambler MW, Roberts GW. Post-traumatic Alzheimer's disease: preponderance of a single plaque type. Neuropath Appl Neurobiol 1991; 17: 69-74.
327. Roberts GW, Gentleman SH, Lynch A et al. β/A4 amyloid protein deposition in the brain after head injury. Lancet 1991; 338: 1422-1423.
328. Nicoll JAR, Roberts GW, Graham DI. Apolipoprotein E e4 allele is associated with deposition of amyloid β-protein following head injury. Nature Medicine 1995; 1: 135-137.
329. Graham DI, Gentleman SM, Lynch A et al. Distribution of β-amyloid protein in the brain following severe head injury. Neuropath Appl Neurobiol 1995; 21: 27-34.
330. Gentleman SM, Nash MJ, Sweeting CJ et al. β-amyloid precursor protein (β-APP) as a marker for axonal injury after head injury. Neurosci Lett 1993; 160: 139-144.
331. Martland HS. Punch drunk. J Am Med Ass 1928; 91: 1103-1107.
332. Courville CB. Punch drunk. Bull Los Angeles Neurol Soc 1962; 27: 160-168.
333. Roberts AH. Brain damage in boxers. London, Pitman Medical Scientific Publication, 1969.
334. Guterman A, Smith RW. Neurological sequelae of boxing. Sports Med 1987; 4: 194-210.
335. Roberts GW, Whitwell HL, Acland PR et al. Dementia in a punch-drunk wife. Lancet 1990; i: 918-919.

336. Roberts GW, Allsop D, Bruton C. The occult aftermath of boxing. J Neurol Neurosurg Psychiatry 1990; 53: 373-378.
337. Tokuda T, Ikeda S, Yanagisawa N et al. Re-examination of ex-boxers' brains using immunohistochemistry with antibodies to amyloid β-protein and tau protein. Acta Neuropathol 1991; 82: 280-285.
338. Roberts GW. Immunocytochemistry of neurofibrillary tangles in dementia pugilistica and Alzheimer's disease: evidence for common genesis. Lancet 1988; ii: 1456-1458.
339. Wisniewski KE, Jervis GA, Moretz RC et al. Alzheimer neurofibrillary tangles in diseases other than senile and presenile dementia. Ann Neurol 1979; 7: 462-465.
340. Dale GE, Leigh PN, Luthert P et al. Neurofibrillary tangles in dementia pugilistica are ubiquitinated. Psychiat 1991; 54: 116-118.
341. Corsellis JAN, Bruton CJ, Freeman-Browne D. The aftermath of boxing. Psychol Med 1973; 3: 270-303.
342. Hof PR, Bouras C, Buee L et al. Differential distribution of neurofibrillary tangles in the cerebral cortex of dementia pugilistica and Alzheimer's disease cases. Acta Neuropathol 1992; 85: 23-30.
343. Geddes JF, Vowles GH, Robinson SFD et al. Neurofibrillary tangles, but not Alzheimer-type pathology, in a young boxer. Neuropath Appl Neurobiol 1996; 22: 12-16.
344. Candy JM, McArthur FK, Oakley AE et al. Aluminum accumulation in relation to senile plaque and neurofibrillary tangle formation in the brains of patients with renal failure. J Neurol Sci 1992; 107: 210-218.
345. Harrington CR, Wischik CM, McArthur FK et al. Alzheimer's disease-like changes in tau protein processing: association with aluminum accumulation in brains of renal dialysis patients. Lancet 1994; 343: 993-997.
346. Burks JS, Alfrey AC, Huddlestone J et al. A fatal encephalopathy in chronic haemodialysis patients. Lancet 1976; i: 764-768.
347. Perl DP, Brody AR. Alzheimer's disease: x-ray spectrometric evidence of aluminum accumulation in neurofibrillary tangle-bearing neurons. Science 1980; 208: 297-299.
348. Candy JM, Oakley AE, Klinowski J et al. Aluminosilicates and senile plaque formation in Alzheimer's disease. Lancet 1986; 1: 354-357.
349. Chafi AH, Hauw J-J, Rancurel G et al. Absence of aluminum in Alzheimer's disease brain tissue: electron microprobe and ion microprobe studies. Neurosci Lett 1991; 123: 61-64.
350. Landsberg JP, McDonald B, Watt JF. Absence of aluminum in neuritic plaque cores in Alzheimer's disease. Nature 1992; 360: 65-68.
351. Masters CL, Simms G, Weinman NA et al. Amyloid plaque core protein in Alzheimer disease and Down syndrome. Proc Natl Acad Sci (USA) 1985; 82: 4245-4249.
352. Tabaton M, Mandybur TI, Perry G et al. The widespread alteration of neurites in Alzheimer's disease may be unrelated to amyloid deposition. Ann Neurol 1989; 26: 771-778.
353. Mandybur TI, Nagpaul AS, Pappas Z et al. Alzheimer neurofibrillary change in subacute sclerosing panencephalitis. Ann Neurol 1977; 1: 103-107.
354. Mandybur TI. The distribution of Alzheimer's neurofibrillary tangles and gliosis in chronic subacute sclerosing panencephalitis. Acta Neuropathol 1990; 80: 307-310.
355. McQuaid S, Allen IV, McMahon J et al. Association of measles virus with neurofibrillary tangles in subacute sclerosing panencephalitis: a combined *in situ* hybridization and immunocytochemical investigation. Neuropath Appl Neurobiol 1994; 20: 103-110.
356. Hirano A, Malamud N, Elizan TS et al. Amyotrophic lateral sclerosis and parkinsonism-dementia complex on Guam. Arch Neurol 1966; 2: 225-232.
357. Geddes JF, Hughes AJ, Lees AJ et al. Pathological overlap in cases of parkinsonism associated with neurofibrillary tangles. A study of postencephalic parkinsonism and comparison with progressive supranuclear palsy and Guamanian parkinsonism-dementia complex. Brain 1993; 116: 281-302.
358. Buee L, Perez-Tur J, Leveugle B et al. Apolipoprotein E in Guamanian amyo-

trophic lateral sclerosis/parkinsonism-dementia complex: genotype analysis and relationships to neuropathological changes. Acta Neuropathol 1996; (In press):.
359. Kiuchi A, Otsuka N, Namba Y et al. Presenile appearance of abundant neurofibrillary tangles without senile plaques in the brain in myotonic dystrophy. Acta Neuropathol 1991; 82: 1-5.
360. Giaccone G, Tagliavini L, Verga L et al. Neurofibrillary tangles of the Indiana kindred of Gerstmann-Straussler-Sheinker disease share antigenic determinants with those of Alzheimer's disease. Brain Res 1990; 530: 325-329.
361. Namba Y, Kawatsu K, Izumi S et al. Neurofibrillary tangles and senile plaques in brain of elderly leprosy patients. Lancet 1992; 340: 978.
362. Cullen KM, Halliday GM. Neurofibrillary tangles in chronic alcoholics. Neuropath Appl Neurobiol 1995; 21: 312-318.
363. Joachim CL, Mori H, Selkoe DJ. Amyloid β-protein deposition in tissues other than brain in Alzheimer's disease. Nature 1989; 341: 226-230.
364. Ikeda M, Shoji M, Yamaguchi H et al. Diagnostic significance of skin immunolabelling with antibody against native cerebral amyloid in Alzheimer's disease. Neurosci Lett 1993; 150: 159-161.
365. Soininen H, Syrjanen S, Heinonen O et al. Amyloid β-protein deposition in skin of patients with dementia. Lancet 1992; 339: 245.
366. Tabaton M, Cammarata S, Mancardi GL et al. Abnormal tau-reactive filaments in olfactory mucosa in biopsy specimens of patients with probable Alzheimer's disease. Neurology 1991; 41: 391-394.
367. Talamo BR, Rudel RA, Kosik KS et al. Pathological changes in olfactory neurons in patients with Alzheimer's disease. Nature 1989; 337: 736-739.
368. Johnston JA, Cowburn RF, Norgren S et al. Increased β amyloid release and levels of amyloid precursor protein (APP) in fibroblast cell lines obtained from family members with the Swedish APP670.671 mutation. FEBS Lett 1994; 354: 274-278.
369. Scheuner D, Eckman C, Jensen M et al. The amyloid β protein deposited in the senile plaques of Alzheimer's disease is increased *in vivo* by the presenilin 1 and 2 and *APP* mutations linked to familial Alzheimer's disease. Nature Medicine 1996; 2: 864-870.
370. Bush AI, Martins RN, Rumble B et al. The amyloid precursor protein of Alzheimer's disease is released by human platelets. J Biol Chem 1990; 265: 15977-15983.
371. Smith RP, Higuchi DA, Broze GJ. Platelet coagulation factor XIa-inhibitor, a form of Alzheimer's amyloid precursor protein. Science 1990; 248: 1126-1128.
372. Beer J, Masters CL, Beyreuther K. Cells from peripheral tissues that exhibit high APP expression are characterized by their high membrane fusion activity. Neurodegeneration 1995; 4: 51-59.
373. Li JC, Kaminskas E. Deficient repair of DNA lesions in Alzheimer's disease fibroblasts. Biochem Biophys Res Commun 1985; 129: 733-738.
374. Bartha E, Szelenyi J, Szilagyi K. Altered lymphocyte acetylcholinesterase activity in patients with senile dementia. Neurosci Lett 1987; 79: 190-194.
375. Kessler JA. Deficiency of a cholinergic differentiating factor in fibroblasts of patients with Alzheimer's disease. Ann Neurol 1987; 21: 95-98.
376. Peterson C, Ratan R, Shelanski M et al. Cytosolic free calcium and cell spreading decrease in fibroblasts from aged and Alzheimer donors. Proc Natl Acad Sci (USA) 1986; 83: 7999-8001.
377. Ueda K, Cole G, Sundsmo M et al. Decreased adhesiveness of Alzheimer's disease fibroblasts: Is amyloid β-protein precursor involved? Ann Neurol 1989; 25: 246-251.
378. Tesco G, Latorica S, Piersanti P et al. Free radical injury in skin cultured fibroblasts from Alzheimer's disease patients. Ann NY Acad Sci 1992; 673: 149-153.
379. Tesco G, Vergelli M, Amaducci L et al. Growth properties of familial Alzheimer skin fibroblasts during *in vitro* aging. Exp Gerontol 1993; 28: 51-58.
380. Zubenko GS, Cohen BM, Crowdon J et al. Platelet membrane fluidity in Alzheimer's disease. Lancet 1984; ii: 235.

381. Zubenko GS, Cohen BM, Reynolds CF et al. Platelet membrane fluidity in Alzheimer's disease and major depression. Am J Psychiatry 1987; 144: 860-868.
382. Hicks N, Brammer MJ, Hymas N et al. Platelet membrane properties in Alzheimer's and multi infarct dementias. Alzheimer's Disease Assoc Disord 1987; 1: 90-97.
383. Piletz JE, Sarasua P, Whitehouse P et al. Intracellular membranes are more fluid in platelets of Alzheimer's disease patients. Neurobiol Aging 1991; 12: 401-406.
384. Duguid JR, De La Paz R, DeGroot J. Magnetic resonance imaging of the midbrain in Parkinson's disease. Ann Neurol 1986; 20: 744-747.
385. Drayer BP, Olanow CW, Burger P et al. Parkinson plus syndrome: Diagnosis using high field MR imaging of the brain. Radiology 1986; 159: 493-498.
386. Olanow CW. An introduction to the free radical hypothesis in Parkinson's disease. Ann Neurol 1992; 32: 52-59.
387. Stern MB, Braffman BH, Skolnick BE et al. Magnetic resonance imaging in Parkinson's disease and parkinsonian syndromes. Neurology 1989; 39: 1524-1526.
388. Huber SJ, Chakeres DW, Paulson GW et al. Magnetic resonance imaging in Parkinson's disease. Arch Neurol 1990; 47: 735-737.
389. Earle KM. Studies in Parkinson's disease including x-ray fluorescent spectroscopy of formalin fixed brain tissue. J Neuropathol Exp Neurol 1968; 27: 1-14.
390. Dexter DT, Wells FR, Lees AJ et al. Increased nigral iron content and alterations in other metal iron occurring in brain in Parkinson's disease. J Neurochem 1989; 52: 1830-1836.
391. Sofic E, Riederer P, Heinsen H et al. Increased iron (III) and total iron content in post mortem substantia nigra of parkinsonian brain. J Neural Transm 1988; 74: 199-205.
392. Hirsch EC, Brandel JP, Galle P et al. Iron and aluminum increase in the substantia nigra of patients with Parkinson's disease: An X-ray microanalysis. J Neurochem 1991; 56: 446-451.
393. Good PF, Olanow CW, Perl DP. Neuromelanin-containing neurons of the substantia nigra accumulate iron and aluminum in Parkinson's disease: A LAMMA study. Brain Res 1992; 593: 343-346.
394. Morris CM, Edwardson JA. Iron histochemistry of the substantia nigra in Parkinson's disease. Neurodegeneration 1994; 3: 277-282.
395. Griffiths PD, Crossman AR. Distribution of iron in the basal ganglia and neocortex in postmortem tissue in Parkinson's disease and Alzheimer's disease. Dementia 1993; 4: 61-65.
396. Nahmias C, Garnett ES, Firnau G et al. Striatal dopamine distribution in parkinsonian patients during life. J Neurol Sci 1985; 69: 223-230.
397. Martin WRW, Stoessl AJ, Adam MJ et al. Positron emission tomography in Parkinson's disease: Glucose and DOPA metabolism. Adv Neurol 1986; 45: 95-98.
398. Brooks DJ, Ibanez V, Sawle GV et al. Differing patterns of striatal 18F-dopa uptake in Parkinson's disease, mutliple system atrophy, and progressive supranuclear palsy. Ann Neurol 1990; 28: 547-555.
399. Bhatt M, Snow BJ, Martin WRW et al. Positron emission tomography suggests that the rate of progression of idiopathic parkinsonism is slow. Ann Neurol 1991; 29: 673-677.
400. Kuhl DE, Metter EJ, Riege WH. Patterns of local cerebral glucose utilization determined in Parkinson's disease by the [18F] fluorodeoxyglucose method. Ann Neurol 1984; 15: 419-424.
401. Peppard RF, Martin WRW, Guttman M et al. Cerebral glucose metabolism in Parkinson's disease and the PD complex of Guam. In: Crossman A, Sambrook M, eds. Neural Mechanisms in Disorders of Movement. London: John Libbey, 1989: 445-452.
402. Gibb WRG, Lees AJ. The relevance of the Lewy body to the pathogenesis of idiopathic Parkinson's disease. J. Neurol Neurosurg Psychiat 1988; 51: 745-752.
403. Gibb WRG, Lees AJ. Anatomy, pigmentation, ventral and dorsal subpopulations of the substantia nigra, and differential cell death in Parkinson's disease. J Neurol Neurosurg Psychiatry 1991; 54: 388-396.

404. Gibb WRG. Idiopathic Parkinson's disease and the Lewy body disorders. Neuropath Appl Neurobiol 1986; 12: 223-234.
405. Fearnley JM, Lees AJ. Aging and Parkinson's disease: Substantia nigra regional selectivity. Brain 1991; 114: 2283-2301.
406. Hirsch E, Graybiel AM, Agid YA. Melanized dopaminergic neurons are differentially susceptible to degeneration in Parkinson's disease. Nature 1988; 334: 345-348.
407. Paulus W, Jellinger K. The neuropathologic basis of different clinical subgroups of Parkinson's disease. J Neuropathol Exp Neurol 1991; 50: 743-755.
408. Forno LS, Norville RL. Ultrastructure of Lewy bodies in the stellate ganglion. Acta Neuropathol 1978; 34: 183-197.
409. Gaspar P, Gray F. Dementia in idiopathic Parkinson's disease: A neuropathological study of 32 cases. Acta Neuropathol 1984; 64: 43-52.
410. Jellinger K. Neuropathological substrates of Alzheimer's disease and Parkinson's disease. J Neurol Transm Suppl 1987; 24: 109-129.
411. Jellinger K. Quantitative changes in some subcortical nuclei in aging, Alzheimer's disease and Parkinson's disease. Neurobiol Aging 1987; 8: 556-561.
412. Jellinger K. Pathology of Parkinson's disease. Changes other than the nigrostriatal pathway. Molec Chem Neuropathol 1991; 14: 153-197.
413. Greenfield JG, Bosanquet FD. The brain stem lesions in Parkinsonism. J Neurol Neurosurg Psychiatry 1953; 16: 213-226.
414. Mann DMA, Yates PO. Pathological basis for neurotransmitter changes in Parkinson's disease. Neuropath Appl Neurobiol 1983; 9: 3-19.
415. Whitehouse PJ, Hedreen JC, White CL et al. Basal forebrain neurons in the dementia of Parkinson's disease. Ann Neurol 1983; 13: 243-248.
416. Ohama E, Ikuta F. Parkinson's disease: Distribution of Lewy bodies and monoamine neuron system. Acta Neuropathol 1976; 34: 311.
417. Braak H, Braak E, Yilmazer D et al. Amygdala pathology in Parkinson's disease. Acta Neuropathol 1994; 88: 493-500.
418. Braak H, Braak E, Yilmazer D de Vos RAI et al. Nigral and extranigral lesions in Parkinson's disease. J Neural Transm 1995; 46: 15-31.
419. Braak H, Braak E. Nuclear configuration and neuronal types of the nucleus niger in the brain of the human adult. Hum Neurobiol 1986; 5: 71-82.
420. Matzuk MM, Saper CB. Preservation of hypothalamic dopaminergic neurons in Parkinson's disease. Ann Neurol 1985; 18: 552-555.
421. Agid Y, Ruberg M, Javoy-Agid F et al. Are dopaminergic neurons selectively vulnerable to Parkinson's disease? Adv Neurol 1993; 60: 148-164.
422. Forno LS, Murphy GM, Eng LF. Immunocytochemical study of Lewy bodies in sympathetic ganglia. Neurodegeneration 1992; 1: 135-144.
423. Oyanagi K, Wakabayashi K, Ohama E et al. Lewy bodies in the lower sacral parasympathetic neurons of a patient with Parkinson's disease. Acta Neuropathol 1990; 80: 558-559.
424. Wakabayashi K, Takahashi H, Ohama E et al. Parkinson's disease: an immunohistochemical study of Lewy body-containing neurons in the enteric nervous system. Acta Neuropathol 1990; 79: 581-583.
425. Bernheimer H, Birkmayer H, Hornykiewicz O et al. Brain dopamine and the syndromes of Parkinson and Huntington: Clinical, morphological and neurochemical correlations. J Neurol Sci 1973; 20: 415-455.
426. McGeer PL, Itagaki S, Akiyama H et al. Rate of cell death in Parkinsonism indicates active neuropathological process. Ann Neurol 1988; 24: 574-576.
427. German DC, Manaye KF, Smith WK et al. Mid brain dopaminergic cell loss in Parkinson's disease: computer visualization. Ann Neurol 1989; 26: 607-614.
428. Halliday GM, McRitchie DA, Cartwright H et al. Midbrain neuropathology in idiopathic Parkinson's disease and diffuse Lewy body disease. J Clin Neurosci 1996; 3: 1-9.
429. Ma SY, Collan Y, Roytta M et al. Cell counts in the substantia nigra: a comparison of single section counts and disector counts in patients with Parkinson's disease

and in controls. Neuropath Appl Neurobiol 1995; 21: 10-17.
430. Foley JM, Baxter D. On the nature of pigment granules in the cells of the locus caeruleus and substantia nigra. J Neuropathol Exp Neurol 1958; 17: 586-598.
431. Bazelon M, Fenichel GM, Randall J. Studies on neuromelanin. I.A melanin system in the human adult brain stem. Neurology 1969; 17: 512-519.
432. Mann DMA, Yates PO. Lipoprotein pigments: Their relationship to aging in the human nervous system. II–The melanin content of pigmented nerve cells. Brain 1974; 97: 489-498.
433. Mann DMA, Yates PO, Barton CM. Variations in melanin content with age in the human substantia nigra. Biochem Exp Biol 1977; 13: 137-139.
434. Mann DMA, Yates PO. The pathogenesis of Parkinson's disease. Arch Neurol 1982; 39: 545-549.
435. Graham DG, Tiffany SM, Bell WR. Auto-oxidation versus covalent binding of quinones as the mechanism of toxicity of dopamine, 6-hydroxydopamine and related compounds towards C1300 neuroblastoma cells in vitro. Molec Pharmacol 1978; 14: 644-653.
436. Mann DMA, Yates PO. The effects of aging on the pigmented nerve cells of the human locus caeruleus and substantia nigra. Acta Neuropathol 1979; 47: 93-97.
437. McGeer PL, McGeer EG, Suzuki JS. Aging and extrapyramidal function. Arch Neurol 1977; 34: 33-35.
438. Hirai S. Aging of the substantia nigra. Adv Neurol Sci 1968; 12: 845-849.
439. Langston JW, Forno LS. The hypothalamus in Parkinson's disease. Ann Neurol 1978; 3: 129-133.
440. Manaye KF, McIntire DD, Mann DMA et al. Locus coeruleus cell loss in the aging human brain: a non-random process. J Comp Neurol 1995; 358: 79-87.
441. Marcyniuk B, Mann DMA. The topography of nerve cell loss from the locus caeruleus in elderly persons. Neurobiol Aging 1989; 10: 5-9.
442. Marcyniuk B, Mann DMA, Yates PO. Topography of nerve cell loss from the locus caeruleus in middle aged persons with Down's syndrome. J Neurol Sci 1988; 83: 15-24.
443. Forno LS, Alvord EC. The pathology of Parkinsonism. In: Recent Advances in Parkinson's disease. Oxford: Blackwell Scientific Publications, 1971: 120-161.
444. Forno LS. Concentric hyaline intraneuronal inclusions of Lewy type in the brains of elderly persons (50 incidental cases); relationship to Parkinsonism. J Amer Geriat Soc 1969; 17: 557-575.
445. Forno LS, Langston JW. Lewy bodies and aging: Relation to Alzheimer's and Parkinson's diseases. Neurodegeneration 1993; 2: 19-24.

CHAPTER 4

Etiological Considerations

4.1. GENETIC FACTORS

4.1.1. Alzheimer's Disease

4.1.1.1. Introduction

It is well recognized[1-7] that the incidence of AD is greater in some families than in others; indeed about 10% of all cases of AD are thought to be inherited. In certain instances the disease may be inherited in what appears to be a classical autosomal dominant fashion, though the genetic basis for this has become apparent only during the past decade. Inheritance seems to be strongest in families where disease onset occurs before 65 years of age, though familial disease in later life may be much more common what has formerly seemed to be the case because of potentially affected individuals dying from unrelated causes before reaching the age at risk or during the early and clinically unrecognized stages of the disease.

Linkage analysis has now shown that AD is associated with four distinct chromosomal loci, these occurring on chromosome 1,[8] chromosome 14,[9-12] chromosome 19,[13] or chromosome 21,[14,15] with mutations (or polymorphisms) in these genes being directly causal or at least, especially in the case of the polymorphisms, conferring a substantially increased risk of developing the disease.

4.1.1.2. The chromosome 21 (APP) gene mutations

The genetic locus on chromosome 21 relates to the gene encoding the amyloid precursor protein (APP). (Fig. 2.5) Two separate disease causing mutations have been found; a double mutation occurring at codons 670/671, [15,16] and a point mutation at codon 717.[14,17-24] The codon 717 mutation involves three separate base changes each replacing the valine at this location by isoleucine, glycine or phenylala-

Sense and Senility: The Neuropathology of the Aged Human Brain, by David M.A. Mann.

nine.[14,17,24] Although this was the first genetic locus associated with AD to be discovered, it is now known that these codon 717 mutations are rare and exist in less than 20 families worldwide; indeed the double mutation at codons 670/671, replacing lysine and methionine by asparagine and leucine respectively, with the codon 670 change being critical, exists in only a single Swedish pedigree.

It is not known with absolute certainty that these APP gene mutations are pathogenic, though this seems highly likely since so far only a single individual bearing such a mutation has not developed clinical disease within a time period equivalent to two standard deviations of the average onset age for all other known family members with that mutation.[25] Furthermore, the APP mutations are found only in cases of AD and in individuals at 50% risk for the disease. Other mutations in the APP gene (at codons 692 and 693) confer an Alzheimer-type pathology and produce the clinical disorder of hereditary cerebral hemorrhage with amyloidosis.[26-29] (Fig. 2.5) However, mutations at codon 665 in a patient with AD,[30] codon 673 in a stroke patient,[31] codon 705 in another patient with AD,[32] codon 705 in some AD and non-AD cases,[33-35] codon 711 in a further patient with AD,[33] codon 713 in both a healthy control[32] and a schizophrenic patient,[36] codon 716 in a control,[37] as well as intron 17 changes[35] are all likely to be non-pathogenic. Haplotype analysis shows that all the AD causing APP mutations represent independent events with none of the reported families being inter-related. Despite their rarity, studies of patients and cell lines bearing these genetic changes have produced striking advances regarding the cause and pathogenesis of the disease, particularly with respect to the amyloidosis. The neuropathology of the APP_{717} mutation is generally typical of severe AD with numerous plaques and tangles but only a mild to moderate amyloid angiopathy.[20,23,38-43] In one family, however, cortical and subcortical Lewy bodies also occur[39,40] and in another much variation in the extent of Aβ deposition (but not in the density of neuritic plaques and tangles) is seen between affected family members.[44] In AD families with the APP_{717} gene mutations a particularly heavy deposition of Aβ, especially $A\beta_{42,}$ has been emphasised[45] whereas the $APP_{670/671}$ mutations appear entirely typical of AD in this respect.[16,45]

Down's syndrome is a much more common form of chromosome 21 related disorder. In about 95% of cases there is an extra full copy of chromosome 21, though in other instances genetic translocation or mosaicism leads to additional chromosomal material involving that part of chromosome 21 obligate for the production of the DS phenotype.[46] Clearly, individuals with a full trisomy will bear, and perhaps even express,[47] an additional copy of the APP gene on chromosome 21 and this is widely thought to be the explanation for the universal development of the pathology of AD among such individuals (see chapter 3).

4.1.1.3. The chromosome 1 and 14 (Presenilin) gene mutations

A more common inherited form of early onset AD is associated with linkage to chromosomes 1 or 14 (particularly). Within the last year the gene on chromosome 14 associated with AD has been identified and this has been termed the presenilin-1 (PS-1) gene.[48-50] To date, at least 27 separate mutations, that segregate with AD in well over 50 families, have been identified in this gene.[48-58] While these mutations are found throughout the PS-1 gene, their distribution is distinctive with most being clustered within exons 5 and 8 of the gene. Changes in similarly conserved regions of an analogous (to PS-1) gene on chromosome 1, termed the presenilin-2 (PS-2) gene, have been linked to AD in certain American kindreds of Volga German ancestry[8] and the mutation responsible in these families has also been identified.[49,59,60] An Italian family has been shown to bear a further mutation in this same PS-2 gene.[60] Again a typical AD pathology is present in cases with both these latter types of mutation though with regard to the PS-1 gene mutation a very heavy formation of plaques and tangles is seen and amyloid angiopathy can be severe.[61-64]

4.1.1.4. The chromosome 19 (ApoE) gene polymorphism

Many instances of late onset familial disease show linkage to chromosome 19[13] where an increased (relative to normal) frequency of the E4 allelic variant of the Apolipoprotein E gene is held to be responsible;[65-79] some cases of early onset familial disease may also be related to an increased possession of the E4 allele.[80-83]

Although, in total, changes at these four genetic loci may be responsible for causing, or predisposing to, AD in many patients it is still uncertain as to what the exact proportion of all cases of AD related to such changes might be. At least half of all cases of AD do not possess an ApoE E4 allele, being mostly of the E3/E3 genotype, and of these only relatively few bear one or other of the mutations on chromosomes 1, 14 and 21. There may, therefore, be further, major genetic risk factors still to be discovered. Some of these might even be directly causal as there are still many early onset families which apparently do not relate to any of the so-far known mutations on chromosomes 1, 14 and 21. Alternatively, other changes may act as important potential risk factors and in this latter context an association between late onset 'sporadic' AD and a common polymorphism in the promoter region of the PS-1 gene has been reported,[84] as have increases in polymorphic changes in the α-1 antichymotrypsin gene[85,86] and in the VLDL receptor gene.[87]

Thus, a substantial number of apparently sporadic cases of disease, with onset in later life, may yet turn out to have a genetic basis, this operating as a risk factor increasing the chance that disease will occur should the individual live long enough. Whether there is truly such an entity as 'sporadic AD', with its inferred environmental (non-genetic) causes, is not known. Perhaps all cases of disease will eventually be shown to be due to the (combined) effects of one or more genetic risk factors which (collectively) increase an individual's susceptibility to the development of pathology.

4.1.2. Genetic Mechanisms of Disease

Despite this abundance of genetic variation associated with AD, the pathology in all cases seems to adhere to a fairly common phenotype. Even in those early onset families associated with mutations on chromosome 1[6,88,89] chromosome 14,[61-64,90] or chromosome 21[16,23,38-43] or in DS (see Chapter 3) the pathology, in the main, seems constant with only relatively minor variations relating to the extent of amyloid angiopathy or plaque density and distribution[45,64,89] or the presence of Lewy bodies[39,40] being recorded. This suggests that all the genetic variants associated with AD feed a common pathogenetic pathway though each may operate at a different level within that pathway while driving the same overall effect. Consequently, much effort has been made in recent times into trying to understand how these particular genetic variations influence the development of the pathological changes of AD, since in these major clues as to the cause and progression of the disease are likely to lie. Transfection studies, in which the mutant gene is introduced into cell lines, studies on the natural biology of cells (fibroblasts) in culture from carriers of the mutation, and transgenic mice bearing some of the disease-associated mutations, together with autopsy studies on the brains of mutation carriers, have all played key parts in furthering our present understanding.

4.1.2.1. APP gene mutations

Initial attempts to understand the effects of the APP mutations involved transgenic mice. Several constructs have been employed (see Higgins and Cordell[91] for review) though in general the results gained so far have been disappointing. Only two models have been able to recapitulate (and then only in part) the pathological changes of AD. One model has been based on a change in the expression of APP_{751} mRNA relative to that of APP_{695},[92-94] an alteration that occurs in human AD.[95] In mice with this construct, amyloid deposits, resembling the early plaques of AD or DS, were seen but cored or neuritic plaques were absent. Although neurones immunoreactive for abnormal tau proteins were present frank NFT were not observed. A more robust amyloid deposition in the form of $A\beta_{42}$ was seen in mice bearing the APP_{717} val→phe mutation,[96] though here no tau positive neuritic or neurofibrillary changes were present.

Studies on transfected cell lines have been more informative. Although the presence of the APP_{717} val→ile and val→gly mutations do not increase the overall amount of Aβ secreted by transfected neuroblastoma cells,[97,98] within that total there is a marked shift towards a preferential production of the longer form, $A\beta_{1-42}$, at the expense of the usually predominant $A\beta._{1-40}$ Because $A\beta_{1-42}$ is known to aggregate spontaneously into fibrils with greater propensity than $A\beta_{1-40}$, once an appropriate concentration has been reached,[99] this favored secretion of $A\beta_{1-42}$ may form the basis of disease in cells bearing this particular mutation. By increasing the local secretion of soluble $A\beta_{1-42}$, a concentration threshold critical to fibrillization may be reached earlier in life. The effect of the mutation may be to promote the catabolic activity of γ-secretase, whose sites of action are close to this mutation, towards the production of Aβ species terminating at ala-42 rather than at val-40. In transgenic mice bearing the APP_{717} val→phe mutation,[96] the Aβ deposited in the tissue is, as expected, in the form of $A\beta._{1-42}$ Such a shift in metabolism, favoring $A\beta_{1-42}$ production and release, should in theory lead to an increased deposition of this peptide species within the brain in cases of human AD bearing this mutation. This does indeed seem to occur in the human APP_{717} mutations[45,100] where vast quantities of $A\beta_{42}$, and notably much more than in 'sporadic' AD, are deposited. This extra Aβ occurs mostly as diffuse (Congo red negative) plaques and the amount of $A\beta_{40}$ deposited is not increased (over values for sporadic AD) to match. Protein chemical studies[101] have likewise detected this increase in $A\beta_{42}$ relative to $A\beta_{40}$ in cases of AD due to APP_{717} val→ile mutation.

On the other hand the mutation at codons 670/671 of the APP gene involves an overall 5-8 fold increase in the total amount of Aβ secreted by neuroblastoma cells, though no disproportionate change in the quantity of $A\beta_{1-42}$ is involved.[97,98,102-104] Incidentally, cell lines bearing the double mutation for codons 670/671 and the codon 717 mutation not only produce more Aβ overall (as would be anticipated given their possession of the APP 670/671 mutation) but within that total the proportion of $A\beta_{1-42}$ relative to $A\beta_{1-40}$ is elevated (in line with possession of the codon 717 mutation).[98] Fibroblasts in culture from carriers of the codon 670/671 mutation also show increased Aβ production.[105]

The presence of the mutation at codons 670/671 is postulated to enhance the cleavage of APP by β-secretase (whose site of action is immediately adjacent to this) at the expense of γ-secretase; γ-secretase activity is unaffected, thereby explaining the lack of change in ratio between Aβ species terminating at val-40 and ala-42.[45,98] Analysis of the trafficking of APP within Madin Darby canine kidney cells[106-108] shows that normal (wild-type) APP is transported to and, following α-secretase activity, is secreted from the cell surface; surplus uncleaved molecules are internalized (recycled) and may be catabolized into Aβ by β-secretase within late endosomes/lysosomes.[109-111] However, mutant (for codons 670/671) APP becomes preferentially diverted into the same secretory vesicles in which wild-type APP is normally cleaved by α-secretase, for catabolism by β-secretase. Hence, following a favored action of β-secretase, at the expense of α-secretase, mutant APP is cleaved so as to enhance Aβ formation, rather than P3 production. This overall increase in production and secretion of both soluble $A\beta_{1-40}$

and $A\beta_{1-42}$ would be expected to elevate local tissue concentrations of both species, though given the greater capacity of the latter for fibrillization, brain deposits preferably containing this $A\beta_{1-42}$ might be anticipated. In cases of human AD due to the $APP_{670/671}$ mutation,[45] $A\beta_{42}$ is indeed the predominant peptide species deposited in the brain but, because this particular Aβ peptide is not preferentially produced relative to $A\beta_{40}$, the proportions of $A\beta_{42}$ and $A\beta_{40}$ are similar to those in sporadic AD.[45] Protein chemical studies,[112] in accord with immunohistochemistry,[45] have also shown the relative amounts of $A\beta_{42}$ and $A\beta_{40}$ in the brain in cases of $APP_{670/671}$ mutation to be similar to those in sporadic AD.

In Down's syndrome the extra copy of the APP gene is presumably transcribed and leads to an increase in APP production.[47] Subsequent catabolism of this (excessive) APP is thought to result in the early deposition of Aβ within brain tissue. Whether APP trafficking and catabolism follows a similar route in DS as in the general population is unknown though because no mutant genes (or proteins) are involved there is no reason to suspect otherwise. Immunohistochemistry[113-117] has shown the early Aβ deposits to contain solely $A\beta_{42}$ perhaps as $A\beta_{1-42}$ or $A\beta_{3-42}$. Indeed, Teller et al[118] have shown that increased levels of $A\beta_{1-42}$ are present in the extracellular fluid of young individuals with DS, even before Aβ deposits become visible (by immunohistochemistry) within the brain tissue. Hence in DS, overexpression of the APP genes[47] may lead to enhanced APP catabolism resulting in an early and excessive formation of $A\beta_{1-42}$; fibrillogenesis will follow when the requisite local concentration of Aβ is reached, this occurring many years in advance of that in the general population. Again, because no actual mutation in the APP gene is present in DS, the ratio between the amount of $A\beta_{42}$ and $A\beta_{40}$ deposited remains normal and similar to that in sporadic AD.[113]

Thus all the mutational changes involving the APP gene seem to bring about the same net effect, namely an early and increased production of $A\beta_{1-42}$, though the routes taken to achieve this end clearly differ according to the particular genetic disturbance present.

4.1.2.2. PS-1 and PS-2 genes

How the various mutations in the PS-1 and PS-2 genes present on chromosomes 14 and 1 respectively, influences the cause of AD is currently not known. The PS-1 gene encodes a 46KD protein that appears to have the characteristics of an integral, rather than a cell surface, membrane protein and which putatively spans the membrane some seven times.[48,49] At least two forms of the protein are produced by alternative splicing of a single gene product, these being designated PS-1 long (PS1-467) and PS-1 short (PS1-463).[48,49,55,119,120] The longer full-length form of 467 amino acids corresponds to the full exon sequence while the shorter form of 463 amino acids lacks a sequence of 12 bases coding for amino acids val-arg-ser-glu (VRSQ). The transcripts for PS1-467/463 appear to be expressed in all body tissues.[48,120,121] Although these transcripts are expressed throughout the brain, greatest levels of expression are seen in the cerebellum and hippocampus and more strongly so in neurones than glial cells.[121,122] A further, alternatively spliced product of 374 amino acids (PS1-374) seems to be only expressed in liver, kidney and spleen.[120] The PS-2 gene codes for a similar-sized (448 amino acids) protein and is also ubiquitously expressed.[59] This again appears to be an integral membrane protein, also spanning the membrane seven times.[59] The expression pattern of the PS-2 gene in the brain parallels that of PS-1, being likewise preferentially expressed in neurones, again most strongly in the cerebellum and hippocampus.[122] In-situ hybridization studies[123] suggest that chromosome 14-linked AD may be associated with a decreased expression of the longer transcript.

Given the general homologies between PS-1 and PS-2 genes–67% homology overall, rising to 84% within putative transmembrane domains–it is likely that the proteins formed by these highly conserved genes serve a similar or, at least, a related function. What this might be and how disturbance of function causes AD is still uncertain. Both the PS-1 and the PS-2 proteins appear to be located in the endoplasmic reticulum and Golgi apparatus[122] and may, therefore, influence the intracellular trafficking and processing (sorting) of newly formed molecules, like APP, through this network en route to their final cellular destination. It seems that full length PS-1 protein is normally post-translationarily processed by a putative enzyme 'presenilinase' to produce a 28KDa amino-terminal fragment and a 18KDa carboxyl terminal fragment.[124,125] Whether this is a step required to produce active forms of the protein or reflect its route of degradation within cells is not known. However, because this cleavage site is close to the exon 8 coded section of the PS-1 gene (where many of the mutations are clustered) it is possible that mutations at this site either favor a change in the susceptibility of PS-1 protein to cleavage by presenilinase or other intracellular enzymes, or that they inhibit this proteinase action. On one hand this might lead to an increased or decreased (respectively) activity on the part of (functional) cleaved PS-1 products or the converse if it is full-length PS-1 that is the active form of the protein. The net result of these changes may be to wrongly direct proteins like APP though the Golgi which in the case of APP may lead to a favored catabolism by β–secretase into Aβ.

Although the effects of using cellular or animal modelling of these mutations on APP metabolism and Aβ deposition have yet to be reported, there is some evidence from humans that both the PS-1 and PS-2 gene mutations, at least, might influence the deposition of Aβ. Fibroblasts from affected individuals produce in culture more Aβ than usual, especially $A\beta_{1-42}$; plasma levels of this are also elevated in carriers of this mutation.[126] Brain tissue levels of Aβ are also much elevated, compared to cases of sporadic AD, with both $A\beta_{42}$ and $A\beta_{40}$ being increased proportionately.[64] In PS-2 gene mutations, $A\beta_{42}$ is again the predominant Aβ peptide deposited in the brain, but the level of both this, and $A\beta_{40}$, is not increased beyond that seen in sporadic AD.[89]

The PS-gene mutations may, therefore, confer an intracellular defect that favors the formation of Aβ (as $A\beta_{1-42}$) perhaps by preferentially diverting newly formed APP along amyloidogenic pathways, involving lysosomes or secretory vesicles, or by failing to ensure that APP is directed along its more usual route of constitutive cleavage by α-secretase. In this respect, because the PS-2 gene mutations occur with later onset and produce a lesser pathology, than the PS-1 gene mutations, the 'gain of dysfunction' imposed by these mutations may be more damaging for changes in the PS-1 protein than it is for ones in the PS-2 protein. Furthermore, the clustering of the mutations (in the PS-1 gene at least) in conserved regions of the gene at sites putatively relating to the transmembrane domains or the hydrophilic loops suggests that the mutations may adversely affect the anchoring or insertion of proteins into the membrane. However, because both PS-1 and PS-2 gene expression is not limited only to brain regions severely involved in the pathology of AD other factors besides a mutated protein must be necessary to produce full disease expression in vulnerable areas.

Polymorphisms in the PS-1 gene[84] may confer a similar, but less aggressive, property as the actual gene mutations, thereby delaying the age at onset into later life and lessening the overall pathological change. Such polymorphic changes might act therefore as important risk factors especially in those (older) individuals with AD where no ApoE E4 allele is present.

Hence, studies of tissues or cell cultures from patients with mutations in chromosomes 1, 14 or 21-linked AD[45,64,89,100,101,105,112] or individuals with DS,[113-118] as well as those which have used cell lines transfected with some of these mutations,[97,98,102-104] or mice transgenic for the same[96] all seem to point towards the same conclusion that $A\beta_{42}$ (and probably $A\beta_{1-42}$) is the critical and predominant Aβ species that is first deposited within the brain in AD. In some way this deposition triggers or promotes other aspects of the pathological cascade that culminates in the widespread formation of plaques and tangles throughout the brain. Carriers of these genetic changes will accumulate, within the extracellular space of the brain, extra amounts of $A\beta_{42}$ which will remain in soluble form until a threshold concentration is reached after which fibrillogenesis will take place. This critical threshold may arise in affected individuals at an earlier time of life than in non-carriers of these changes, with the pathological process commencing, and clinical change ensuing, according to the rate of build up of soluble $A\beta_{42}$. This in turn will depend upon competing rates of formation and clearance from the tissue. The variable ages of onset associated with each of the mutations presumably relate to the dynamics of this process with some favoring $A\beta_{42}$ production, or preventing clearance of the same, more than others.

If this is so, other genetic changes which favor the development of AD, such as the possession of the apolipoprotein E E4 allele or α-1 antichymotrypsin polymorphisms might likewise affect either the formation of soluble Aβ ($A\beta_{42}$), promote its fibrillization, or enhance its stability in the tissue once fibril formation has occurred. These might act independently or in concert with the mutations on chromosomes 1, 14 and 21 to influence the rate or 'aggressiveness' of the process in these latter patients.

4.1.2.3. Apolipoprotein E gene

Over the past few years overwhelming evidence has emerged linking late-onset forms of familial and sporadic AD to the possession of at least one copy of the E4 allelic variant of the ApoE gene.[65-79] Persons with such an allele show a dose-dependant lowering of the age of onset of disease whereas possession of the E2 allele is claimed to have the reverse effect, delaying age at onset, relative to E3 homozygotes.[78,127-129] Although many cases of early onset sporadic or familial AD are not associated with an increase in the frequency of the E4 allele, especially in those familial cases related either to APP or PS-1 gene mutations,[66,83,129-131] an elevation of E4 allele frequency does seem to occur in other early onset families especially if certain family members are affected later in life.[80-83] Age at onset effects, with E2 and E4, similar to those in late onset AD have been noted in patients with chromosome 21-linked AD (i.e. $APP_{670/671}$ and APP_{717} mutations)[130,131] but apparently not so in chromosome 14-linked disease,[129] nor interestingly in HCHWA-D due to APP_{692} or APP_{693} mutations.[132] Indeed, in the single instance of a patient with the APP_{717} val→ile mutation remaining clinically unaffected two standard deviations beyond the mean age at onset for other affected family members, the ApoE genotype was E2/E3 whereas all other affected individuals had an E3/E4 genotype.[25] In DS, however, although the overall E4 allelic frequency is normal,[66,133-135] possession of an E4 allele is associated with an earlier death, with those persons having an E2 allele living longest.[133] This protective (longevity) effect of the E2 allele is also seen in healthy, elderly old people living up to and beyond 100 years of age.[136]

This relationship between cases of late onset AD and the possession of an ApoE E4 allele may not, however, hold in those cases of AD with onset very late in life (i.e. after 75 years of age). In these subjects the ApoE E4 allele frequency seems not to be significantly raised[127,137-140] and many individuals in this age range remain cognitively preserved despite carrying one or even two copies of the E4 allele. Hence, possession of the E4 allele may exert its maximal effects between ages of 55 and 75 years, tailing off beyond this point. The reason for this is not clear though it may be that in younger persons the ApoE E4 allele is acting in concert with other (environmental or genetic) risk factors to produce an overall susceptibility to disease. At late age, in the absence or lessening of such factors, the ApoE E4 effect may not on its own be strong enough to dictate AD.

Nonetheless, in all etiological forms of AD, all types of Aβ containing plaques, whether they be diffuse or cored, cerebral or cerebellar, come to contain the ApoE protein.[44,72,117,141-144] ApoE protein binds directly to the Aβ fibril, and it has been suggested that the E4 isoform might bind more avidly to Aβ than the E3 isoform[68,145-147] though not all investigators agree on this.[148,149] Accordingly it has been proposed that the E4 protein may promote the deposition of Aβ, an argument reinforced by observations that the E4 allele frequency is similarly (to AD) elevated in other late onset disorders where Aβ deposition also occurs, such as CLBD[77,150-153] or in survivors of acute head injury,[154,155] but remains normal in Creutzfeldt-Jakob disease,[66,156] motor neurone disease,[157,158] frontal lobe dementia,[159] or Huntington's disease,[75] disorders where widespread Aβ deposition does not generally take place. Although early reports in AD,[44,72,160-162] DS[163] or normal brain aging[160] suggested that the overall amyloid burden of the brain, in terms of plaque number or the amount of tissue occupied, was increased in carriers of an E4 allele compared to those without, later studies have not confirmed these findings either in AD[115,164-167] or DS.[168] Further, the use of carboxy-terminal specific antibodies to Aβ has shown that while the total $A\beta_{42}$ load in the brain does not vary in AD[115,165] or DS[168] according to the presence of an E4 allele, the amount of $A\beta_{40}$ does increase, at least in AD,[115,165] in a dose-dependent way in line with the copy number of E4 alleles present. $A\beta_{40}$ is associated principally with the later occurring cored type of plaque, rather than the early $A\beta_{42}$ containing diffuse plaques.[100,113] Hence, the presence of ApoE E4 protein may not differentially (compared to other ApoE isoforms including E2) influence the amount or molecular form of Aβ that is initially deposited as $A\beta_{42}$, though it may affect the subsequent evolution or maturation of such plaques, into ones containing $A\beta_{40}$, this occurring quicker or to a greater extent when E4 protein is present. Hence the presence of E4 protein in plaques may hasten their transformation from a (relatively) benign $A\beta_{42}$ containing diffuse form into a malignant $A\beta_{40}$ containing cored type.[115,169] This evidence implies that possession of the E4 allele, and production of the E4 protein, confers a more 'aggressive' pathological course to the illness, thereby lowering the age at onset in AD[127] and causing an earlier death in DS.[133] The finding[170] of a maximal loss of choline acetyltransferase activity from the frontal cortex of AD patients with two copies of the E4 allele, compared to those with one copy, or no copies, is consistent with such a viewpoint. In possessors of an E4 allele the build up of damage by plaques and tangles may occur at a faster rate thereby leading to an earlier crossing of the "threshold" to clinical change (i.e. lowering of onset age). Duration of disease also seems to be less in carriers of E4 alleles[115]–this again being consistent with a more aggressive course. How the presence of E4 protein within plaques might lead to their more rapid maturation is not clear. Because cored, $A\beta_{40}$ containing plaques are preferentially associated with microglial cells compared to diffuse, $A\beta_{42}$ containing ones,[114,171] it is possible that E4 protein is chemotactically more attractive to such cells than are E3 and E2 isoforms; heightened activity on the part of microglial cells might hasten their transformation. Thus although generically, ApoE protein might facilitate Aβ fibril formation in vitro,[146,147] this effect does not seem to be isoform-specific within tissues–all three major forms seem equally adept. However, when the ApoE E4 isoform is bound to Aβ a more rapid maturation of such deposits seems to occur, carrying with it the more malignant aspects of plaque formation. It is perhaps in this latter capacity that the isoform specific effects of E4 are manifest.

Alternatively, because the E4 isoform can bind only weakly, if at all, (compared to E3 and E2) to tau protein this might have the effect of destabilizing the cell's microtubule system and facilitating NFT formation.[172] The findings that neurones, in vitro, are unable to form microtubules in the presence of E4 protein, compared to E3 protein[173] and that phosphorylation of tau at Ser[262] (a site critical for successful microtubule assembly) inhibits the normal binding by E3 protein[174] would support such an argument. However, this explanation of the isoform specific effect of E4 in AD seems unlikely since, as argued earlier, in several studies no differences in the extent of tau pathology have been noted across genotype groups.[44,72,164,166] Furthermore, in CLBD, where the ApoE allelic frequency is raised similarly to that in AD,[77,150-153] NFT are sparse or absent and levels of pathological tau are low.[153]

ApoE is a 34 KDa glycoprotein that acts as a ligand for lipoproteins facilitating the movement of cholesterol across cell membranes via the LDL receptor.[175] Although synthesized primarily in the liver, ApoE is also manufactured in the CNS with ApoE mRNA being expressed predominantly by astrocytes;[176-178] in the peripheral nervous system it is expressed by Schwann cells,[179,180] particularly in response to injury. ApoE mRNA is not expressed by neurones[176,177] though they seemingly contain ApoE protein,[68,181] presumably having taken this up via their LDL receptor from the extracellular space after its release from glial cells.[72]

ApoE may play an important role during the normal development of the CNS, or following injury where its expression is upregulated in relationship to neuronal repair and the redistribution of lipids for myelin synthesis.[179,180] Furthermore, ApoE influences the extent of neurite outgrowth in vitro, with the usual E3 isoform promoting outgrowth while the E4 isoform retards this.[182] Levels of ApoE protein are lower in the brains of E4 allele carriers with AD than in those persons with AD who are homozygous for the E3 allele.[183] Hence, (one of) the major roles for ApoE may be in phospholipid homeostasis and the maintenance of (nerve) cell membranes. The E4 protein may thus exert its effect in AD through defective repair mechanisms operating particularly at synapse level. In this context the finding of a reduced number of synapses and a marked disruption of the dendritic cytoskeleton in ApoE deficient (knock-out) mice and a failure to compensate for removal of entorhinal projection[184] add weight to the hypothesis that human carriers of E4 alleles with AD would be less able to restrain the march of pathology, or to compensate for its effects, to the same extent as E3 or E2 carriers. In the former the disease would become more manifest and progress quicker though perhaps for other genetic or environmental reasons, such persons might have been in any case "predestined" to have developed the disease. In motor neurone disease ApoE allelic frequencies are normal overall,[157,158] though those patients with an ApoE E4 allele seem more likely to present with a bulbar palsy–i.e. a more aggressive form of the illness–than those without. This again might point towards a role for ApoE in disease restriction with E4 carriers suffering more in this respect. The lowering of age at death in DS[133] and the earlier age of onset in chromosome 21 linked AD[130,131] in E4 patients would also be consistent with this role for ApoE.

It is clear, therefore, that ApoE is a protein with probably many and diverse roles in terms of CNS function. The exact part it might play in determining the clinical or pathological characteristics of AD remains to be elucidated. It may have several, complementary sites of action, some of which may apply equally to all cases of AD while in others isoform-dependent processes might be selective for certain subgroups of patients.

From what has been said so far, possession of one of the mutations on chromosomes 1, 14 or 21, or trisomy of chromosome 21, will (almost) inevitably lead to the development of AD at some time of life. Possession of the ApoE E4 allele will also increase your chances of developing disease, though this is by no means certain, even for persons who are homozygous for this. An interaction with other, as yet unknown genes, may be required for the full expression of disease. Yet at least half of all cases of AD, only some of whom relate to early onset familial disease, do not possess any of the so-far known AD-causing mutations on chromosomes 1, 14 or 21 nor do they possess an ApoE E4 allele. This implies that there are major, as well as perhaps other minor, genetic risk factors yet to be discovered that favor the development of AD. Other (common) polymorphic changes, perhaps on chromosomes 1 or 14,[84] or genetic variations in some of the amyloid associated molecules such as ACT[85,86] or the VLDL receptor[87] might be involved and these may interact with other genetic changes (e.g. ApoE allele) to influence the rate or formation of the pathological changes. The full range of genetic susceptibility in AD remains to be ascertained.

4.1.2.4. Tau gene

Although tau (hyper) phosphorylation and PHF formation are fundamental aspects of the pathology of AD, no mutational or polymorphic changes in the tau gene, specific to the disease, have yet been identified. The tau protein produced in AD is seemingly normal, at least as far as its amino-acid sequence is concerned, and the pathology seems to result entirely from post-translational modifications (see Chapter 2). While an overexpression of the tau gene may occur, in some neurones, towards the end stages of the illness,[185] this is likely to reflect a compensatory response on the part of such surviving neurones to loss of neighboring, or complementary acting, cells.

4.1.2.5. APLP gene

The APLPs, APLP1 and APLP2, bear strong homology to APP, and share a similar cellular distribution,[186-188] though they lack the gene sequence corresponding to Aβ.[186,187,189] Because APLPs can be processed in similar fashion to APP, within the Golgi apparatus, [188,190] it remains possible that changes in these particular proteins could play a part in the disease process, perhaps by interferring with normal APP metabolism in a way which predisposes towards the catabolism of the latter into Aβ (e.g. by competing with APP for α-secretase mediated processing). However, no alterations in APLP processing have as yet been demonstrated in AD, nor have any mutations in the APLP genes been detected. A role for these proteins in the pathogenesis of AD therefore still remains open.

4.1.2.6. Other genetic changes

In accordance with the hypothesis that environmental (toxic) factors might play a role in the development of AD, especially in later life, genetic changes predisposing towards an ineffective prevention of the damaging effects of free radical mediated reactions have been sought. In this respect a possibly increased frequency of polymorphisms in the CYP2D6 (debrisoquine hydroxylase) gene in AD, associated with the cytochrome P450 detoxification system, conferring a "poor metabolizer" status has been ruled out;[191] findings consistent with observations in AD of a normal activity of this enzyme.[192] Nonetheless, an increased frequency of polymorphisms conferring a poor metabolizer phenotype has been detected in patients with CLBD,[191,193] where ApoE E4 allele frequencies are also elevated.[78,150-154] However, since no changes in the CYP2D6 gene occur in AD without Lewy bodies[191,193] this latter polymorphic change may be important more with respect to the pathogenesis of Lewy body formation rather than of Alzheimer-type changes. Similarly increased frequencies of CYP2D6 polymorphisms in PD itself[191,194] strengthen such a conclusion. Mutations in mitochondrial DNA have been reported in some patients with AD[195] though whether these are of etiological or functional significance remains uncertain.

If, on the other hand, the full range of genetic risk factors for AD is now essentially known, it has to be concluded that there are indeed important (and again perhaps largely unrecognized) environmental factors that may directly cause the disease or act in concert with certain of the already recognized genetic changes to precipitate the same.

4.1.2. Parkinson's Disease

Traditionally, PD has been considered to be mostly a sporadic disease of later life[196-198] though an apparent family history, suggesting inheritance, of the disease has occasionally been reported.[199-202] The genetic basis for these latter cases is unknown and while it does not relate to changes in obvious candidate genes such as the tyrosine hydroxylase gene,[203] a greater than usual frequency of one or more of the (common) polymorphisms within the debrisoquine hydroxylase gene (CYP2D6) has been reported in apparently sporadic cases of PD.[194] As mentioned, the protein encoded by this gene (debrisoquine hydroxylase) forms part of the cytochrome P450 system which detoxifies xenobiotic compounds. Such polymorphisms are thought to confer a 'poor metabolizer' phenotype which may either render possessors at greater risk of cell damage from extraneous inorganic and organic compounds, some of which may participate in free radical-type reactions involving cells of the substantia nigra (e.g. MPTP), or produce an endogenous toxin that may act accordingly. Hence, genetic variations of this kind may confer a disproportionate vulnerability upon certain nerve cell types (e.g. pigmented neurones) which perhaps because of their particular metabolic profile (catecholamine oxidation) may be naturally prone to the damaging effects of reactive oxygen species. The finding of point mutations or deletions in the mitochondrial DNA in some patients with PD[204,205] (but see refs. 206, 207) lends some support to this argument. However, it remains uncertain whether these mitochondrial gene changes in PD occur more frequently than in aging alone,[208-210] nor is it clear whether such (extra) changes (if present) actually lead to a reduction in the number of organelles or the cell's capacity for oxidative metabolism.[210]

4.2. Environmental Factors

4.2.1. Introduction

In the past, the absence of any apparent previous family history of the disorder in question was generally taken to indicate that disease had arisen sporadically and, by implication, factors other than those of a genetic nature were causally responsible. However, despite there being so many living patients with dementia or Parkinsonism where an obvious previous family history of the condition is lacking, it has not as yet proved possible to unequivocally relate disorders such as AD or PD to any single environmental factor that might be so ubiquitous as to account for such a widespread prevalence. Thus, the possible role of environmental factors in the pathogenesis of neurodegenerative disease has lost much favor over the past few years. This has been compounded most recently by the discovery of genetic risk factors such as the ApoE E4 allele, possession of which increases the chances of developing AD in many, but importantly not all, elderly individuals.[65-83] Genetic factors like the E4 allele do not usually produce a clear-cut mode of inheritance of disease, in contrast to the point mutations like those on chromosomes 1, 14 and 21 in AD[15-24,48-60] where development of disease in carriers of that defective gene is virtually predictable and follows an autosomal dominant course of transmission. Therefore, inheritance of genetic risk factors, like the ApoE E4 allele, increases susceptibility to AD, but does not guarantee that this will occur; many elderly persons with E3/E4, or even E4/E4, genotypes do not become demented no matter how long they live.[127,137-140] In this way disease can "skip" generations, or appear "sporadically" depending solely upon parental genotypes and the manner of genetic recombination in their offspring. The presence of such genetic risk factors further undermines the possibility that environmental factors might play a significant role in causing neurodegenerative disease.

Nonetheless, clear associations between certain environmental factors and the presence of the pathology of AD and PD, or at least of changes strongly reminiscent of these disorders have been recorded suggesting that in particular circumstances these can take on important causative roles.

4.2.2. Alzheimer's Disease

4.2.2.1. Head Injury

As described in Chapter 3, pathological changes resembling those of AD occur in the brains of head injured persons, these resulting either from trauma substained following road traffic or other accident, or from that which is occasioned deliberately during professional or amateur boxing. In such individuals substantial numbers of amyloid plaques, usually of a diffuse type, can occur widely throughout the cerebral cortex, these in relationship to accidental trauma sometimes appearing within a matter of days following the injury.[211-213] These plaques carry the same molecular characteristics (i.e. are composed of $A\beta_{42}$) as the diffuse plaques of AD and DS.[212] NFT occur in boxers[214] (but not in the accidentally head injured[213]) though not usually in the same topographical regions as in AD; indeed in some instances NFT may even occur when Aβ deposits are absent.[215,216] Such trauma seems to result in an upregulation of the APP gene with APP levels increasing in certain vulnerable cell types.[217] Some of this additional APP may become diverted along amyloidogenic pathways leading to local increases in $A\beta_{42}$ concentration in the extracellular fluid, these eventually becoming of sufficient magnitude as to precipitate fibrillization and plaque formation. The increased incidence of the ApoE E4 allele in such head injured patients showing Aβ deposition[212] might have been anticipated since this protein is widely believed to play a role in promoting Aβ deposition.

Although instances of head injury like these partially recapitulate the pathological characteristics of AD there are notable departures in both the form, and the distribution, of the lesions that sets them apart from this latter disorder.[215] Nevertheless, perhaps the best evidence for a role for head injury in the causation of AD comes from those isolated case histories[218-220] where a single blow to the head in the absence of any other significant lifestyle event has seemingly led to a dementing disorder that is both clinically and pathologically typical of AD. Perhaps in these instances the trauma has acted as a trigger 'activating' the pathological cascade of AD. Whether such persons would, in the fullness of time and because of particular genetic factors, in any case have developed AD, even if the traumatic event had never occurred, cannot be determined, though this is possible. Hence, traumatic brain injury can, at least partially, provoke tissue reactions leading to Aβ deposition and NFT formation in affected regions in many individuals, though only in isolated circumstances does this seem to result in the full course of pathological events culminating in an entity clinically and histopathologically indistinguishable from AD.

4.2.2.2. Aluminum

That aluminum is neurotoxic is undisputed.[221] A relationship between aluminum (intoxication) and the pathology of AD has long been claimed, dating back to 1976 when Crapper and colleagues detected elevated gross tissue levels of aluminum in the brains of patients dying with AD[222] and showed this to accumulate in the nucleus of nerve cells.[223] Data of this kind based on gross tissue analyses have proved contentious[221] and although some subsequent, and more sophisticated, analyses have likewise claimed increases in aluminum (in plaques)[224-226] others have failed to confirm this.[227-229] Perl and Brody[230] reported increased aluminum in NFT bearing cells, relative to their non-NFT bearing neighbours, as did Good et al[231], though yet again this has not always been confirmed.[228,232]

Epidemiological studies relating local concentrations of aluminum in drinking water to the prevalence of dementia (and by implication to that of AD) have also proved conflictory with some reports claiming an association[233-237] but others not;[238,239] no clear cut conclusion regarding any risk from drinking water is yet possible.[240]

Animal studies have not helped to resolve this issue. Early work showing that injection of aluminum salts into brain could induce a filamentous change similar to NFT in neurones[241] raised much expectation, though enthusiasm was dampened when subsequent analyses revealed these fibrillary structures to differ ultrastructurally from PHF being composed of 10nm neurofilaments.[242] However, recent work employing cell (neuroblastoma) cultures treated with aluminum salts has shown that these express certain epitopes (e.g. for phosphorylated tau) associated with AD.[243,244] Nevertheless, it has yet to be demonstrated that such tau changes actually mimic PHF formation; phosphorylation of tau in this way may simply reflect one of many "stresses" that such cells may undergo when exposed to aluminum; progression to PHF formation need not necessarily follow.

Perhaps the best evidence for a role for aluminum in the pathogenesis of AD has come from studies of those patients who in former times inadvertantly suffered chronic exposure to high circulating levels of this during renal dialysis. Such persons developed a (reversible) encephalopathy known as dialysis dementia and in many instances plaque-like structures composed of Aβ were later seen at autopsy.[245] Although the presence of NFT in this disorder is uncertain[245-248] levels of pathological tau suggestive of PHF formation are increased in some patients.[248]

A clear role for aluminum in the pathogenesis of AD remains unproven though it is popular opinion that this is an unlikely cause of the disease.

4.2.2.3. Viruses

A possible association between a prior viral infection of the brain and the occurrence of AD has long been sought. Because of its propensity for infection of the temporal lobe, herpes simplex virus (HSV) has remained an attractive candidate. Acute infection of the temporal lobe by HSV is achieved through neurone to neurone spread of virus from the trigeminal ganglion and other sites of latent infection.[249] However, once spread to the brain the virus may become latent within neurones only to be come reactivated at a later date. Chronic rounds of recurrent reactivation and spread through the CNS might lead to a progressive neuronal degeneration triggering the pathology of AD or exacerbating the ongoing damage of AD within already affected individuals. Early evaluations of serum or cerebrospinal fluid HSV-1 antibody levels suggested that demented patients with AD had higher titers than controls[250] though direct evidence for the presence of HSV-1 in the brains of patients with AD has been sparse. HSV-1 epitopes have been detected by immunohistochemistry in some patients with AD,[251-253] as has HSV genome (DNA) or message (RNA),[254-260] though in these and other studies no such positive findings were recorded in far many other individuals.[253,256-262] However, methodologies such as Southern blotting, in-situ hybridization or immunohistochemistry are plagued by low sensitivity; DNA hybridization can detect only one copy of each genome present within a cell and RNA hybridization, up to 10^4 copies per cell. Nonetheless, the finding of HSV message in neurones of the trigeminal ganglion of 81% patients with AD, compared to 47% of controls,[262] together with observations that NFT, apparently identical to those in AD, occur in the brains of persons suffering from reactivated measles virus in SSPE[263,264] or in postencephalitic Parkinsonism[265] has sustained interest in the possibility that latent HSV may participate in, or even initiate, the pathogenetic cascade of AD upon reactivation.

Recent work, using the vastly more sensitive polymerase chain reaction has confirmed the high incidence of HSV-1 genome in the brains of patients with AD,[266,267] this being present in more than 75% of patients. However, a similar prevalence of HSV-1 was found in control subjects. Furthermore, the pattern of distribution of HSV throughout the brain did not reveal any preferential concentration within the temporal lobe with the cerebellum and striatum being just as likely to be infected.[267] Nonetheless, it remains possible that AD is associated with a more frequent reactivation of the virus, even though its actual presence within the tissue may be no more common than in control subjects. Such episodes of reactivation may be more 'relevant' early in the course of the illness than in end-stages of disease and further study is required to clarify any association between the onset of disease and that of viral reactivation.

4.2.3. Parkinson's Disease

Since most cases of PD appear to arise sporadically in the absence of any obvious family history of the disorder and because of a discordance for the disorder in many twin studies, it has been presumed that the disease mechanism is largely outside the influence of genetic control.[197,198,268] Four main mechanisms have been proposed to explain the cause of the degeneration and loss of nerve cells from the substantia nigra, locus caeruleus and other brain regions. Firstly, cell damage may result from the direct action on dopaminergic neurones of environmental or endogenous toxins like as 1-methyl, 4-phenyl, 1236 tetrahydropyridine (MPTP).[269-272] Secondly, excess excitatory amino acids may induce neurotoxic changes in nigral cells.[273] Thirdly, deficiencies in a neuronotrophic factor, such as brain derived neurotrophic factor, may adversely affect the survival of nigral cells.[274] Lastly, free radical mediated oxidative stress may lead to the degeneration of nigral cells.[275] Of these, the concepts of an environmental toxin, akin to MPTP, if not this molecule itself, or free-radical mediated reactions have gained most support; indeed the mechanism of any putative intoxication may proceed through just such a free radical mediated reaction.

4.2.3.1. PD and MPTP

The involvement of MPTP with human PD has a colorful history. It was first discovered serendipitously as a result of a chemist's attempt to synthesize a meperidine analogue, a "synthetic heroin",[269] in which the compound was produced as a contaminatory by-product. The accident came to light with the observation of a number of young addicts who had developed a symptomatology identical to human PD within days or weeks of injection of this synthetic "heroin".[269,270,276] Investigation of these, and subsequent cases, led to the identification of MPTP as the culprit and since then much basic research has ensued through which the mode of action of this substance has become defined.

Models of PD in primates and rodents are now widely available using the selective toxin MPTP. These have confirmed that MPTP injections produce a highly selective neurotoxicity involving nerve cells of the substantia nigra, though the molecule is itself non-toxic. Toxicity is conferred through the metabolism of MPTP into the cation 1-methyl, 4-phenylpyridine (MPP+), via the intermediate 1-methyl, 4-phenyl 2,3 dihydropyridine (MPDP+). Metabolism of MPTP is performed by glial cells (astrocytes) using the enzyme monoamine oxidase B.[277] MPP+ selectively enters dopaminergic cells via the dopamine uptake receptor.[278] Once in the cell MPP+ will bind to neuromelanin granules thereby facilitating its accumulation within the cell. MPP+ is thought to penetrate the mitochondrion by an energy dependent system for concentrating lipophilic cations[279] where it inhibits the activity of the enzyme NADH ubiquinone reductase (complex I) of the electron transport chain[280-282] and by doing so reduces levels of ATP production. MPP+ also inhibits another mitochondrial enzyme, α ketoglutaric acid dehydrogenase (α KGDH),[283] thereby further interfering with ATP production via inhibition of the Krebs cycle. At the same time MPP+ may participate in reactions that generate free radicals, particularly hydroxyl radicals, these in turn further increasing oxidative stress. How MPP+ inhibits complex I is not clear though it may bind to it at, or near, the rotenone site.[282] While the initial binding of MPP+ to complex I appears weak and reversible, over time this becomes stronger leading to a progressive and irreversible decline in enzyme activity.[282] Such changes cause a poisoning of the respiration chain and eventually cell death. In older MPTP treated animals eosinophilic inclusions, reminiscent of Lewy bodies, have been found in the same brain regions where they occur in human PD.[284] Other selective neurotoxins such as 6-hydroxydopa[285,286] or manganese[287] which destroy the substantia nigra may similarly participate in free radical mediated reactions.

However, a weakness of the hypothesis that MPTP, or a compound similar to this, either widely present in the environment or endogenously produced in the brain or elsewhere in the body, might cause human PD comes from observations that in experimental animals, as well as in MPTP-induced human parkinsonism, the noradrenergic locus caeruleus is less damaged and the serotonergic raphe unaffected.[281,284] Such findings are in clear contrast to human PD where the former cell group, especially, is decimated.[288-291] Hence, while MPTP (and 6-hydroxydopa) neurotoxicity clearly provides a good model to study the clinical effects of, and to develop potential treatments for, human parkinsonism its relevance to the pathogenesis of PD itself is less certain. Nevertheless, deficiencies of complex I, and to a lesser extent complexes II and III, specific to the substantia nigra, have been reported in PD[292-296] as have alterations in αKGDH activity.[294,297] Such losses have been localized, using immunohistochemistry, to the pigmented nerve cells of the substantia nigra.[298] These changes in complex I (and other complexes) should, as with MPTP-induced neurotoxicity, result in an increased production of reactive oxygen species which, if unscavenged, will foster cell damage, perhaps especially to any residual complex I enzyme.[282] Moreover, inhibition of complex I will lead to diminished ATP production and this may result in the release of stored dopamine from secretory vesicles (storage here is energy dependent); catabolism of this by monoamine oxidase will further increase free radical production. The cause of the complex I deficiency in PD is not known and may represent an inherited defect in mitochondrial DNA. Alternatively, unchecked or unrepaired age-associated damage to the mitochondrial DNA (on which the respiratory proteins are coded) may be responsible. In this context observations of increased frequencies of point mutations (but see refs. 206, 207) or gene deletions[205] in the mitochondrial DNA of patients with PD may be important.

Further indirect evidence in support of the hypothesis that MPTP, or a substance like this, may be responsible for PD comes from observations that certain isoquinoline derivatives structurally related to MPTP and MPP+ can also damage the substantia nigra in rodents[299] and in some primates[300] (but see ref. 301) and are likewise neurotoxic to PC12 cells and ventral mesencephalic cells in culture.[302-305] Again these probably act upon mitochondrial complex 1 and αKGDH activities.[306,307] Such compounds, which are naturally produced in the brain through condensation reactions involving catecholamines and aldehydes, may also generate reactive oxygen species and isoquinolinium cations through autooxidation.[308] The latter, as with MPP+, can act as substrates for the dopamine re-uptake system. It is possible, therefore, that substances like isoquinolines, which may be formed endogenously within the brain in PD or perhaps metabolized into such from an environmental precursor, can damage the substantia nigra in a manner akin to MPTP following a high acute, or a chronic low, dose exposure.

4.2.3.2. PD, Free Radicals and Oxidative Stress

Pigmented cells of the substantia nigra are potentially highly vulnerable to free radical mediated damage. Their neurotransmitter dopamine is catabolized (oxidized) enzymatically by monoamine oxidase, and non-enzymatically by auto-oxidation; quinones, hydrogen peroxide and oxyradicals are all by-products of these reactions. Many biochemical and histochemical studies on postmortem brains from patients with PD suggest an involvement of free radical mediated oxidative stress leading to degeneration of nigral nerve cells (see ref. 309). Oxidative stress is normally reduced or prevented by two enzymes, catalase and glutathione peroxidase. The level of both of these enzymes in the substantia nigra is reduced in PD[275,310] suggesting a decreased capacity to combat free radical injury. Glutathione, which scavenges oxygen radicals, is also decreased.[311-315] and basal lipid peroxidation is increased.[316-318] The levels of manganese dependent superoxide dismutase[319,320] and its mRNA[321] are both elevated in PD and increases in 8-hydroxy-2deoxyguanosine, a product of free-radical mediated DNA damage, also occur[322] again suggesting increased oxidative stress.

Increased amounts of iron occur within the substantia nigra in patients with PD[313,323-326] being accompanied by an elevation in its valence state as the more toxic Fe^{3+}. Free iron, especially Fe^{3+}, is neurotoxic as it promotes the formation, via the Fenton reaction, of free radicals from hydrogen peroxide, particularly the hydroxyl radical.[327] The excess iron present within the substantia nigra in PD could be located either in the neuromelanin pigment of surviving cells,[231,328-331] in Lewy bodies,[332,333] or in a safely stored form (as ferritin) in microglial cells.[326,333] This would have been scavenged from degenerated neurones since neuromelanin in normal individuals is known to contain iron[231,328-331] and microglial cells are widely present in the substantia nigra in PD.[334] However, some may still occur as increased free iron, available to take part in free radical type reactions. However, because an accumulation of iron in brain tissue occurs in other neurodegenerative diseases it is possible that the elevated iron in the substantia nigra in PD merely represents a secondary "safe" storage in microglial cells, as ferritin, this having been scavenged from degenerating neurones. Nevertheless, this latter molecule has been reported as unchanged or even reduced in amount in PD.[324] Thus, the presence of high levels of iron, of which some might be in a potentially reactive form, and microglial cells, with their peroxidative capabilities, could place surviving cells of the substantia nigra at further risk of free radical damage. Such changes might prolong or exacerbate the basic underlying cause of the neurodegeneration and set in place a vicious cycle of nigral cell death, iron accumulation and neurotoxicity. Other sources of free radicals could come from the auto-oxidation of dopamine (and noradrenaline in cells of the locus caeruleus).[335] Loss of the most heavily pigmented cells[336] from the substantia nigra in PD, in which auto-oxidative processes may be greatest, would be consistent with such a suggestion. Hence, there is much evidence to suggest that the substantia nigra does indeed suffer oxidative stress in PD. The sources of such free radicals seem multiple; increased free iron, reduced anti-oxidant capacity, lipid peroxidation, catecholamine auto-oxidation and mitochondrial defects may all play a part, separately or convergingly, in producing an adverse climate of changes that damages these vulnerable cells and leads to their degeneration. While this hypothesis may satisfactorily explain the degeneration of dopaminergic nigral cells in PD, and in part that of the noradrenergic cells of the locus caeruleus and the serotonergic cells of the raphe (sources of free radicals in these latter cell types perhaps relating to catecholamine metabolism or pigmentation), it cannot readily account for the involvement of non-pigmented, non-aminergic neurones like those of the nucleus basalis of Meynert which are also seriously damaged in the disorder.[337,338] Alternatively, exposure to damaging agents may have occurred much earlier in life producing a sub-clinical damage in the substantia nigra and other brain regions. 'Natural' aging may then take over in later life to complete the process of nigral and other cell loss begun much earlier, eventually producing clinical symptomatology. In this context longitudinal observations of MPTP-treated subjects with subclinical dopamine damage, or other subjects who are chronic amphetamine abusers (amphetamines can act as dopaminergic neurotoxins) will be important, since presymptomatic damage in these subjects may become unmasked in later life as physiological aging changes in the substantia nigra commence and gain momentum.[272]

4.2.3.3. PD and viruses

The possibility of a viral factor in the development of PD has long been proposed, ever since cases of clinical parkinsonism relating to prior epidemics of encephalitis lethargica appeared in the 1930-1950s.[339] However, since 1960 no new cases of postencephalitic parkinsonism have appeared and no new instances of encephalitis lethargica have not been reported since 1935.[340] By now all patients affected in this way will have died and the disorder is thus solely of historical interest. Nonetheless, the fact that viruses can damage the substantia nigra and the absence of a proven environmental cause for PD has meant that research into a viral origin for the disorder has continued. Viruses such as measles or herpes simplex have been implicated, though tangible evidence for an involvement of prior infection by either of these and subsequent development of PD is lacking.

4.2.3.4. Summary

In summary therefore, the etiology of PD remains enigmatic though evidence for an 'environmental' cause is strong. Diverse factors, some truly environmental, others which may be internally modified environmental agents, or even endogenously produced metabolites, could be responsible. A common thread shared by all of these seems to be their ability to facilitate the formation of free radicals and to induce oxidative stress. However, for full expression of the disease, as distinct from cases of incidental Lewy body disease in which the damage may not progress or do so only very slowly, additional genetic susceptibility factors may be required which might set the scene of a poor metabolic cell environment in which such potentially harmful molecules fail to be detoxified and rendered harmless. Toxic (environmental) factors together with a genetic susceptibility may operate to bring about the disorder; the balance of each may vary much from individual to individual and this should be considered when seeking any single biochemical or genetic marker for the disorder. Hence, environmental factors could impact upon the substantia nigra in late life, "finishing off" the job that aging may have started many years earlier.

4.2.4. Conclusion

In conclusion, the development of AD is likely to be largely, or even wholly, determined by genetic factors which drive or modulate steps in the pathological cascade culminating in the formation and accumulation of the characterizing destructive lesions of the disease–namely the plaques and tangles–which damage nerve cells and produce clinical dysfunction. There is no firm evidence at present that outside (environmental) factors can produce AD per se, though risks like head injury or exposure to toxins like aluminum might precipitate, or hasten the course of, illness in persons who for genetic reasons would in any case have been predestined to develop the disease in all good time. In PD, genetic factors are much less clear cut, but there is some evidence that these might confer a susceptibility which could exacerbate or facilitate environmentally or endogenously derived free radical mediated reactions which decimate vulnerable cell types such as the substantia nigra.

The features that distinguish clinically demented persons, with the typical histopathological features of AD, from other non-demented or mildly demented persons whose brains contain structurally similar changes, lie either with the relative amount of pathology that has developed or the distribution of such changes. In the non-demented or mildly demented subject such changes may be absent or less severe, and when present may be "concentrated" in key regions of the temporal lobe. These kinds of observations seem to place individuals with varying degrees of clinical disability upon a sliding scale of pathological change with the severity and the topography of tissue damage widening, and the functional disturbance increasing, as the disease progresses. Likewise, the histopathological features that typify clinical PD can be found in lesser amounts in some non-Parkinsonian elderly subjects, again suggesting that clinically affected and unaffected individuals might exist on a continuum of pathological change with clinical disability appearing once a certain threshold of 'tolerance' to damage has been crossed.

In the last chaper, it is argued that in each of these clinically affected and unaffected scenarios the same pathological processes and mechanisms, leading to the appearance of destructive lesions, are in place though the ways in which they are triggered or promoted over time may differentiate between them. The real distinctions between "aging" and "disease" may be not so much with pathological variation but with etiological differences, principally genetic factors. The nature and balance of these latter risks may ultimately determine the amount of tissue damage that takes place, the timescale over which this occurs and the amount of clinical (functional) derangement that ensues. Distinctions should not perhaps be drawn in terms of 'normal' or 'pathological' aging, but rather in terms of 'successful' or 'unsuccessful' aging.

References

1. Heston LL, Mastri AR, Anderson VE et al. Dementia of the Alzheimer type. Arch Gen Psychiat 1981; 38: 1085-1090.
2. Nee LE, Polinsky RJ, Eldridge R et al. A family with histologically confirmed Alzheimer's disease. Arch Neurol 1983; 40: 203-208.
3. Heyman A, Wilkinson WE, Hurwitz BJ et al. Alzheimer's disease: Genetic aspects and associated clinical disorders. Ann Neurol 1983; 14: 507-515.
4. Breitner JCS, Silverman JM, Mohs RC et al. Familial aggregation in Alzheimer's disease: comparison of risk among relatives of early- and late-onset cases, and among male and female relatives in successive generations. Neurology 1988; 38: 207-212.
5. Huff FJ, Auerbach J, Chakravarti A et al. Risk of dementia in relatives of patients with Alzheimer's disease. Neurology 1988; 38: 786-790.
6. Bird TD, Sumi SM, Nemens EJ et al. Phenotypic heterogeneity in familial Alzheimer's disease: A study of 24 kindreds. Ann Neurol 1989; 25: 12-25.
7. Mayeux R, Stern Y, Spanton S. Heterogeneity in dementia of the Alzheimer type: evidence of subgroups. Neurology 1985; 35: 453-461.
8. Levy-Lahad E, Wijsman EM, Nemens E et al. A familial Alzheimer's disease locus on chromosome 1. Science 1995; 269: 970-973.
9. Schellenberg GD, Bird TD, Wijsman E et al. Genetic linkage evidence for a familial Alzheimer's disease locus on chromosome 14. Science 1992; 258: 668-671.
10. Mullan M, Houlden H, Windelspecht M et al. A locus for familial early onset Alzheimer's disease on the long arm of chromosome 14, proximal to α-antichymotrypsin. Nature Genet 1992; 2: 340-342.
11. Van Broeckhoven C, Backhovens H, Cruts M et al. Mapping of a gene predisposing to early onset Alzheimer's disease to chromosome 14q 24.3. Nature Genet 1992; 2: 335-339.
12. St George-Hyslop PH, Haines J, Rogaev E et al. Genetic evidence for a novel familial Alzheimer's disease locus on chromosome 14. Nature Genet 1992; 2: 330-334.
13. Schellenberg GD, Boehinke M, Wijsman EM et al. Genetic association and linkage analysis of the apolipoprotein CII locus and familial Alzheimer's disease. Ann Neurol 1992; 31: 223-227.
14. Goate A, Chartier-Harlin MC, Mullan M et al. Segregation of a missense mutation in the amyloid precursor gene with familial Alzheimer's disease. Nature 1991; 349: 704-706.

15. Mullan M, Crawford F, Axelman K et al. A pathogenic mutation for probable Alzheimer's disease in the APP gene at the N-terminus of β-amyloid. Nature Genet 1992; 1: 345-347.
16. Lannfelt L, Bogdanovic N, Appelgren M et al. Amyloid precursor protein mutation causes Alzheimer's disease in a Swedish family. Neurosci Lett 1994; 168: 254-256.
17. Chartier-Harlin MC, Crawford F, Houlden H et al. Early onset Alzheimer's disease caused by mutations at codon 717 of the β-amyloid precursor protein gene. Nature 1991; 353: 844-845.
18. Naruse S, Igarashi S, Aoki K et al. Missense mutation val→ile in exon 17 of amyloid precursor protein gene in Japanese familial Alzheimer's disease. Lancet 1991; 337: 978-979.
19. Yoshizawa T, Komatsuzaki Y, Iwamoto H et al. Screening of the mis-sense mutation producing the 717 val→ile substitution in the amyloid precursor protein in Japanese familial and sporadic Alzheimer's disease. J Neurol Sci 1993; 117: 12-15.
20. Karlinsky H, Vaula G, Haines JL et al. Molecular and prospective phenotypic characterization of a pedigree with familial Alzheimer's disease and a missense mutation in codon 717 of the β-amyloid precursor protein gene. Neurology 1992; 42: 1445-1453.
21. Fidani L, Rooke K, Chartier-Harlin MC et al. Screening for mutations in the open reading frame and promoter of the β-amyloid precursor protein gene in familial Alzheimer's disease: identification of a further family with APP_{717} Val→Ile. Hum Molec Genet 1992; 13: 165-168.
22. Murrell J, Farlow M, Ghetti B et al. A mutation in the amyloid precursor protein associated with hereditary Alzheimer's disease. Science 1991; 254: 97-99.
23. Mullan M, Tsuji S, Miki T et al. Clinical comparison of Alzheimer's disease in pedigrees with the codon 717 val→ile mutation in the amyloid precursor protein gene. Neurobiol Aging 1993; 14: 407-419.
24. Kennedy AM, Newman S, McCaddon A et al. Familial Alzheimer's disease. Brain 1993; 116: 309-324.
25. St George-Hyslop P, Crapper-McLachlan D, Tuda T et al. Alzheimer's disease and possible gene interaction. Science 1994; 263: 537.
26. Van Broeckhoven C, Haan J, Bakker E et al. Amyloid β protein precursor gene and hereditary cerebral hemorrhage with amyloidosis (Dutch). Science 1990; 248: 1120-1128.
27. Levy E, Carman MD, Fernandez-Madrid IJ et al. Mutation of the Alzheimer's disease amyloid gene in hereditary cerebral haemorrhage, Dutch type. Science 1990; 248: 1124-1126.
28. Van Duinen SG, Castano EM, Prelli F et al. Hereditary cerebral haemorrhage with amyloidosis in patients of the Dutch origin is related to Alzheimer disease. Proc Natl Acad Sci (USA) 1987; 84: 5991-5994.
29. Hendriks L, Van Duijn CM, Cras P et al. Presenile dementia and cerebral haemorrhage linked to a mutation at codon 692 of the β-amyloid precursor protein gene. Nature Genet 1992; 1: 218-221.
30. Peacock ML, Murman DL, Sima AAF et al. Novel amyloid precursor protein gene mutation (codon 665^{Asp}) in a patient with late-onset Alzheimer's disease. Ann Neurol 1994; 35: 432-438.
31. Peacock ML, Warren JT, Roses AD et al. Novel polymorphism in A4-region of amyloid precursor protein gene in a patient without Alzheimer's disease. Neurology 1993; 43: 1254-1256.
32. Forsell C, Lannfelt L. Amyloid precursor protein mutation at codon 713 (Ala→Val) does not cause schizophrenia: non-pathogenic variant found at codon 705 (silent). Neurosci Lett 1995; 184: 90-93.
33. Adroer R, Lopez-Acedo C, Olivia R et al. A novel silent variant at codon 711 and a variant at codon 708 of the APP sequence detected in Spanish Alzheimer and control cases. Neurosci Lett 1993; 150: 33-34.
34. Balbin M, Abrahamson M, Gustafson L et al. A novel mutation in the β-protein coding region of the amyloid β-protein precursor (APP) gene. Hum Genet 1992; 89: 580-582.
35. Kamino K, Orr HT, Payami H et al. Linkage and mutational analysis of familial

Alzheimer disease kindreds for the APP gene region. Am J Hum Genet 1992; 51: 998-1014.
36. Jones CT, Morris S, Yates CM et al. Mutation in codon 713 of the β-amyloid precursor protein gene presenting with schizophrenia. Nature Genet 1992; 1: 306-309.
37. Zubenko GS, Farr J, Stiffler JS et al. Clinically-silent mutation in the putative iron-responsive element in exon 17 of the β-amyloid precursor protein gene. J Neuropathol Exp Neurol 1992; 51: 459-463.
38. Ghetti B, Murrell J, Benson MD et al. Spectrum of amyloid β-protein immunoreactivity in hereditary Alzheimer's disease with a guanine to thymine missense change at position 1924 of the APP gene. Brain Res 1992; 571: 133-139.
39. Lantos PL, Luthert PJ, Hanger D et al. Familial Alzheimer's disease with the amyloid precursor protein position 717 mutation and sporadic Alzheimer's disease have the same cytoskeletal pathology. Neurosci Lett 1992; 137: 221-224.
40. Lantos PL, Ovenstone IMK, Johnson J et al. Lewy bodies in the brain of two members of a family with the 717 (val to ile) mutation of the amyloid precursor protein gene. Neurosci Lett 1994; 172: 77-79.
41. Mann DMA, Jones D, Snowden JS et al. Pathological changes in the brain of a patient with familial Alzheimer's disease having a missense mutation at codon 717 in the amyloid precursor protein gene. Neurosci Lett 1992; 137: 225-228.
42. Hanger DP, Mann DMA, Neary D et al. Tau pathology in a case of familial Alzheimer's disease with a valine to glycine mutation at position 717 in the amyloid precursor protein. Neurosci Lett 1992; 145: 178-180.
43. Cairns NJ, Chadwick A, Lantos PL et al. βA4 protein deposition in familial Alzheimer's disease with the mutation in codon 717 of the β/A4 amyloid precursor protein gene and sporadic Alzheimer's disease. Neurosci Lett 1993; 149: 137-140.
44. Schmechel D, Saunders AM, Strittmatter WJ et al. Increased amyloid β peptide deposition in cerebral cortex as a consequence of apolipoprotein E genotype in late-onset Alzheimer disease. Proc Natl Acad Sci (USA) 1993; 90: 9649-9653.
45. Mann DMA, Iwatsubo T, Ihara Y et al. Predominant deposition of Aβ42(43) in plaques in cases of Alzheimer's disease and hereditary cerebral haemorrhage associated with mutations in the amyloid precursor protein gene. Am J Pathol 1996; 148: 1257-1266.
46. Hook EB. Down's syndrome: Its frequency in human populations and some factors pertinent to variations in rats. In: de la Cruz FF, Gerald PS, eds. Trisomy 21 (Down's syndrome): Research Perspectives. Baltimore: Baltimore University Park Press, 1981: 3-68.
47. Rumble B, Retallack R, Hilbich C et al. Amyloid (A4) protein and its precursor in Down's syndrome and Alzheimer's disease. N Engl J Med 1989; 320: 1446-1452.
48. Sherrington R, Rogaev EI, Liang Y et al. Cloning of a novel gene bearing missense mutations in early onset familial Alzheimer's disease. Nature 1995; 375: 754-760.
49. Clark RF, Hutton M, Fuldner R et al. The structure of the presenilin 1 (S182) gene and identification of six novel mutations in early onset AD families. Nature Genet 1995; 11: 219-222.
50. Wasco W, Pettingell WP, Jondro PD et al. Familial Alzheimer's chromosome 14 mutations. Nature Medicine 1995; 1: 848.
51. Perez-Tur J, Frodlich S, Prihar G et al. A mutation in Alzheimer's disease destroying a splice acceptor site in the presenilin-1 gene. NeuroReport 1996; 7: 204-207.
52. Perez-Tur J, Croxton R, Wright C et al. A further presenilin 1 mutation in the exon 8 cluster in familial Alzheimer's disease. Neurodegeneration 1996; 5: 207-212.
53. Sorbi S, Nacmias B, Forleo P et al. Missense mutation of S182 gene in Italian families with early-onset Alzheimer's disease. Lancet 1995; 346: 439-440.
54. Campion D, Flaman J-M, Brice A et al. Mutations of the presenilin 1 gene in families with early-onset Alzheimer's disease. Hum Molec Genet 1995; 4: 2373-2377.
55. Cruts M, Backhovens H, Wang S-Y et al. Molecular genetic analysis of familial

early-onset Alzheimer's disease linked to chromosome 14q 24.3. Hum Molec Genet 1995; 4: 2363-2371.
56. Tanahashi H, Mitsunaga Y, Takahashi K et al. Missense mutation of S182 gene in Japanese familial Alzheimer's disease. Lancet 1995; 346: 440.
57. Hutton M, Busfield F, Wragg M et al. Complete analysis of the presenilin 1 gene in early onset Alzheimer's disease. NeuroReport 1996; 7: 801-805.
58. Botova K, Vitek M, Mitsuda H et al. Mutation analysis of presenilin 1 gene in Alzheimer's disease. Lancet 1996; 347: 130-131.
59. Levy-Lahad E, Wasco W, Poorkaj P et al. Candidate gene for the chromosome 1 familial Alzheimer's disease locus. Science ; 269: 973-977.
60. Rogaev EI, Sherrington R, Rogaeva EA. Familial Alzheimer's disease in kindreds with missense mutations in a gene on chromosome 1 related to the Alzheimer's disease type 3 gene. Nature 1995; 376: 775-778.
61. Martin JJ, Gheuens J, Bruyland K et al. Early onset Alzheimer's disease in 2 large Belgian families. Neurology 1991; 41: 62-68.
62. Lampe TH, Bird TD, Nochlin D et al. Phenotype of chromosome 14-linked familial Alzheimer's disease in a large kindred. Ann Neurol 1994; 36: 368-378.
63. Haltia M, Viitanen M, Sulkava R et al. Chromosome 14-encoded Alzheimer's disease: genetic and clinicopathological description. Ann Neurol 1994; 36: 362-367.
64. Mann DMA, Iwatsubo T, Cairns NJ et al. Amyloid (Aβ) deposition in chromosome 14-linked Alzheimer's disease: predominance of $A\beta_{42(43)}$. Ann Neurol 1996; 40: 149-156.
65. Corder EH, Saunders AM, Strittmatter WJ et al. Gene dose of apolipoprotein E Type 4 allele and the risk of Alzheimer's disease in late onset families. Science 1993; 261: 921-923.
66. Saunders AM, Schmader K, Breitner JCS et al. Apolipoprotein E E4 allele distribution in late onset Alzheimer's disease and in other amyloid forming diseases. Lancet 1993; 342: 710-711.
67. Saunders AM, Strittmatter WJ, Schmechel D et al. Association of Apolipoprotein E4 with late-onset familial and sporadic Alzheimer's disease. Neurology 1993; 43: 1467-1472.
68. Strittmatter WJ, Saunders AM, Schmechel D et al. Apolipoprotein E: high-avidity binding to β-amyloid and increased frequency of type 4 allelle in late-onset familial Alzheimer's disease. Proc Natl Acad Sci (USA) 1993; 90: 1977-1981.
69. Poirier J, Davignon J, Bouthillier D et al. Apolipoprotein polymorphism and Alzheimer's disease. Lancet 1993; 342: 697-699.
70. Payami H, Kaye J, Heston LL et al. Apolipoprotein E genotype and Alzheimer's disease. Lancet 1993; 342: 738.
71. Noguchi S, Murakami K, Yamada N. Apolipoprotein E genotype and Alzheimer's disease. Lancet 1993; 342: 737.
72. Rebeck GW, Reiter JS, Strickland DK et al. Apolipoprotein E in sporadic Alzheimer's disease: Allelic variation and receptor interactions. Neuron 1993; 11: 575-580.
73. Houlden H, Crook R, Duff K et al. Confirmation that the apolipoprotein E4 allele is associated with late onset familial Alzheimer's disease. Neurodegeneration 1993; 2: 283-288.
74. Mayeux R, Stern Y, Ottman R et al. The Apolipoprotein E4 allele in patients with Alzheimer's disease. Ann Neurol 1993; 34: 752-754.
75. Pickering-Brown SM, Roberts D, Owen F et al. Apolipoprotein E4 alleles and non-Alzheimer forms of dementia. Neurodegeneration 1994; 3: 95-96.
76. Sorbi S, Nacmias B, Forleo P et al. ApoE allele frequencies in Italian familial and sporadic Alzheimer's disease. Neurosci Lett 1994; 177: 100-102.
77. St Clair D, Norrman J, Perry R et al. Apolipoprotein E4 allele frequency in patients with Lewy body dementia, Alzheimer's disease and age matched controls. Neurosci Lett 1994; 176: 45-46.
78. West HL, Rebeck GW, Hyman BT. Frequency of the apolipoprotein E, E2 allele is diminished in sporadic Alzheimer disease. Neurosci Lett 1994; 175: 46-48.

79. Zubenko GS, Stiffler S, Stabler S et al. Association of the apolipoprotein E e4 allele with clinical subtypes of autopsy-confirmed Alzheimer's disease. Am J Med Genet 1994; 54: 199-205.
80. van Duijn CM, de Knijff P, Cruts M et al. Apolipoprotein E4 allele in a population-based study of early-onset Alzheimer's disease. Nature Genet 1994; 7: 74-78.
81. Chartier-Harlin MC, Parfitt M, Legrain S et al. Apolipoprotein E, E4 allele as a major risk factor for sporadic early and late-onset form of Alzheimer's disease. Hum Molec Genet 1994; 3: 569-574.
82. Okuizumi K, Onodera O, Tanaka H et al. ApoE-e4 and early onset Alzheimer's disease. Nature Genet 1994; 7: 10-11.
83. Perez-Tur J, Campion D, Martinez M et al. Evidence for apolipoprotein E e4 association in early-onset Alzheimer's patients with late-onset relatives. Am J Med Genet 1995; 60: 550-553.
84. Wragg M, Hutton M, Talbot C et al. Genetic association between an intronic polymorphism in the presenilin-1 gene and late onset Alzheimer's disease. Lancet 1996; 347: 509-512.
85. Talbot CJ, Houlden H, Craddock N et al. Polymorphism in AACT gene may lower age of onset of Alzheimer's disease. NeuroReport 1996; 7: 534-536.
86. Kamboh MI, Sanghera DK, Ferrell RE et al. APOE e4-associated Alzheimer's disease risk is modified by α-antichymotrypsin polymorphism. Nature Genet 1995; 10: 486-488.
87. Okuizumi K, Onodera O, Namba Y et al. Genetic association of the very low density lipoprotein (VLDL) receptor gene with sporadic Alzheimer's disease. Nature Genet 1995; 11: 207-209.
88. Bird TD, Lampe TH, Nemens EJ et al. Familial Alzheimer's disease in American descendants of the Volga Germans: probable genetic founder effect. Ann Neurol 1988; 23: 25-31.
89. Mann DMA, Iwatsubo T, Nochlin D et al. Amyloid (Aβ) protein deposition in chromosome 1 linked Alzheimer's disease–the Volga German kindreds. Ann Neurol 1996; (In press):.
90. Kennedy AM, Newman SK, Frackowiak RSJ et al. Chromosome 14 linked familial Alzheimer's disease: A clinico-pathological study of a single pedigree. Brain 1995; 118: 185-205.
91. Higgins LS, Cordell B. Genetically engineered animal models of human neurodegenerative diseases. Neurodegeneration 1995; 4: 117-129.
92. Quon D, Wang Y, Catalano R et al. Formation of β amyloid protein in brains of transgenic mice. Nature 1991; 352: 239-441.
93. Higgins LS, Holtzman DM, Rabin J et al. Transgenic mouse brain histopathology resembles early Alzheimer's disease. Ann Neurol 1994; 35: 598-607.
94. Higgins LS, Rodems JM, Catalano C et al. Early Alzheimer's disease-like histopathology increases in frequency with age in mice transgenic for β-APP751. Proc Natl Acad Sci (USA) 1995; 92: 4402-4406.
95. Johnson SA, McNeil T, Cordell B et al. Relationship of neuronal APP751/APP695 in mRNA ratios to neuritic plaque density in Alzheimer's disease. Science 1990; 248: 854-857.
96. Games D, Adams D, Alessandrini R et al. Alzheimer-type neuropathology in transgenic mice overexpressing V717F β amyloid precursor protein. Nature 1995; 373: 523-527.
97. Cai X-D, Golde TE, Younkin SG. Release of excess amyloid β protein from a mutant amyloid β protein precursor. Science 1993; 259: 514-516.
98. Suzuki N, Cheung TT, Cai X-D et al. An increased percentage of long amyloid β protein secreted by familial amyloid β protein precursor (βAPP717) mutants. Science 1994; 264: 1336-1340.
99. Jarrett JT, Berger EP, Lansbury PT. The carboxy terminus of the β-amyloid protein is critical for the seeding of amyloid formation. Implications for the pathogenesis of Alzheimer's disease. Biochemistry 1993; 32: 4693-4697.
100. Iwatsubo T, Odaka N, Suzuki N et al. Visualization of Aβ42(43)-positive and Aβ40-positive senile plaques with end-specific Aβ monclonal antibodies: Evidence that

an initially deposited species is Aβ1-42(43). Neuron 1994; 13: 45-53.
101. Tamaoka A, Odaka A, Ishibashi Y et al. APP717 mis-sense mutation affects the ratio of amyloid β protein species (Aβ1-42/43 and Aβ1-40) in familial Alzheimer's disease brain. J Biol Chem 1994; 269: 32721-32724.
102. Citron M, Oltersdorf T, Haass C et al. Mutation of the β-amyloid precursor protein in familial Alzheimer's disease increases β-protein production. Nature 1992; 360: 672-674.
103. Dovey HF, Suomesaari-Chrysler S, Lieberburg I et al. Cells with a familial Alzheimer's disease mutation produce authentic β-peptide. NeuroReport 1993; 4: 1039-1042.
104. Felsenstein KM, Hunihan LW, Roberts SB. Altered cleavage and secretion of a recombinant β-APP bearing the Swedish familial Alzheimer's disease mutation. Nature Genet 1994; 6: 251-256.
105. Johnston JA, Cowburn RF, Norgren S et al. Increased β amyloid release and levels of amyloid precursor protein (APP) in fibroblast cell lines obtained from family members with the Swedish APP670.671 mutation. FEBS Lett 1994; 354: 274-278.
106. Lo ACY, Haass C, Wagner SL et al. Metabolism of the "Swedish" amyloid precursor protein variant in Madin-Darby canine kidney cells. J Biol Chem 1994; 269: 30966-30973.
107. De Strooper B et al. Basolateral secretion of amyloid precursor protein in Madin-Darby canine kidney cells is disturbed by alterations of intracellular pH and by introducing a mutation associated with familial Alzheimer's disease. J Biol Chem 1995; 270: 4058-4065.
108. Haass C, Lemere CA, Capell A et al. The Swedish mutation causes early-onset Alzheimer's disease by β-secretase cleavage within the secretory pathway. Nature Medicine 1995; 1: 1291-1296.
109. Haass C, Koo EH, Mellon A et al. Targeting of cell-surface β-amyloid precursor protein to lysosomes : alternative processing into amyloid bearing fragments. Nature 1992; 357: 500-502.
110. Estus S, Golde TE, Kunishita T et al. Potentially amyloidogenic carboxy-terminal derivatives of the amyloid protein precursor. Science 1992; 255: 726-728.
111. Golde TE, Estus S, Younkin LH et al. Processing of the amyloid protein precursor to potentially amyloidogenic derivatives. Science 1992; 255: 728-730.
112. Naslund J, Schierhorn A, Hellman U et al. Relative abundance of Alzheimer Aβ amyloid peptide variants in Alzheimer disease and normal aging. Proc Natl Acad Sci (USA) 1994; 91: 8378-8382.
113. Iwatsubo T, Mann DMA, Odaka A et al. Amyloid β protein (Aβ) deposition: Aβ42(43) precedes Aβ40 in Down syndrome. Ann Neurol 1995; 37: 294-299.
114. Mann DMA, Iwatsubo T, Fukumoto H et al. Microglial cells and amyloid β protein (Aβ) deposition; association with Aβ40 containing plaques. Acta Neuropath 1995; 90: 472-477.
115. Mann DMA, Iwatsubo T, Pickering-Brown SM et al. Preferential deposition of amyloid β protein (Aβ) in the form $A\beta_{40}$ in Alzheimer's disease is associated with a gene dosage effect at the Apolipoprotein E E4 allele. Neurosci Lett 1996; in press.
116. Kida E, Wisniewski KE, Wisniewski HM. Early amyloid-β deposits show different immunoreactivity to the amino- and carboxy-terminal regions of β-peptide in both Alzheimer's disease and Down's syndrome brain. Neurosci Lett 1995; 193: 1-4.
117. Lemere CA, Blusztajn JK, Yamaguchi H et al. Sequence of deposition of heterogeneous amyloid β-peptides and APO E in Down syndrome: Implications for initial events in amyloid plaque formation. Neurobiol Dis 1996; 3: 16-32.
118. Teller JK, Russo C, de Busk LM et al. Presence of soluble amyloid β peptide precedes amyloid plaque formation in Down's syndrome. Nature Medicine 1996; 2: 93-95.
119. Anwar R, Moynihan TP, Ardley H et al. Molecular analysis of the presenilin 1 (S182) gene in 'sporadic' cases of Alzheimer's disease: identification and characterization of unusual splice variants. J Neurochem 1996; 66: 1774-1777.

120. Sahara N, Yahagi Y-I, Takagi H et al. Identification and characterization of presenilin I-467, I-463 and I-374. FEBS Lett 1996; 381: 7-11.
121. Suzuki T, Nishiyama K, Murayama S et al. Regional and cellular presenilin I gene expression in human and rat tissues. Biochem Biophys Res Commun 1996; 219: 708-713.
122. Kovacs DM, Fausett HJ, Page KJ et al. Alzheimer-associated presenilins 1 and 2: Neuronal expression in brain and localization to intracellular membranes in mammalian cells. Nature Medicine 1996; 2: 224-229.
123. Barton AJL, Crook BW, Karran EH et al. Alteration in brain presenilin 1 mRNA expression in early onset familial Alzheimer's disease. Neurodegeneration 1996; 5: 213-218.
124. Ward RV, Davis JB, Gray CW et al. Presenilin-1 is processed into two major cleavage products in neuronal cell lines. Neurodegeneration 1996; (In press):.
125. Thinakaran G, Borchelt DR, Lee MK et al. Endoproteolysis of presenilin 1 and accumulation of processed derivatives in vivo. Neuron 1996; 17: 181-190.
126. Scheuner D, Eckman C, Jensen M et al. The amyloid β protein deposited in the senile plaques of Alzheimer's disease is increased *in vivo* by the presenilin 1 and 2 and *APP* mutations linked to familial Alzheimer's disease. Nature Medicine 1996; 2: 864-870.
127. Corder EH, Saunders AM, Risch NJ et al. Protective effect of apolipoprotein E type 2 allele decreases risk of late onset Alzheimer's disease. Nature Genet 1994; 7: 180-184.
128. Talbot C, Lendon C, Craddock N et al. Protection against Alzheimer's disease with apoE e2. Lancet 1994; 343: 1432-1433.
129. Van Broeckhoven C, Backhovens H, Cruts M et al. APOE genotype does not modulate age of onset in families with chromosome 14 encoded Alzheimer's disease. Neurosci Lett 1994; 169: 179-180.
130. Hardy J, Houlden H, Collinge J et al. Apolipoprotein E genotype and Alzheimer's disease. Lancet 1993; 342: 737-738.
131. Nacmias B, Latorraca S, Piersanti P et al. ApoE genotype and familial Alzheimer's disease: a possible influence on age of onset in APP717 Val→Ile mutated families. Neurosci Lett 1995; 183: 1-3.
132. Haan J, van Broeckhoven C, van Duijn CM et al. The apolipoprotein E e4 allele does not influence the clinical expression of the amyloid precursor protein gene codon 693 or 692 mutations. Ann Neurol 1994; 36: 434-437.
133. Royston MC, Mann DMA, Pickering-Brown S et al. ApoE2 allele promotes longevity and protects patients with Down's syndrome from the development of dementia. NeuroReport 1994; 5: 2583-2585.
134. Wisniewski T, Morelli L, Wegiel J et al. The influence of Apolipoprotein E isotypes on Alzheimer's disease pathology in 40 cases of Down's syndrome. Ann Neurol 1995; 37: 136-138.
135. van Gool WA, Evenhuis HM, van Duijn CM. A case-control study of apolipoprotein E genotypes in Alzheimer's disease associated with Down's syndrome. Ann Neurol 1995; 38: 225-230.
136. Schachter F, Faure-Delanef L, Guenot F et al. Genetic associations with human longevity at the APOE and ACE loci. Nature Genet 1994; 6: 29-31.
137. Lannfelt L, Lilius L, Nastase M et al. Lack of association between apolipoprotein E allele e4 and sporadic Alzheimer's disease. Neurosci Lett 1994; 169: 175-178.
138. Henderson AS, Easteal S, Jorm AF et al. Apolipoprotein E allele e4, dementia, and cognitive decline in a population sample. Lancet 1995; 346: 1887-1890.
139. Sobel E, Louhija J, Sulkava R et al. Lack of association of the apolipoprotein E allele e4 with late-onset Alzheimer's disease in Finnish centenarians. Neurology 1995; 45: 903-907.
140. Corder EH, Basun H, Lannfelt L et al. Attenuation of apolipoprotein E e4 allele gene dose in late age. Lancet 1996; 347: 542.
141. Namba Y, Tomonaga M, Kawasaki H et al. Apolipoprotein E immunoreactivity in cerebral amyloid deposits and neurofibrillary tangles in Alzheimer's disease and Kuru plaque amyloid in Creutzfeldt-Jakob disease. Brain Res 1991; 541: 163-166.

142. Kida E, Golabek AA, Wisniewski T et al. Regional differences in apolipoprotein E immunoreactivity in diffuse plaques in Alzheimer's disease brain. Neurosci Lett 1994; 167: 73-76.
143. Wisniewski T, Frangione B. Apolipoprotein E: a pathological chaperone protein in patients with cerebral and systemic amyloid. Neurosci Lett 1992; 135: 235-238.
144. Davies CA, Mann DMA. Co-localization of apolipoprotein E and amyloid β protein in Down's syndrome. Ann Neurol 1996; (in press):.
145. Strittmatter WJ, Weisgraber KH, Huang DY et al. Binding of human apolipoprotein E to synthetic amyloid β peptide: isoform specific effects and implications for late-onset Alzheimer disease. Proc Natl Acad Sci (USA) ; 90: 8098-8102.
146. Ma J, Yee A, Brewer B et al. Amyloid-associated proteins of α-1 antichymotrypsin and apolipoprotein E promote assembly of Alzheimer β protein into filaments. Nature 1994; 372: 92-94.
147. Sanan DA, Weisgraber KH, Russell SJ et al. Apolipoprotein E associates with β amyloid peptide of Alzheimer's disease to form novel monofibrils. J Clin Invest 1994; 94: 860-869.
148. LaDu MJ, Falduto MT, Manelli AM et al. Isoform-specific binding of apolipoprotein E to β-amyloid. J Biol Chem 1994; 269: 23403-23406.
149. LaDu MJ, Pederson TM, Frail DE et al. Purification of apolipoprotein E attenuates isoform-specific binding to β-amyloid. J Biol Chem 1995; 270: 9039-9042.
150. Benjamin R, Leake A, Edwardson JA et al. Apolipoprotein E genes in Lewy body and Parkinson's disease. Lancet 1994; 343: 1565.
151. Pickering-Brown S, Mann DMA, Bourke JP et al. Apolipoprotein E4 and Alzheimer's disease pathology in Lewy body disease and in other β-amyloid forming diseases. Lancet 1994; 343: 1155.
152. Galasko D, Saitoh T, Xia Y et al. The apolipoprotein E allele E4 is over-represented in patients with the Lewy body variant of Alzheimer's disease. Neurology 1994; 44: 1950-1951.
153. Harrington CR, Louwagie J, Rossau R et al. Influence of apolipoprotein E genotype on senile dementia of the Alzheimer and Lewy body types. Am J Pathol 1994; 145: 1472-1484.
154. Nicoll JAR, Roberts GW, Graham DI. Apolipoprotein E e4 allele is associated with deposition of amyloid β-protein following head injury. Nature Medicine 1995; 1: 135-137.
155. Mayeux R, Ottman R, Maestre G et al. Synergistic effects of traumatic head injury and apolipoprotein-epsilon 4 in patients with Alzheimer's disease. Neurology 1995; 45: 555-557.
156. Pickering-Brown S, Mann DMA, Owen F et al. Allelic variations in Apolipoprotein E and prion protein genotype related to plaque formation and age of onset in sporadic Creutzfeldt-Jakob disease. Neurosci Lett 1994; 187: 127-129.
157. Al-Chalabi A, Enayat ZE, Bakker MC et al. Association of apolipoprotein E e4 allele with bulbar-onset motor neuron disease. Lancet 1996; 347: 159-160.
158. Mui S, Rebeck W, McKenna-Yasek D et al. Apolipoprotein E e4 allele is not associated with earlier age at onset in amyotrophic lateral sclerosis. Ann Neurol 1995; 38: 460-463.
159. Pickering-Brown SM, Siddons M, Mann DMA et al. Apolipoprotein E allelic frequencies in patients with lobar atrophy. Neurosci Lett 1995; 188: 205-207.
160. Berr C, Hauw J-J, Delaere P et al. Apolipoprotein E allele E4 is linked to increased deposition of the amyloid β-peptide (A-β) in cases with or without Alzheimer's disease. Neurosci Lett 1994; 178: 221-224.
161. Ohm TG, Kirca M, Bohl J et al. Apolipoprotein E polymorphism influences not only cerebral senile plaque load but also Alzheimer-type neurofibrillary tangle formation. Neuroscience 1995; 66: 585-587.
162. Polvikoski T, Sulkava R, Haltia M et al. Apolipoprotein E, dementia, and cortical deposition of β-amyloid protein. N Engl J Med 1995; 333: 1242-1247.
163. Hyman B, West HL, Rebeck GW et al. Neuropathological changes in Down's syndrome, hippocampal formation: effect of age and apolipoprotein E genotype. Arch Neurol 1995; 52: 373-378.

164. Benjamin R, Leake A, Ince PG et al. Effects of apolipoprotein E genotype on cortical neuropathology in senile dementia of the Lewy body type and Alzheimer's disease. Neurodegeneration 1995; 4: 443-448.
165. Gearing M, Mori H, Mirra SS. Aβ peptide length and apolipoprotein E genotype in Alzheimer's disease. Ann Neurol 1996; 39: 395-399.
166. Heinonen O, Lehtovirta M, Soininen H et al. Alzheimer pathology of patients carrying apolipoprotein E E4 allele. Neurobiol Aging 1995; 16: 505-513.
167. Itoh Y, Yamada M. Apolipoprotein E and the neuropathology of dementia. N Engl J Med 1996; 334: 599-600.
168. Mann DMA, Pickering-Brown SM, Siddons MA et al. The extent of amyloid deposition in brain in patients with Down's syndrome does not depend upon the apolipoprotein E genotype. Neurosci Lett 1995; 196: 105-108.
169. Wisniewski HM, Wegiel J, Kotula L. Some neuropathological aspects of Alzheimer's disease and its relevance to other disciplines. Neuropath Appl Neurobiol 1996; 22: 3-11.
170. Soininen H, Kosunen O, Helisalmi S et al. A severe loss of choline acetyltransferase in the frontal cortex of Alzheimer patients carrying apolipoprotein e4 allele. Neurosci Lett 1995; 187: 79-82.
171. Fukumoto H, Asami-Odaka A, Suzuki N et al. Association of Aβ40 positive senile plaques with microglia cells in the brains of patients with Alzheimer's disease and non-demented aged individuals. Neurodegeneration 1996; 5: 13-17.
172. Strittmatter WJ, Weisgraber K, Goedert M et al. Hypothesis: Microtubule instability and paired helical filament formation in the Alzheimer disease brain are related to Apolipoprotein E genotype. Exp Neurol 1994; 125: 163-171.
173. Nathan BP, Chang K-C, Bellosta S et al. The inhibitory effect of apolipoprotein E4 on neurite outgrowth is associated with microtubule depolymerization. J Biol Chem 1995; 270: 19791-19799.
174. Huang DY, Weisgraber KH, Goedert M et al. ApoE3 binding to tau tandem repeat I is abolished by tau serine262 phosphorylation. Neurosci Lett 1995; 192: 1-4.
175. Mahley RW. Apolipoprotein E: cholesterol transport protein with expanding role in cell biology. Science 1988; 240: 622-630.
176. Diedrich JF, Minnigan H, Carp RI et al. Neuropathological changes in Scrapie and Alzheimer's disease are associated with increased expression of apolipoprotein E and cathepsin D in astrocytes. J Virol 1991; 65: 4759-4768.
177. Poirier J, Hess M, May PC et al. Astrocytic apolipoprotein E mRNA and GFAP mRNA in hippocampus after entorhinal cortex lesioning. Mol Brain Res 1991; 11: 97-106.
178. Pitas RE, Boyles JK, Lee SH et al. Astrocytes synthesize apolipoprotein E and metabolize apolipoprotein E-containing lipoproteins. Biochim Biophys Acta 1987; 917: 148-161.
179. Skene JHP, Shooter EM. Denervated sheath cells secrete a new protein after nerve injury. Proc Natl Acad Sci (USA) 1983; 80: 4169-4173.
180. Muller HW, Ignatius MJ, Hangen DH et al. Expression of specific sheath cell proteins during peripheral nerve growth and regeneration in mammals. J Cell Biol 1986; 102: 393-402.
181. Han S-H, Hulette C, Saunders AM et al. Apolipoprotein E is present in hippocampal neurons without neurofibrillary tangles in Alzheimer's disease and in age-matched controls. Exp Neurol 1994; 128: 13-26.
182. Nathan BP, Bellosta S, Sanan DA et al. Differential effects of apolipoproteins E3 and E4 on neuronal growth in vitro. Science 1994; 264: 850-852.
183. Bertrand P, Poirier J, Oda T et al. Association of apolipoprotein E genotype with brain levels of apolipoprotein E and apolipoprotein J (clusterin) in Alzheimer's disease. Mol Brain Res 1995; 33: 174-178.
184. Masliah E, Mallory M, Alford M et al. Abnormal synaptic regeneration in APP695 transgenic and apoE knockout mice. In: Iqbal K, Mortimer JA, Winblad B et al., eds. Research Advances in Alzheimer's Disease and Related Disorders. John Wiley and Sons, 1995: 405-414.
185. Barton AJL, Harrison PJ, Najlerahim A et al. Increased tau messenger RNA in Alzheimer's disease hippocampus. Am J Pathol 1990; 137: 497-502.

186. Wasco W, Bupp K, Magendantz M et al. Identification of a mouse brain cDNA that encodes a protein related to the Alzheimer disease-associated amyloid β precursor. Proc Natl Acad Sci (USA) 1992; 89: 10758-10762.
187. Wasco W, Gurubhagavatula S, Paradis M d et al. Isolation and characterization of APLP2 encoding a homologue of the Alzheimer's associated amyloid β protein precursor. Nature Genet 1993; 5: 95-99.
188. Webster M-T, Groome N, Francis PT et al. A novel protein, amyloid precursor-like protein 2, is present in human brain, cerebrospinal fluid and conditioned media. Biochem J 1995; 310: 95-99.
189. Sprecher CA, Grant FJ, Grimm G et al. Molecular cloning of the cDNA for a human amyloid precursor protein homolog: evidence for a multigene family. Biochemistry 1993; 32: 4481-4486.
190. Slunt HH, Thinakaran G, Von Koch C et al. Expression of a ubiquitous, cross-reactive homologue of the mouse β-amyloid precursor protein (APP). J Biol Chem 1994; 269: 2637-2644.
191. Rempfer R, Crook R, Houlden H et al. Parkinson's disease, but not Alzheimer's disease, Lewy body variant associated with mutant alleles at cytochrome P450 gene. Lancet 1994; 344: 815.
192. Benitez J, Barquero MS, Coria F et al. Oxidative polymorphism of debrisoquine is not related to the risk of Alzheimer's disease. J Neurol Sci 1993; 117: 8-11.
193. Saitoh T, Xia Y, Chen X et al. The CYP2D6B mutant allele is overrepresented in the Lewy body variant of Alzheimer's disease. Ann Neurol 1995; 37: 110-112.
194. Smith CAD, Gough AC, Leigh PN et al. Debrisoquine hydroxylase gene polymorphism and susceptibility to Parkinson's disease. Lancet 1992; 339: 1375-1377.
195. Lin FH, Lin R, Wisniewski HM et al. Detection of point mutations in codon 331 of mitochondrial NADH dehydrogenase subunit 2 in Alzheimer's brains. Biochem Biophys Res Commun 1991; 182: 238-246.
196. Jenner P, Schapira AHV, Marsden CD. New insights into the cause of Parkinson's disease. Neurology 1992; 42: 2241-2250.
197. Ward CD, Duvoisin RC, Ince SE et al. Parkinson's disease in 65 pairs of twins and in a set of quadruplets. Neurology 1983; 33: 815-824.
198. Duvoisin R. Genetics of Parkinson's disease. Adv Neurol 1986; 45: 307-312.
199. Perry TL, Wright JM, Berry K et al. Dominantly inherited apathy, central hypoventilation, and Parkinson's syndrome. Neurology 1990; 40: 1882-1887.
200. Johnson WG. Genetic susceptibility to Parkinson disease. Neurology 1991; 41: 82-87.
201. Golbe LI, Iorio G, Bonavita V et al. A large kindred with autosomal dominant Parkinson disease. Ann Neurol 1990; 27: 276-282.
202. Maraganore DM, Harding AE, Marsden CD. A clinical and genetic study of Parkinson disease. Mov Disord 1991; 6: 205-211.
203. Tanaka H, Ishikawa A, Ginns EL et al. Linkage analysis of juvenile parkinsonism to tyrosine hydroxylase. Neurology 1991; 41: 719-722.
204. Ozawa T, Tanaka M, Ino H et al. Distinct clustering of point mutations in mitochondrial DNA among patients with mitochondrial encephalomyopathies and with Parkinson's disease. Biochem Biophys Res Commun 1991; 176: 938-946.
205. Ikebe S, Tanaka M, Ohno K et al. Increase of deleted mitochondrial DNA in the striatum in Parkinson's disease and senescence. Biochem Biophys Res Commun 1988; 170: 1044-1048.
206. Lestienne P, Nelson I, Riederer P et al. Mitochondrial DNA in postmortem brain from patients with Parkinson's disease. J Neurochem 1991; 57: 1809-1991.
207. Lestienne P, Nelson I, Riederer P et al. Normal mitochondrial genome in brain from patients with Parkinson's disease and complex I defect. J Neurochem 1990; 55: 1810-1812.
208. Corral-Debrinski M, Horton T, Lott MT et al. Mitochondrial DNA deletions in human brain: regional variability and increase with advanced age. Nature Genet 1992; 2: 324-329.
209. Mecocci P, MacGarvey U, Kaufman AE et al. Oxidative damage to mitochondrial DNA shows marked age-dependent increases in

human brain. Ann Neurol 1993; 34: 609-616.
210. Filburn CR, Edris W, Tamatani M et al. Mitochondrial electron transport chain activities and DNA deletions in regions of the rat brain. Mech Aging Dev 1996; 87: 35-46.
211. Roberts GW, Gentleman SM, Lynch A et al. β/A4 amyloid protein deposition in the brain after head injury. Lancet 1991; 338: 1422-1423.
212. Nicoll JAR, Roberts GW, Graham DI. Apolipoprotein E e4 allele is associated with deposition of amyloid β-protein following head injury. Nature Medicine 1995; 1: 135-137.
213. Graham DI, Gentleman SM, Lynch A et al. Distribution of β-amyloid protein in the brain following severe head injury. Neuropath Appl Neurobiol 1995; 21: 27-34.
214. Roberts GW, Allsop D, Bruton C. The occult aftermath of boxing. J Neurol Neurosurg Psychiatry 1990; 53: 373-378.
215. Hof PR, Bouras C, Buee L et al. Differential distribution of neurofibrillary tangles in the cerebral cortex of dementia pugilistica and Alzheimer's disease cases. Acta Neuropathol 1992; 85: 23-30.
216. Geddes JF, Vowles GH, Robinson SFD et al. Neurofibrillary tangles, but not Alzheimer-type pathology, in a young boxer. Neuropath Appl Neurobiol 1996; 22: 12-16.
217. Gentleman SM, Nash MJ, Sweeting CJ et al. β-amyloid precursor protein (β-APP) as a marker for axonal injury after head injury. Neurosci Lett 1993; 160: 139-144.
218. Corsellis JAN, Brierley JB. Observations on the pathology of insidious dementia following head injury. J Ment Sci 1959; 105: 714-720.
219. Rudelli R, Strom JO, Welch PT et al. Posttraumatic premature Alzheimer's disease: Neuropathologic findings and pathogenetic considerations. Arch Neurol 1982; 39: 570-575.
220. Clinton J, Ambler MW, Roberts GW. Post-traumatic Alzheimer's disease: preponderance of a single plaque type. Neuropath Appl Neurobiol 1991; 17: 69-74.
221. McLachlan DR. Inorganic neurotoxins in dementia caused by neurodegeneration. In: Calne DB, ed. Neurodegenerative Diseases. Philadelphia, London, Toronto, Montreal, Sydney, Tokyo: WB Saunders Company, 1994: 241-249.
222. Crapper DR, Krishnan SS, Quittkat S. Aluminum, neurofibrillary degeneration and Alzheimer's disease. Brain 1976; 99: 67-80.
223. Lukiw WJ, Kruck TPA, Krishnan B et al. Nuclear compartmentalization of aluminum in Alzheimer's disease. Neurobiol Aging 1992; 13: 115-121.
224. Candy JM, Oakley AE, Klinowski J et al. Aluminosilicates and senile plaque formation in Alzheimer's disease. Lancet 1986; 1: 354-357.
225. Nikaido T, Austin JH, Trueb L et al. Studies in aging of the brain. II Microchemical analyses of the nervous system in Alzheimer patients. Arch Neurol 1972; 27: 549-554.
226. Masters CL, Simms G, Weinman NA et al. Amyloid plaque core protein in Alzheimer disease and Down syndrome. Proc Natl Acad Sci (USA) 1985; 82: 4245-4249.
227. Chafi AH, Hauw J-J, Rancurel G et al. Absence of aluminum in Alzheimer's disease brain tissue: electron microprobe and ion microprobe studies. Neurosci Lett 1991; 123: 61-64.
228. Jacobs RW, Duong T, Jones RE et al. A reexamination of aluminum in Alzheimer's disease: analysis by energy dispersive x-ray microprobe and flameless atomic absorption spectrophotometry. Can J Neurol Sci 1989; 16: 498-503.
229. Landsberg JP, McDonald B, Watt JF. Absence of aluminum in neuritic plaque cores in Alzheimer's disease. Nature 1992; 360: 65-68.
230. Perl DP, Brody AR. Alzheimer's disease: x-ray spectrometric evidence of aluminum accumulation in neurofibrillary tangle-bearing neurons. Science 1980; 208: 297-299.
231. Good PF, Olanow CW, Perl DP. Neuromelanin-containing neurons of the substantia nigra accumulate iron and aluminum in Parkinson's disease: A LAMMA study. Brain Res 1992; 593: 343-346.
232. Lovell MA, Ehmann WD, Markesbery WR. Laser microprobe analysis of brain alumi-

num in Alzheimer's disease. Ann Neurol 1993; 33: 36-42.
233. Martyn CN, Osmond C, Edwardson JA et al. Geographic relation between Alzheimer's disease and aluminum in drinking water. Lancet 1989; i: 59-62.
234. Neri LC, Hewitt D. Aluminum, Alzheimer's disease and drinking water. Lancet 1991; 338: 592-593.
235. Forbes WF, Hayward LM, Agwani N. Dementia, aluminum, and fluoride. Lancet 1991; 338: 1592-1593.
236. Vogt T. Water quality and health–Study of a possible relationship between aluminum in drinking water and dementia (Abstract). Oslo, Central Bureau of Statistics of Norway, 1986.
237. Michel P, Commenges D, Dartigues JF et al. Study of the relationship between aluminum concentrations in drinking water and risk of Alzheimer's disease. In: Iqbal K, McLaughlan DRC, Winblad B et al., eds. Alzheimer's Disease Basic Mechanisms, Diagnosis and Therapeutic Strategies. Chichester: J Wiley & Sons, 1991: 387-389.
238. Wood DJ, Cooper C, Stevens J et al. Bone mass and dementia in hip fracture in patients from areas with different aluminum concentrations in water supplies. Age Aging 1988; 17: 415-419.
239. Wettstein A, Aeppli J, Gautschi K et al. Failure to find a relationship between mnestic skills of octogenarians and aluminum in drinking water. Arch Occup Environ Health 1991; 63: 97-103.
240. Doll R. Alzheimer's disease and environmental aluminum. Age Aging 1993; 22: 138-153.
241. Klazo I, Wisniewski HM, Streicher E. Experimental production of neurofibrillary degeneration: I. Light microscopic observations. J Neuropathol Exp Neurol 1965; 24: 187-199.
242. Terry RD, Pena C. Experimental production of neurofibrillary degeneration: 2. Electron microscopy, phosphatase histochemistry and electron probe analysis. J Neuropathol Exp Neurol 1965; 24: 200-210.
243. Guy S, Jones D, Mann DMA et al. Neuroblastoma cells treated with aluminum-EDTA express an epitope associated with Alzheimer's disease neurofibrillary tangles. Neurosci Lett 1991; 121: 166-168.
244. Mesco ER, Kachen C, Timiras PS. Effects of aluminum on tau proteins in human neuroblastoma cells. Molec Chem Neuropathol 1991; 14: 199-212.
245. Candy JM, McArthur FK, Oakley AE et al. Aluminum accumulation in relation to senile plaque and neurofibrillary tangle formation in the brains of patients with renal failure. J Neurol Sci 1992; 107: 210-218.
246. Alfrey AC, LeGendre GR, Kaehny WD. The dialysis encephalopathy syndrome: possible aluminum intoxication. N Engl J Med 1976; 294: 184-188.
247. Burks JS, Alfrey AC, Huddlestone J et al. A fatal encephalopathy in chronic haemodialysis patients. Lancet 1976; i: 764-768.
248. Harrington CR, Wischik CM, McArthur FK et al. Alzheimer's disease-like changes in tau protein processing: association with aluminum accumulation in brains of renal dialysis patients. Lancet 1994; 343: 993-997.
249. Stroop WG, Rock D, Fraser NW. Localization of herpes simplex virus in the trigeminal and olfactory systems in the mouse central nervous system during acute and latent infections by *in situ* hybridization. Lab Invest 1984; 51: 27-38.
250. Libikova H, Pogady J, Wiedermann V et al. Search for herpetic antibodies in the cerebrospinal fluid in senile dementia and mental retardation. Acta Virologica 1975; 19: 494-495.
251. Mann DMA, Yates PO, Davies JS et al. Viruses and Alzheimer's disease. J Neurol Neurosurg Psychiatry 1982; 45: 759-760.
252. Esiri MM. Viruses and Alzheimer's disease. J Neurol Neurosurg Psychiatry 1982; 45: 759.
253. Pogo BGT, Casals J, Elizan TS. A study of viral genomes and antigens in brains of patients with Alzheimer's disease. Brain 1987; 110: 907-915.
254. Sequiera LW, Jennings LC, Carrasco LH et al. Detection of herpes simplex viral genome in brain tissue. Lancet 1979; ii: 608-612.

255. Fraser NW, Lawrence WC, Wroblewska Z et al. Herpes simplex virus type 1 DNA in human brain tissue. Proc Natl Acad Sci (USA) 1981; 78: 6461-6465.
256. Taylor GR, Crow TJ, Markakis DA et al. Herpes simplex virus and Alzheimer's disease: a seach for virus DNA by spot hybridisation. J Neurol Neurosurg Psychiatry 1984; 47: 1061-1065.
257. Efstathiou S, Minson AC, Field HJ et al. Detection of herpes simplex virus-specific DNA sequences in latently infected mice and in humans. J Virol 1986; 57: 446-455.
258. Croen KD, Ostrove JM, Dragovic LJ et al. Latent herpes simplex virus in human trigeminal ganglia. Detection of an immediate early gene "anti-sense" transcript by *in situ* hybridization. N Engl J Med 1987; 317: 1427-1432.
259. Gordon YJ, Johnson B, Romanowski E et al. RNA complementary to herpes simplex virus type 1 ICP0 gene demonstrated in neurons of human trigeminal ganglia. J Virol 1988; 62: 1832-1835.
260. Steiner I, Spivack JG, O'Boyle DR et al. Latent herpes simplex virus type 1 transcription in human trigeminal ganglia. J Virol 1988; 62: 3493-3496.
261. Middleton PJ, Petric M, Kozak M et al. Herpes simplex viral genome and senile and presenile dementias of Alzheimer and Pick. Lancet 1980; i: 1038.
262. Dealty AM, Haase AT, Fewster PH et al. Human herpes virus infections and Alzheimer's disease. Neuropath Appl Neurobiol 1990; 16: 213-223.
263. Wisniewski KE, Jervis GA, Moretz RC et al. Alzheimer neurofibrillary tangles in diseases other than senile and presenile dementia. Ann Neurol 1979; 7: 462-465.
264. Tabaton M, Mandybur TI, Perry G et al. The widespread alteration of neurites in Alzheimer's disease may be unrelated to amyloid deposition. Ann Neurol 1989; 26: 771-778.
265. Geddes JF, Hughes AJ, Lees AJ et al. Pathological overlap in cases of parkinsonism associated with neurofibrillary tangles. A study of postencephalic parkinsonism and comparison with progressive supranuclear palsy and Guamanian parkinsonism-dementia complex. Brain 1993; 116: 281-302.
266. Jamieson GA, Maitland NJ, Wilcock GK et al. Herpes simplex virus type 1 DNA is present in specific regions of brain from aged people with and without senile dementia of the Alzheimer type. J Pathol 1992; 167: 365-368.
267. Bertrand P, Guillaume D, Hellauer K et al. Distribution of herpes simplex virus type 1 DNA in selected areas of normal and Alzheimer's disease brains: A PCR study. Neurodegeneration 1993; 2: 201-208.
268. Marsden CD. Parkinson's disease in twins. J Neurol Neurosurg Psychiatry 1987; 50: 105-106.
269. Langston JW, Ballard P, Tetrud JW et al. Chronic parkinsonism in humans due to a product of meperidine analogue synthesis. Science 1983; 219: 979-980.
270. Burns RS, Chiueh CC, Markey SP et al. A primate model of parkinsonism: selective destruction of dopaminergic neurons in the pars compacta of the substantia nigra by N-methyl-4-phenyl-1,2,3,6-tetrahydropyridine. Proc Natl Acad Sci (USA) 1983; 80: 4546-4550.
271. Davis GC, Williams AC, Markey SP et al. Chronic parkinsonism secondary to intravenous injection of meperidine analgue. Psychiat Res 1979; 1: 249-254.
272. Calne DB, Langston JW. Aetiology of Parkinson's disease. Lancet 1983; ii: 1457-1459.
273. Spencer PS, Nunn PB, Hugon J et al. Guam amyotrophic lateral sclerosis-parkinsonism-dementia linked to a plant excitant neurotoxin. Science 1987; 237: 517-522.
274. Hyman C, Hofer M, Barde YA et al. BDNF is a neurotrophic factor for dopaminergic neurons of the substantia nigra. Nature 1991; 350: 230-233.
275. Ambani LM, Van Woert MH, Murphy S. Brain peroxidase and catalase in Parkinson's disease. Arch Neurol 1975; 32: 114-118.
276. Ballard PA, Tetrud JW, Langston JW. Permanent human parkinsonism due to 1-methyl-4-phenyl-1,2,3,6-tetrahydropyridine (MPTP): Seven cases. Neurology 1985; 35: 949-956.

277. Salach JI, Singer TP, Castagnoli N et al. Oxidation of the neurotoxic amine MPTP by monoamine oxidases A and B and suicide inactivtion of the enzymes by MPTP. Biochem Biophys Res Commun 1984; 125: 831-835.
278. Javitch JA, D'Amato RJ, Strittmatter SM et al. Parkinsonism-inducing neurotoxin, N-methyl-4-phenyl-1,2,3,6-tetrahydropyridine: uptake of the metabolite N-methyl-4-phenylpyridinium by dopamine neurons explains selective toxicity. Proc Natl Acad Sci (USA) 1985; 82: 2173-2177.
279. Ramsay RR, Dadger J, Trevor AJ et al. Energy-driven uptake of MPP+ by brain mitochondria mediates the neurotoxicity of MPTP. Life Sci 1986; 39: 581-588.
280. Nicklas WJ, Vyas I, Heikkila RE. Inhibition of NADH-linked oxidation in brain mitochondria by 1-methyl-4-phenylpyridinium, a metabolite of 1-methyl-4-phenyl-1,2,3,6-tetrahydropyridine. Life Sci 1985; 36: 2503-2508.
281. Nicklas WJ, Youngster SK, Kindt MV et al. MPTP, MPP+ and mitochondrial function. Life Sci 1987; 40: 721-729.
282. Cleeter MJW, Cooper JM, Schapira AHV. Irreversible inhibition of mitochondrial complex I by 1-methyl-4-phenylpyridinium–evidence for free radical involvement. J Neurochem 1992; 58: 786-789.
283. Mizuno Y, Saitoh T, Sone N. Inhibition of mitochondrial α-ketoglutarate dehydrogenase by 1-methyl-4-phenylpyridinium ion. Biochem Biophys Res Commun 1987; 143: 971-976.
284. Forno LS, Langston JW, DeLanney LE et al. Locus ceruleus lesions and eosinophilic inclusions in MPTP-treated monkeys. Ann Neurol 1986; 20: 449-455.
285. Heikkila RE, Cohen G. Inhibition of biogenic amine uptake by hydrogen peroxide: A mechanism for toxic effects of 6-hydroxydopamine. Science 1971; 172: 1257-1258.
286. Graham DG, Tiffany SM, Bell WR. Auto-oxidation versus covalent binding of quinones as the mechanism of toxicity of dopamine, 6-hydroxydopamine and related compounds towards C1300 neuroblastoma cells in vitro. Molec Pharmacol 1978; 14: 644-653.
287. Archibald FS, Tyree C. Manganese poisoning and the attack of trivalent manganese upon catecholamines. Arch Biochem Biophys 1987; 256: 638-650.
288. Greenfield JG, Bosanquet FD. The brain stem lesions in Parkinsonism. J Neurol Neurosurg Psychiatry 1953; 16: 213-226.
289. Forno LS, Langston JW. Lewy bodies and aging: Relation to Alzheimer's and Parkinson's diseases. Neurodegeneration 1993; 2: 19-24.
290. Gaspar P, Gray F. Dementia in idiopathic Parkinson's disease: A neuropathological study of 32 cases. Acta Neuropathol 1984; 64: 43-52.
291. Mann DMA, Yates PO, Hawkes J. The pathology of the human locus caeruleus. Clin Neuropathol 1983; 2: 1-7.
292. Schapira AHV, Cooper JM, Dexter D. Mitochondrial complex I deficiency in Parkinson's disease. Lancet 1989; i: 1269.
293. Schapira AHV, Cooper JM, Dexter D et al. Mitochondrial complex 1 deficiency in Parkinson's disease. J. Neurochem 1990; 54: 823-827.
294. Schapira AHV, Mann VM, Cooper JM et al. Anatomic and disease specificity of NADH CoQ1 reductase (complex I) deficiency in Parkinson's disease. J Neurochem 1990; 55: 2142-2145.
295. Mizuno Y, Ohta S, Tanaka M. Deficiencies in complex I subunits of the respiratory chain in Parkinson's disease. Biochem Biophys Res Commun 1989; 163: 1450-1455.
296. Parker WD, Boyson SJ, Parks JK. Abnormalities of the electron transport chain in idiopathic Parkinson's disease. Ann Neurol 1989; 26: 719-723.
297. Mizuno Y, Matuda S, Yoshino H et al. An immunohistochemical study on α-ketoglutarate dehydrogenase complex in Parkinson's disease. Ann Neurol 1994; 35: 204-210.
298. Hattori N, Tanaka M, Ozawa T et al. Immunohistochemical studies on complexes I, II, III and IV of mitochondria in Parkinson's disease. Ann Neurol 1991; 30: 563-571.
299. McNaught KStP, Thull U, Carrupt P-A et al. Nigral cell loss produced by infusion of isoquinoline derivatives structurally related

to 1-methyl-4-phenyl-1,2,3,6-tetrahydropyridine. Neurodegeneration 1996; 5: 265-270.
300. Yoshida M, Niwa T, Nagatsu T. Parkinsonism in monkeys produced by chronic administration of an endogenous substance of the brain, tetrahydroisoquinoline: the behavioural and biochemical changes. Neurosci Lett 1990; 119: 109-113.
301. Yoshida M, Ogawa M, Suzuki K et al. Parkinsonism produced by tetrahydroisoquinoline (TIQ) or the analogues. In: Narabayashi H, Yanagisawa Y, Mizuno Y, eds. Advances in Neurology. New York: Raven Press Limited, 1993: 207-211.
302. McNaught KStP, Altomare C, Cellamare S et al. Inhibition of α-ketoglutarate dehydrogenase by isoquinoline derivatives structurally related to 1-methyl-4-phenyl-1,2,3,6-tetrahydropyridine (MPTP). NeuroReport 1995; 6: 1105-1108.
303. Nishi K, Mochizuki H, Furukawa Y et al. Neurotoxic effects of 1-methyl-4-phenylpyridinium (MPP+) and tetrahydroisoquinoline derivatives on dopaminergic neurons in ventral mesencephalic striatal co-culture. Neurodegeneration 1994; 3: 33-42.
304. Saporito M, Heikkila RE, Youngster SK et al. Dopaminergic neurotoxicity of 1-methyl-4-phenylpyridinium analogs in cultured neurons: relationship to the dopamine uptake system and inhibition of mitochondrial respiration. J Pharmac Exp Ther 1992; 260: 1400-1409.
305. Maruyama W, Nakahara D, Ota M et al. N-methylation of dopamine-derived 6,7-dihydroxy-1,2,3,4-tetrahydroisoquinoline, (R)-salsolinol, in rat brains: In vivo microdialysis study. J Neurochem 1992; 59: 395-400.
306. McNaught KStP, Thull U, Carrupt PA et al. Inhibition of complex I by isoquinoline derivatives structurally related to 1-methyl-4-phenyl-1,2,3,6-tetrahydropyridine (MPTP). Biochem Pharmacol 1995; 50: 1903-1911.
307. McNaught KStP, Thull U, Carrupt PA et al. Effects of isoquinoline derivatives structurally related to 1-methyl-4-phenyl-1,2,3,6-tetrahydropyridine (MPTP) on mitochondrial respiration. Biochem Pharmacol 1996; 51: 1503-1511.
308. Maruyama W, Dostert W, Naoi M. Dopamine-derived 1-methyl-6,7-dihydroxyisoquinoline as hydroxy radical promoters and scavengers: In vivo and in vitro studies. J Neurochem 1995; 64: 2635-2643.
309. Fahn S, Cohen G. The oxidant stress hypothesis in Parkinson's disease: evidence supporting it. Ann Neurol 1992; 32: 804-812.
310. Kish SJ, Morito C, Hornykiewicz O. Glutathione peroxidase activity in Parkinson's disease. Neurosci Lett 1985; 58: 343-346.
311. Perry TL, Godin DV, Hansen S. Parkinson's disease: a disorder due to nigral glutathione deficiency? Neurosci Lett 1982; 33: 305-310.
312. Perry TL, Yong VW. Idiopathic Parkinson's disease, progressive supranuclear palsy and gluathione metabolism in the substantia nigra of patients. Neurosci Lett 1986; 67: 269-274.
313. Riederer P, Sofic E, Rausch WD et al. Transitional metals, ferritin, glutathione, ascorbic acid in parkinsonian brains. J Neurochem 1989; 52: 515-526.
314. Sofic E, Lange KW, Jellinger K et al. Reduced and oxidized glutathione in the substantia nigra of patients with Parkinson's disease. Neurosci Lett 1992; 142: 128-130.
315. Sian J, Dexter DT, Lees AJ et al. Alterations in glutathione levels in Parkinson's disease and other neurodegenerative disorders affecting basal ganglia. Ann Neurol 1994; 36: 348-355.
316. Dexter DT, Carter CJ, Agid F et al. Lipid peroxidation as a cause of nigral cell death in Parkinson's disease. Lancet 1986; ii: 639-640.
317. Dexter DT, Carter CJ, Wells FR et al. Basal lipid peroxidation is increased in Parkinson's disease. J Neurochem 1989; 52: 381-389.
318. Dexter DT, Holley AE, Flitter WD et al. Increased levels of lipid hydroperoxides in the parkinsonian substantia nigra: An HPLC and ESR study. Mov Disord 1994; 9: 92-97.
319. Marttila RJ, Lorentz H, Rinne UK. Oxygen toxicity protecting enzymes in Parkinson's disease. Increase of superoxide dismutase-like activities in the substantia

nigra and basal nucleus. J Neurol Sci 1988; 86: 321-331.

320. Saggu H, Cooksey J, Dexter D et al. A selective increase in particulate superoxide dismutase activity in parkinsonian substantia nigra. J Neurochem 1989; 53: 692-697.
321. Ceballos I, Lafon M, Javoy-Agid F et al. Superoxide dismutase and Parkinson's disease. Lancet 1990; 335: 1035-1036.
322. Sanchez-Ramos JR, Overvik E, Ames BN. A marker of oxyradical-mediated DNA damage (8-hydroxy-2'-deoxyguanosine) is increased in the nigro-striatum in Parkinson's disease brain. Neurodegeneration 1994; 3: 197-204.
323. Dexter DT, Wells FR, Agid F et al. Increased nigral iron content in post mortem Parkinsonian brain. Lancet 1987; ii: 1219-1220.
324. Dexter DT, Carayon A, Javoy-Agid F et al. Alterations in the levels of iron, ferritin and other trace metals in Parkinson's disease and other neurodegenerative diseases affecting the basal ganglia. Brain 1991; 114: 1953-1975.
325. Sofic E, Paulus W, Jellinger K et al. Selective increase of iron in substantia nigra zona compacta of Parkinsonian brains. J Neurochem 1991; 56: 978-982.
326. Morris CM, Edwardson JA. Iron histochemistry of the substantia nigra in Parkinson's disease. Neurodegeneration 1994; 3: 277-282.
327. Olanow CW. An introduction to the free radical hypothesis in Parkinson's disease. Ann Neurol 1992; 32: 52-59.
328. Perl DP, Good PF. Comparative techniques for determining cellular iron distribution in brain tissues. Ann Neurol 1992; 32: 576-581.
329. Jellinger K, Kienzl E, Paulus W et al. Presence of iron in melanized dopamine neurons in Parkinson's disease. J Neurochem 1992; 59: 1168-1171.
330. Zecca L, Pietra R, Goj C et al. Iron and other metals in neuromelanin, substantia nigra and putamen of human brain. J Neurochem 1994; 62: 1097-1101.
331. Enochs WS, Nilges MJ, Swartz HM. Purified human neuromelanin, synthetic dopamine melanin as a potential model pigment, and the normal human substantia nigra: Characterization by electron paramagnetic resonance spectroscopy. J Neurochem 1992; 61: 68-79.
332. Hirsch EC, Brandel JP, Galle P et al. Iron and aluminum increase in the substantia nigra of patients with Parkinson's disease: An X-ray microanalysis. J Neurochem 1991; 56: 446-451.
333. Jellinger K, Paulus W, Grundke-Iqbal U et al. Brain iron and ferritin in Parkinson's and Alzheimer's diseases. J Neurol Transm. Park Dis Dement Sec 1990; 2: 327-340.
334. McGeer PL, Itagaki S, Akiyama H et al. Rate of cell death in Parkinsonism indicates active neuropathological process. Ann Neurol 1988; 24: 574-576.
335. Graham DG. On the origin and significance of neuromelanin. Arch Path Lab Med 1979; 103: 359-362.
336. Fearnley JM, Lees AJ. Aging and Parkinson's disease: Substantia nigra regional selectivity. Brain 1991; 114: 2283-2301.
337. Arendt T, Bigl V, Arendt A et al. Loss of neurones in the nucleus basalis of Meynert in Alzheimer's disease, paralysis agitans and Korsakoff's disease. Acta Neuropathol 1983; 61: 101-108.
338. Rogers JD, Brogan D, Mirra SS. The nucleus basalis of Meynert in neurological disease: A quantitative morphological study. Ann Neurol 1985; 17: 163-170.
339. Hoehn MD, Yahr MM. Parkinsonism: Onset, progression and mortality. Neurology 1967; 17: 427-442.
340. Duvoisin RC, Yahr MD. Encephalitis and parkinsonism. Arch Neurol 1965; 12: 227-239

CHAPTER 5

Relationships Between Aging and Neurodegenerative Disease

5.0. INTRODUCTION

Simple comparisons between the catalogues of damage that occur in the 'normal' elderly and in patients with the common neurodegenerative disorders of old age (AD and PD) show much overlap. For example, in AD the larger pyramidal cells of the cerebral cortex (especially those of the frontal, temporal and cingulate gyri) and the hippocampus (CA1 and subiculum), amygdala (cortical and medial nuclei), suprachiasmatic nucleus, nucleus basalis complex, locus caeruleus and ventral tegmentum are all seriously decimated, yet these same cell types are affected, though to a lesser extent, in normally aged individuals. In PD the gross loss of the pigmented cells of the substantia nigra and, again, cells of the locus caeruleus and nucleus basalis contrasts similarly with the lesser involvement in 'normal' aging. However, this is not always the rule with, for example, Purkinje cells in the cerebellum[1] and neurones of the sexually dimorphic nucleus[2] being nerve cell types that are clearly affected with aging but do not apparently suffer additional damage in either AD or PD. Some regions, such as the cranial nerve nuclei, the mammillary bodies, the olivary and pontine nuclei, the dentate nucleus, paraventricular and supraoptic nuclei[3] seem to resist equally the effects of aging and both of these forms of neurodegeneration. Many of the regressive changes taking place within surviving cells in areas of cell loss (i.e. dendritic reduction, perikaryal shrinkage, loss of nucleic acid) also seem magnified in AD and PD.

There is further evidence of pathological overlap between old age and neurodegenerative disease. NFT, typical of AD, are present in many elderly non-demented persons though these are mostly few in number and usually topographically restricted to certain regions of the hippocampus (CA1 and subiculum), entorhinal cortex and amygdala.[4-7] It is perhaps without coincidence that these same three regions show an early affliction by NFT in persons with DS,[8-11] these individuals eventually developing, entirely and predictably, the pathological changes of AD if they live into middle age and beyond.[12-16] The Aβ deposits that constitute the focus of plaque formation in AD[17-20] occur widely in many non-demented elderly people.[17,18,21-25] Indeed, the prevalence and intensity of formation of such deposits increases with age so that a very large proportion of people over 70 years of age show some Aβ somewhere in their brains. Moreover, Aβ nearly always has the same morphological ('diffuse' plaques),[17-24] and chemical[20,25] characteristics as that seen in younger individuals with DS.[10,26-29] Furthermore, some elderly patients have been described who, while clinically demented, show much Aβ within the cerebral cortex yet few NFT are seen, these being common only in the hippocampus, especially the parahippocampal gyrus (entorhinal cortex).[30,31] Hence, there are clear parallels in which certain patients, with much cortical Aβ but few and restricted NFT, form similar pathological cohorts within each of the normally aged, the demented (AD) and the (non-demented) DS groups. Likewise, the Lewy body, pathognomonic of Parkinson's disease, is commonly seen in cells of the substantia nigra in non-Parkinsonian old people,[32] the prevalence of this particular change rising with age.[33-35]

5.1. PATHOLOGY IN THE NORMAL ELDERLY–PRECLINICAL DISEASE?

The incidence and prevalence of clinical AD rises with age; estimates of the number of old people affected clinically by AD vary but seem to average 10-15% of all people over 65 years[36] though as many as 47% of those over 85 years of age may be affected.[37] PD is similarly age-related. Hence, the high clinical prevalence rates for both these disorders among elderly people, and the knowledge that in PD a 50% cell loss from the substantia nigra and, at least, a similar depletion of striatal dopamine are necessary to produce symptomatic parkinsonism (how much pathology can be 'tolerated' in AD before clinical change occurs is not known) argue that within the general population there must be many persons who can be described clinically as within 'normal limits', but who actually contain some disease changes in their brains but at levels presymptomatic for these disorders. Such persons, if they were to live long enough, would presumably continue to accrue pathology eventually sufficient to cause clinical expression of the disease. The critical question, therefore, is to what extent are we justified in correlating this likelihood of a large number of preclinical, asymptomatic individuals in the population at large with the frequently reported high prevalence rates of Alzheimer or Parkinson-type pathology within so-called unaffected or normal individuals? In other words, how far can we claim that all persons who, for example, show cortical deposits of Aβ with (or perhaps even without) NFT in the hippocampus and other regions, but in amounts less than would customarily be equated with AD, are actually suffering from early or incipient disease? Similarly, do all persons who show Lewy or 'pale' bodies within some cells of the substantia nigra have early PD? These contentious issues need urgent clarification. Since it is not possible to perform 'longitudinal' pathological studies on human brain tissues we cannot, with certainty, differentiate, from individuals seen at autopsy to have a limited Alzheimer or Parkinson-type pathology, those who would have developed a full disease expression from those who would not had they lived longer. Yet, we do have some clues, e.g., in DS where the chances of developing AD in later life are entirely predictable. In younger DS patients the evolving pathology closely resembles that seen in many non-demented old persons. To establish whether such affected old people do indeed have early disease is important, not simply for the sake of providing a diagnostic label nor even for the identification of cases of early disease for the devising or implementation of therapeutic strategies, but for its philosophical implications for aging research in general.

5.2. IMPLICATIONS FOR AGING RESEARCH

If all persons with minimal Alzheimer or Parkinson-type pathology do indeed represent cases of early disease, then much of the aging research that has so far gone on in humans requires re-evaluation. In most instances, such research was conducted on patients chosen 'at random' or 'as available', providing they did not meet minimal clinical criteria for psychiatric or neurological disease. Hence, in many, or perhaps even in most, of the foregone studies on human 'aging', subjects making up the 'normal' elderly cohort would in all likelihood be 'contaminated' with some old people with preclinical disease. Inclusion of such patients, along with the limited pathology they would already have accrued, will obviously bias data sets and predispose towards an achievement of statistical significance when correlating with age, for example, the number of neurones in the substantia nigra, or the density of pyramidal cells in the temporal cortex. In only a few studies on aging were patients screened for Alzheimer or Parkinson-type changes.[38] In other studies, the presence of such minimal changes was recognized but was found acceptable within the applied definition of the 'normal' elderly individuals.[6,39,40] However, most studies where screening for pathology had occurred predated the modern immunohistochemical (e.g. α-β-protein, α-tau or α-ubiquitin immunostaining) or silver (e.g. methenamine silver, Gallyas or Campbell stain) methods now widely used for the sensitive and specific detection of Aβ deposits, NFT and Lewy bodies; it is therefore uncertain as to how effective these exclusion criteria were. Despite this, in at least one study, where rigorous and effective screening for Alzheimer-type changes was made,[38] no evidence for an age-related decline in pyramidal cell density within the cerebral cortex was found. It is perhaps now necessary to reconsider what we feel to be true concerning aging, as it affects the human nervous system at least.

Some clinical studies[41,42] have attempted this. In these, sensitive (psychometric) screening tests were applied to potential recruits and, on the basis of performance using WAIS or mini-mental status, all but the top performers were rejected. In this way the 'customary' findings in PET scans of a decrease with age in cerebral cortical glucose utilization,[43] or striatal fluorodopa metabolism,[44] were negated.[41,42] While it is probable that most, or even all, of these 'top performers' would be free from the overt pathological changes associated with clinical AD or PD, and possibly minimal change too, it cannot be necessarily inferred that all of those rejected would have contained some degree of Alzheimer or Parkinson-type pathology within their brains, though perhaps many would have done so.

5.3. THE CONCEPT OF NORMALITY

The major problem lies within the definition, in a form that is generally acceptable and appropriate, of exactly what is meant by 'normal' when applied to the elderly population. Dictionaries provide numerous definitions of the word 'normal'. Most imply the meaning 'usual', and by this terminology those many elderly persons with minimal pathology in their brains could be included as 'normal' since clearly findings like these in such persons are indeed 'usual'. This same problem crops up outside the nervous system. Atherosclerosis occurs within the aorta and large systemic arteries of most middle-aged and elderly persons and its prevalence is age-related. However, is minimal atherosclerosis considered to be 'normal'? Similarly, with osteoporosis. The fact that a finding is commonplace, does not necessarily make it 'normal'.

In the context of biological aging the definition of 'normal' should perhaps exclude the presence in the brain, or other tissue, of pathological changes, however slight, that can be associated with definite diseases of the nervous or other system. Persons falling into such a category might be described as having aged 'successfully'[45] since they will have escaped entirely the common neurological disorders of later life. Hence, it is necessary to clearly differentiate between tissue changes that might be defined as those of 'usual aging' as opposed to those of 'successful aging'; a true 'normality' with respect to the changes of aging would be equated with the latter. To date, the effects of 'usual aging' in old people have generally been studied. Although application of such rigorous distinctions may lead to the selection of a 'supernormal' group of individuals by doing this one is not trying to define brain changes that may adequately describe the elderly population as a whole but solely attempting to discriminate, within this broad level of change, that (perhaps constant) component, which may be due to the passage of time alone (i.e. the process of aging), from that which is variably super-added due to the (minor or early) effects of the major diseases of later life. It is not the 'fault' of the 'process of aging' if many, or even most, individuals become 'tarnished' with disease in later life. How common such 'super persons' might be is unclear. In relationship to Aβ deposition about 15-40% of persons over 65 years do not show any Aβ deposits whatsoever in their brains[21-23] whereas 90% do not show Lewy bodies within the nigra;[32-35] obviously some of those persons without Lewy bodies may have Aβ deposits and vice versa, since in no study have both been looked for at the same time.

Therefore, at present, it is not possible to be sure as to what extent, in humans, nerve cells are damaged and lost simply as a result of the passage of time alone. It might be possible to consider the human cerebellum as a model for aging, since here no Aβ deposits nor NFT occur with aging (although the former do often occur in AD)[17,46] nor are Lewy bodies ever seen at this site. Hence, it is possible that the Purkinje cell loss from the cerebellum that occurs during life (some 2.5% per decade)[1] marks the extent of change in the rest of the brain. Yet Purkinje cells are extremely sensitive to the effects of hypoxia or alcohol[47] and the basilar/vertebral arteries are particularly prone to atherosclerosis. Therefore some (or even all) of this cell loss with age might reflect the effects of cardiovascular or cerebrovascular disease, or alcohol intoxication rather than those of aging per se.

Furthermore, it is also entirely possible that these 'static' markers of the infrastructure of the brain do not reflect the efficiency with which it is used. More dynamic measures are needed and here changes in nucleic acids or protein metabolism or oxygen uptake and glucose metabolism with age may take on more relevance. It may be the case that with the passage of time major problems do not necessarily arise with the 'hardware', but with the efficiency with which the 'software' operates. Nevertheless a 'weakened system' either generally throughout the brain, or perhaps especially compromised in specific vulnerable regions, may pave the way for additional deleterious changes that predispose towards or themselves precipitate the neurodegenerative diseases of old age. In this way aging may act as a 'springboard' to disease in later life.

5.4. RELEVANCE OF ANIMAL STUDIES

Although some aged higher primates,[48-51] dogs[52-54] and bears[51] may accumulate small quantities of Aβ or NFT, rodents and other (lower) mammals seemingly do not show such changes, perhaps because of their differing composition of the APP gene[55] In any case none of these animal species are prone to spontaneously develop the neurodegenerative diseases of old age in humans. Therefore, clues to the effects of the passage of time on the central nervous system in an environment free from confounding disease and which could have general application to humans might come from animal studies (see Coleman and

Flood;[56] Rogers et al[57]). However, the considerable variations in findings between species and even within different strains of the same species permits only the broadest of comparisons to be made. The limited data so far gained from the examination of (relatively few) higher primates may be more pertinent to humans than the greater wealth of rodent data, which in many instances, has contradicted itself through a lack of recognition that prolonged developmental phases (up to 12 months) may occur in some rat strains, inappropriate chronological definitions of what is 'young' or 'aged', or a failure to let (or manipulate through dietary restriction) animals reach an age old enough to make them comparable with the oldest of humans.[58]

5.5. AGING AND DISEASE—A CONTINUUM

In the past pathologists, cell biologists or biochemists have sought to differentiate patients with AD from the non-demented elderly, the Parkinsonian from the non-Parkinsonian subjects, either in terms of the amount of tissue change on a notional scale, or (preferably) on the basis of the presence of a precise molecular or structural marker in the tissue of one group (the diseased) and its absence in the other. Likewise clinical distinctions have been drawn according to points on a quantitative sliding scale of disability or in terms of particular neuropsychological features specific to each category of cognitive performance.

This premise is based upon the traditionalist view that the two categories (i.e. demented, non-demented; Parkinson, non-Parkinson) represent distinct clinical and pathological entities. Yet as we have seen there are real problems in defining a normally aged population, both clinically and pathologically. In practice, a clinically normal group will include persons who are completely free of pathology along with those who will show a variable degree of change, but always at a level below that critical for the generation of clinical symptoms. Overtly diseased patients present little problem in detecting either a pathological cause or the clinical consequences. Perhaps three groups should be defined:

(i) The normally aged without any clinical or pathological evidence of disease.
(ii) Those with (minimal) pathological change but still capable of functioning clinically within normal range.
(iii) Those with overt clinical and pathological change consistent with disease.

If these criteria are to be adopted then clinical cohorts of normality would encompass categories (i) and (ii) whereas pathological cohorts would be restricted to category (i). Clinical cohorts of disease would encompass category (iii) whereas pathological cohorts would include categories (ii) and (iii). In this way clinical studies will always suffer from the inability to exclude patients with sub-clinical or pre-clinical disease until the subtle contribution that the small amount of ongoing pathology might make can be assessed and taken into account. On the other hand pathological studies will suffer since the precise cut-off point, in terms of the minimal amount of pathology required to dictate the beginnings of clinical symptoms, cannot be determined as patients are unlikely to die exactly at this 'threshold'; this in any case will vary greatly according to an individuals' level of functional reserve. Hence, most pathological studies will be based on cases of established disease where early or initial indices of change have disappeared under a mountain of accrued damage and where many alterations (particularly those of a chemical nature) may reflect or be partially obscured by compensatory or adaptive responses.

5.6. GENETIC SUSCEPTIBILITY

One way forward may be to move away from classifications based on 'disease' and 'normality' towards ones which acknowledge the presence of evolving subthreshold disease changes. This would not involve use of the familiar end-stage markers of pathological cascade processes (i.e. plaques, tangles, Lewy bodies), which in any case cut across traditional diagnostic groups, but engage etiological factors that make the development of such changes likely or even predictable, or which determine the extent to which such changes might progress. These etiological factors refer to the genetic variations that confer varying levels of susceptibility to damaging changes in cell and tissue function. Each and everyone of us comprises a rich mix of genetic variation some of which may be especially good and promote a conservation of cell and tissue function, some of which may be not so good and cause or facilitate damage, even on a scale eventually leading to disease.

In AD, for example, some cases are associated with specific variations (mutations), in genes on chromosomes 1, 14 and 21, which almost invariably cause the disease to develop in bearers of such changes. Likewise in DS an additional chromosome 21 bears similar fruit. However, in other cases of disease, not associated with such precise mutational changes, there is an enrichment of one or more (usually) fairly common gene variations (polymorphisms), possession of which seems to increase the statistical likelihood that disease will occur. Such variations do not cause the disease solely by their presence (e.g. APOE E4 allele in AD, CYP2D allele in PD). In non-demented individuals, or non-Parkinsonian subjects the disease causing mutations are obviously not present and critical genetic risk factors are low in frequency. For example the APOE E4 allele frequency is at least 2-3 times lower in non-demented subjects than in those with AD, this falling even further in the very elderly. In this way the elderly population may be grouped as (i) successfully aged; (ii) usually (normally), and partially successfully, aged; (iii) unsuccessfully aged. In the first group genetic characteristics would favor the preservation of tissue integrity and clinical change would be absent. In the second group, and this might include the majority of elderly individuals, a variable amount of unfavorable genetic variation would be present and this might favor the formation of only a limited amount of pathology, this perhaps accruing over a prolonged time period. Clinical change would be absent or minimal. In the third group, either specific disease causing mutations would be present or the balance of genetic risk factors would be so extreme as to lead inexorably to a progress in pathology producing overt disease with clinical symptoms. Hence, according to our genetic framework relatively few of us will age really well (successfully), reaching perhaps the 9th or 10th decade without apparent brain damage nor trace of physical or cognitive deterioration, or both. A substantial minority will age poorly (unsuccessfully) succumbing to overt nervous system disease, sometimes by middle age but usually later. The great majority of us will age partially successfully, retaining for the most part our physical and mental faculties late into life, in spite of a certain amount of damage that will undoubtedly have taken place. This accrual of pathology may progress at such a slow rate that sufficient damage to cause disease might only emerge well beyond the normal life expectancy, with death from other (unrelated) systemic illness intervening long before this time is reached. Within this scenario all grades of pathological change (ranging from nil to abundant) and concomitant clinical dysfunction are possible depending on the balance of genetic risk carried. Environmental factors however are not completely excluded, though the major role of these might perhaps be to bring out, at an earlier time of life, those changes which would in any case have occurred later in life because of genetic predisposition. How such factors might operate is unclear. They might, for example, tune directly into the pathological cascades (e.g. head injury and APP) or they may lessen resistance to disease (i.e. reduce functional reserve) such that the pathological threshold to clinical disease is lowered, symptoms then being brought on by a lower amount of additional tissue damage.

5.7. CONCLUDING REMARKS

If aging (successfully or partially so) and neurodegeneration (unsuccessful aging) form a continuum and if the real distinctions between these lie solely, or at least principally, with a relative balance and strength of genetic factors that each predispose towards the activation and progress of pathological cascade processes, then the quest to find clinical or tissue markers capable of differentiating between age and disease is not likely to be attainable. It will be our own genetic profile that will ultimately decide the outcome with possession of bad genes meaning disease, possession of good genes, none. Most of us will have an assortment of good and bad genes but only when all these predisposing factors are known and the relative risks each convey can be assessed and balanced against each other will it become possible to distinguish between who has and or who does not have disease (irrespective of clinical or tissue status), or perhaps more importantly to determine who is not affected but who would in future develop disease. Environmental factors may still play a part, though perhaps at most these might 'simply' provide 'fine tuning' to the system maybe bringing forward, or even delaying, the onset of change, or increasing the severity or rate of progression of change, in persons who would in any case, because of their genetic profile, have developed disease at some stage of their life.

Are the genes that dictate the process of aging (i.e. successful aging) the 'good side' of those bad genes that promote pathology. This is possible though unlikely. These 'health promoting' genes probably fulfill housekeeping roles maintaining the integrity of the genome and ensuring a faithful and efficient transcription and tranlation of the message, thereby keeping up the flow of proteins capable of performing well in their prescribed roles. Failures here with time, these perhaps stemming from a degraded mechanism for maintaining the DNA, may result in partial cell dysfunction and even some death and could act as a trigger upon which genetic changes promoting pathology might gain momentum. If so, this might explain the time dependency of such pathology. In early and middle age the CNS remains fundamentally fit and healthy, well able to cope with many of the adverse genetic risk factors, only falling foul of those with sufficient severity (e.g. mutations on chromosomes 1, 14 and 21). In later life, the system might not be so well-blessed and adverse genetic tendencies can then begin to take effect and exact their toll on this weakened infrastructure; disease may come about or be prevented (restricted) according to the particular mix of genetic factors present.

It is therefore most likely that both the processes of aging and those underlying neurodegeneration are fundamentally under genetic control with separate, though complementary, genes dictating each. Both processes can operate separately; the latter potentially conferring a disease state may not operate at all in some healthy elderly individuals. In most of us both processes will be ongoing to a greater or lesser extent as we get older, the balance of these ultimately determining how well or how poorly we function in later life with respect to our physical capabilities and our mental faculties.

If this argument is correct then we are left with the inevitable conclusion that prognostications regarding the potential mental well-being of individuals in later life will depend more upon a person's possession, or otherwise, of particular health promoting or disease determining genes rather than any particular lifestyle they might occupy. Clearly, poor diet and hygiene, excessive alcohol consumption, cigarette smoking, drug habituation and occupational hazards are all risky factors that will damage and impair organs and tissues and compromise physical health at any time of life, let alone during old age. Yet these would seem to have little relevance in terms of influencing brain function and changes in lifestyle may do little to lessen the likelihood of "acquiring" neurodegenerative disease. Care of our bodies may prolong physical life, yet here again major killers like cardiovascular disease, carcinoma and diabetes are likely to be caused, at least partially, by inherited genetic risk factors.

At present, therefore, is seems that little can be done to reduce the risk of neurodegenerative disease in later life. A "healthy lifestyle" is not the answer. Nervous system disease will only be prevented or attenuated by intervention at gene or protein level thereby modifying or negating the deleterious tissue effects these might convey. The identification of these genetic risk factors and an understanding of how they operate must be the priority of future research since epidemiology predicts a global expansion of the proportion of elderly persons in the world's population and the burgeoning health care problems they will bring. That even more of us than at present will grow very old seems inevitable; that many of us should pay the price for this in terms of neurological or psychiatric disease should not be considered such an inevitability. The success of future research will be to see the threat of this, like that of the other major systemic illnesses afflicting mankind, is removed from the horizon.

References

1. Hall TC, Miller AKH, Corsellis JAN. Variations in the human Purkinje cell population according to age and sex. Neuropath Appl Neurobiol 1975; 1: 267-292.
2. Swaab DF, Fliers E, Partiman TS. The suprachiasmatic nucleus of the human brain in relation to sex, age and senile dementia. Brain Res 1985; 342: 37-44.
3. Goudsmit E, Hopman MA, Fliers E et al. The supraoptic and paraventricular nuclei of the human hypothalamus in relation to sex, age and Alzheimer's disease. Neurobiol Aging 1990; 11: 529-536.
4. Tomlinson BE, Blessed G, Roth M. Observations on the brains of non-demented old people. J Neurol Sci 1968; 7: 331-356.
5. Ball MJ. Neuronal loss, neurofibrillary tangles and granulovacuolar degeneration in the hippocampus with aging and dementia. A quantitative study. Acta Neuropathol 1977; 37: 111-118.
6. Mann DMA, Yates PO, Marcyniuk B. Some morphometric observations on the cerebral cortex and hippocampus in presenile Alzheimer's disease, senile dementia of Alzheimer type and Down's syndrome in middle age. J Neurol Sci 1985; 69: 139-159.
7. Mann DMA, Tucker CM, Yates PO. The topographic distribution of senile plaques and neurofibrillary tangles in the brains of non-demented persons of different ages. Neuropath Appl Neurobiol 1987; 13: 123-139.
8. Mann DMA, Esiri MM. The site of the earliest lesions of Alzheimer's disease. N Engl J Med 1988; 318: 789-790.
9. Mann DMA, Esiri MM. Regional acquisition of plaques and tangles in Down's syndrome patients under 50 years of age. J Neurol Sci 1989; 89: 169-179.
10. Mann DMA, Brown AMT, Prinja D et al. An analysis of the morphology of senile plaques in Down's syndrome patients of different ages using immunocytochemical and lectin histochemical methods. Neuropath Appl Neurobiol 1989; 15: 317-329.
11. Mann DMA, Prinja D, Davies CA et al. Immunocytochemical profile of neurofibrillary tangles in Down's syndrome patients of different ages. J Neurol Sci 1989; 92: 247-260.
12. Mann DMA. The pathological association between Down syndrome and Alzheimer disease. Mech Aging Dev 1988; 43: 99-136.
13. Mann DMA. Alzheimer's disease and Down's syndrome. Histopath 1988; 13: 125-138.
14. Whalley A. The dementia of Down's syndrome and its relevance to aetiological studies of Alzheimer's disease. Ann NY Acad Sci 1982; 396: 39-53.
15. Oliver C, Holland AJ. Down's syndrome and Alzheimer's disease: a review. Psychol Med 1986; 16: 307-322.
16. Wisniewski HM, Rabe A. Discrepancy between Alzheimer type neuropathology and dementia in persons with Down's syndrome. Ann NY Acad Sci 1986; 477: 247-259.
17. Ikeda S-I, Allsop D, Glenner GG. The morphology and distribution of plaque and related deposits in the brains of Alzheimer's disease and control cases: an immunohistochemical study using amyloid β protein antibody. Lab Invest 1989; 60: 113-122.

18. Ogomori K, Kitamoto T, Tateishi J et al. β amyloid protein is widely distributed in the central nervous system of patients with Alzheimer's disease. Am J Pathol 1989; 134: 243-251.
19. Bugiani O, Giaccone G, Frangione B et al. Alzheimer patients: preamyloid deposits are more widely distributed than senile plaques throughout the central nervous system. Neurosci Lett 1989; 103: 262-268.
20. Iwatsubo T, Odaka N, Suzuki N et al. Visualization of Aβ42(43)-positive and Aβ40-positive senile plaques with end-specific Aβ monclonal antibodies: Evidence that an initially deposited species is Aβ1-42(43). Neuron 1994; 13: 45-53.
21. Mann DMA, Brown AMT, Prinja D et al. A morphological analysis of senile plaques in the brains of non-demented persons of different ages using silver, immunocytochemical and lectin histochemical staining techniques. Neuropath Appl Neurobiol 1990; 16: 17-25.
22. Davies L, Wolska B, Hilbich C et al. β4 amyloid protein deposition and the diagnosis of Alzheimer's disease : prevalence in aged brains determined by immunocytochemistry compared with conventional neuropathologic techniques. Neurology 1988; 38: 1688-1693.
23. Price JL, Davis PB, Morris JC et al. The distribution of plaques, tangles and related immunohistochemical markers in healthy aging and Alzheimer's disease. Neurobiol Aging 1991; 12: 295-312.
24. Ohgami T, Kitamoto T, Shin R-W et al. Increased senile plaques without microglia in Alzheimer's disease. Acta Neuropathol 1991; 81: 242-247.
25. Fukumoto H, Asami-Odaka A, Suzuki N et al. Amyloid β protein (Aβ) deposition in normal aging has the same characteristics as that in Alzheimer's disease: predominance of Aβ42(43) and association of Aβ40 with cored plaques. Am J Pathol 1996; 148: 259-265.
26. Rumble B, Retallack R, Hilbich C et al. Amyloid (A4) protein and its precursor in Down's syndrome and Alzheimer's disease. N Engl J Med 1989; 320: 1446-1452.
27. Iwatsubo T, Mann DMA, Odaka A et al. Amyloid β protein (Aβ) deposition: Aβ42(43) precedes Aβ40 in Down syndrome. Ann Neurol 1995; 37: 294-299.
28. Lemere CA, Blusztajn JK, Yamaguchi H et al. Sequence of deposition of heterogeneous amyloid β-peptides and APO E in Down syndrome: Implications for initial events in amyloid plaque formation. Neurobiol Dis 1996; 3: 16-32.
29. Kida E, Wisniewski KE, Wisniewski HM. Early amyloid-β deposits show different immunoreactivity to the amino- and carboxy-terminal regions of β-peptide in both Alzheimer's disease and Down's syndrome brain. Neurosci Lett 1995; 193: 1-4.
30. Braak H, Braak E. Neurofibrillary changes confined to the entorhinal region and an abundance of cortical amyloid in cases of presenile and senile dementia. Acta Neuropathol 1990; 80: 479-486.
31. Bouras C, Hof PR, Morrison JH. Neurofibrillary tangle densities in the hippocampal formation in a non-demented population define subgroups of patients with differential early pathologic changes. Neurosci Lett 1993; 153: 131-135.
32. Gibb WRG. Idiopathic Parkinson's disease and the Lewy body disorders. Neuropath Appl Neurobiol 1986; 12: 223-234.
33. Gibb WRG, Lees AJ. The relevance of the Lewy body to the pathogenesis of idiopathic Parkinson's disease. J. Neurol Neurosurg Psychiat 1988; 51: 745-752.
34. Forno LS, Langston JW. Lewy bodies and aging: relation to Alzheimer's and Parkinson's diseases. Neurodegeneration 1993; 2: 19-24.
35. Forno LS. Concentric hyaline intraneuronal inclusions of Lewy type in the brains of elderly persons (50 incidental cases); relationship to Parkinsonism. J Amer Geriat Soc 1969; 17: 557-575.
36. Katzman R. Alzheimer's disease. N Engl J Med 1986; 314: 964-973.
37. Evans DA, Funkenstein H, Albert MS et al. Prevalence of Alzheimer's disease in a community population of older persons. JAMA 1989; 262: 2551-2556.
38. Terry RD, De Teresa R, Hansen LA. Neocortical cell counts in normal human adult aging. Ann Neurol 1987; 21: 530-539.

39. Terry RD, Peck A, De Teresa R et al. Some morphometric aspects of the brain in senile dementia of the Alzheimer type. Ann Neurol 1981; 10: 184-192.
40. West MJ. Regionally specific loss of neurons in the aging human hippocampus. Neurobiol Aging 1993; 14: 287-293.
41. Duara R, Margolin RA, Robertson-Tschabo EA et al. Cerebral glucose utilization as measured with positron emission tomography in 21 resting healthy men between the ages of 21 and 83 years. Brain 1983; 106: 761-775.
42. Sawle GV, Colebatch JG, Shah A et al. Striatal function in normal aging: implications for Parkinson's disease. Ann Neurol 1990; 28: 799-804.
43. Kuhl DE, Metter EJ, Riege WH et al. Effects of human aging on patterns of local cerebral glucose utilization determined by the (18 F) flurodeoxyglucose method. J Cereb Blood Flow Metab 1982; 2: 163-171.
44. Martin WRW, Palmer MR, Patlak CS et al. Nigrostriatal function in humans studied with positron emission topography. Ann Neurol 1989; 26: 535-542.
45. Calne D, Calne JS. Normality and disease. Can J Neurol Sci 1988; 14: 3-14.
46. Mann DMA, Jones D, Prinja D et al. The prevalence of amyloid (A4) protein deposits within the cerebral and cerebellar cortex in Down's syndrome and Alzheimer's disease. Acta Neuropathol 1990; 80: 318-327.
47. Torvik A, Torp S, Lindboe CF. Atrophy of the cerebellar vermis in aging. A morphometric and histologic study. J Neurol Sci 1986; 76: 283-294.
48. Selkoe DJ, Bell DS, Podlisny MB et al. Conservation of brain amyloid proteins in aged mammals and humans with Alzheimer's disease. Science 1987; 235: 873-877.
49. Walker LC, Kitt CA, Schwam E et al. Senile plaques in aged squirrel monkeys. Neurobiol Aging 1987; 8: 291-296.
50. Struble RG, Price DL, Cork LC et al. Senile plaques in cortex of aged normal monkeys. Brain Res 1985; 361: 267-275.
51. Cork LC, Powers RE, Selkoe DJ et al. Neurofibrillary tangles and senile plaques in aged bears. J Neuropathol Exp Neurol 1988; 47: 629-641.
52. Giaccone G, Verga L, Finazzi M et al. Cerebral preamyloid deposits and congophilic angiopathy in aged dogs. Neurosci Lett 1990; 114: 178-183.
53. Wisniewski T, Lalowski M, Bobik M et al. Amyloid β1-42 deposits do not lead to Alzheimer's neuritic plaques in aged dogs. Biochem J 1996; 313: 575-580.
54. Wegiel J, Wisniewski HM, Dziewiatkowski J et al. Fibrillar and non-fibrillary amyloid in the brain of aged dogs. In: Iqbal K, Mortimer JA, Winblad B et al., eds. Research Advances in Alzheimer's Disease and Related Disorders. John Wiley & Sons Ltd, 1995: 703-707.
55. Volloch V. Possisble mechanism for resistance to Alzheimer's disease (AD) in mice suggests new approach to generate a mouse model for sporadic AD and may explain familial resistance to AD in man. Neurodegeneration (In press).
56. Coleman PD, Flood DG. Neurone numbers and dendritic extent in normal aging and Alzheimer's disease. Neurobiol Aging 1987; 8: 521-545.
57. Rogers J, Magistretti PJ, Bolis LC. Animal models for aging research. Neurobiol Aging 1991; 12: 619-701.
58. Coleman PD, Finch C, Joseph J. The need for multiple time points in aging studies. Neurobiol Aging 1990; 11: 1-2.

INDEX

I

L

M

N

O

P

R

S

T

U

V

W

MIX
Papier aus verantwortungsvollen Quellen
Paper from responsible sources
FSC® C105338

If you have any concerns about our products,
you can contact us on
ProductSafety@springernature.com

In case Publisher is established outside the EU,
the EU authorized representative is:
Springer Nature Customer Service Center GmbH
Europaplatz 3, 69115 Heidelberg, Germany

Printed by Libri Plureos GmbH
in Hamburg, Germany